ENERGY POLICIES, POLITICS AND PRICES

ENHANCED OIL RECOVERY

METHODS, ECONOMIC BENEFITS AND IMPACTS ON THE ENVIRONMENT

ENERGY POLICIES, POLITICS AND PRICES

Additional books in this series can be found on Nova's website under the Series tab.

Additional e-books in this series can be found on Nova's website under the e-book tab.

ENERGY POLICIES, POLITICS AND PRICES

ENHANCED OIL RECOVERY

METHODS, ECONOMIC BENEFITS AND IMPACTS ON THE ENVIRONMENT

ALICIA KNIGHT
EDITOR

nova publishers
New York

Library of Congress Cataloging-in-Publication Data

Enhanced oil recovery (Knight)
Enhanced oil recovery : methods, economic benefits and impacts on the environment / editors, Alicia Knight.
pages cm. -- (Energy policies, politics and prices)
Includes index.
ISBN 978-1-63463-917-0 (hardcover)
1. Secondary recovery of oil. I. Knight, Alicia, 1976- II. Title.
TN871.37.E54 2014
622'.3381--dc23

2014050205

Published by Nova Science Publishers, Inc. † New York

CONTENTS

PREFACE

Significant quantities of oil can be extracted using Enhanced Oil Recovery (EOR) methods. One of the main methods of this type is CO_2-EOR, which also has a positive impact on the environment as it results in the practically permanent storage of CO_2, the main greenhouse gas (GHG). When its use is economically viable, the technology for extracting oil using CO_2 (CO_2-EOR) enables significant quantities of oil to be obtained and at the same time, provides storage for large volumes of CO_2. It thus brings in both financial profits and environmental benefits if the oil extracted is used instead of crude oil obtained by conventional methods and does not entail any increase in consumption. This book discusses the different methods of EOR, as well as all of the economic benefits and impacts it has on the environment.

Chapter 1 - Significant quantities of oil can be extracted using Enhanced Oil Recovery (EOR) methods. One of the main methods of this type is CO_2-EOR, which also has a positive impact on the environment as it results in the practically permanent storage of CO_2, the main greenhouse gas (GHG). The demand brought on by the increase in facilities with CO_2-EOR can also entail additional benefits for CO_2-producing industries, provided that the margins are positive. In recent years oil prices have led to an increase in the number of CO_2-EOR facilities, which means that in the case of reservoirs with favorable characteristics the revenue obtained exceeds the costs incurred, e.g. those of carbon capture and storage (CCS) and transport. The cost of CO_2 is also a basic element in the economics of these facilities: It depends on whether the gas originates naturally or from industry, and on the distance to the point of injection. Moreover, in an economic analysis of investments of this type all risks must be taken into account, including those involving commodity prices. The management options available must also be considered, such as the possibility of temporarily halting oil production if prices fall below a certain threshold level but may recover, or of closing down the operation entirely if it is considered that there is little chance of making a profit in the future or when the maximum amount that can profitably be extracted has been reached.

This chapter examines the extent to which this technology is used and looks at economic and environmental issues, especially the expected profit in line with current and future market oil prices, costs and risks. The options open to facility managers are also examined from an economic viewpoint.

Chapter 2 - This commentary deals with the notable contributions to enhanced oil recovery (EOR), particularly of large chemical companies. In general, EOR encompasses a bundle of different methods for the recovery of additional amounts of original oil in place –

e.g. by injection of gas, steam, microbes or chemicals. Especially the latter method is focused on the application of polymers and surfactants and is an attractive option to improve the economics of petroleum production (chemical EOR is good for brown fields and for even green fields). As the development of a reservoir can be very challenging, an interdisciplinary team should be formed and work together to achieve the successful design and implementation of an EOR program. Such an approach allows the combination of the different kinds of needed expertise and avoids problems on the long run (e.g. lack of supply of material or undesired performance of material in the field).

The expertise, in particular, of large chemical companies experienced in polymer and surfactant production is reflected by following points:

- Selection of relevant raw materials, which are available in the desired amounts.
- Secure a reliable feedstock supply by either being backward integrated or by having long term supply contracts in place.
- Capabilities to scale-up manufacturing from lab to industrial scale production without changing product quality (from couple of 100 kg for single-well test to 100000 to for full field implementation).
- Utilize state of the art production processes to ensure cost-effectiveness.
- Leverage global production footprint to guarantee supply reliability and cost effectiveness.
- Benefit from a global supply chain network and expertise to supply large quantities of chemicals even to remote locations.
- Manage regulation needs (e.g. REACH).
- Contribute own R&D resources to develop new compounds with an improved performance profile or to identify more cost-effective alternative solutions.
- Another important factor, which should not be neglected, is the financial strength of a global chemical company. As a full field implementation can last several years and requires significant amounts of chemicals, a strong chemical partner can cope with investment needs to fulfill even a strong growth in chemical demands. This fact guaranties long-term supply of the needed volume of chemicals in desired quality.

Chapter 3 - Enhanced oil recovery (EOR) becomes increasingly important to the petroleum industry. SC-CO_2 has been successful applied to a large extent under some favorable reservoir conditions. In this chapter, mutual interactions between each of three different crude oils and SC-CO_2 are experimentally studied. First, the solubility of CO_2 in ultra-heavy oil is conducted to determine the viscosity reducing behavior of SC-CO_2. Second, the interfacial tension between CO_2 and crude oil is measured as a function of the temperature and pressure. Third, CO_2 core displacement experiments are conducted to investigate the production characteristics of SC-CO2 flooding. It is found that the CO_2 solubility in the ultra-heavy oil increases as the pressure increases, but the increasing amplitude becomes increasingly smaller. CO_2 has an obvious effect on reducing the viscosity of ultra-heavy oil; CO2 is able to reduce the viscosity by 99.64%. As the pressure increases, the equilibrium interfacial tension of the CO_2-crude oil system decreases. When the pressure is greater than 3.0MPa, the influence of temperature on the interfacial tension becomes increasingly obvious; as the temperature increases, the interfacial tension of CO2-crude oil becomes greater. As the pressure increases, the extent of CO_2 flooding enhancing oil recovery increases. Compared to the asphaltene content of crude oil, the content of the oil produced is little different when the injection pressure is smaller. When the injection pressure is large, the asphaltene content of the oil produced significantly decreases because precipitation of asphaltene occurs, and the precipitation adsorbs and deposits into the porous media. During the CO_2 flooding, the oil production rate will suddenly rise before gas breakthrough occurs, which is followed by a phase of rapid decline in the oil production rate. The curve of oil production rate looks similar to an inverted “V”. The maximum oil production rate can be 3.09~9.18 times that of the average oil production rate and can be 1.52~3.39 times of the gas breakthrough oil production rate. There are two conditions that can be used to predict the gas breakthrough in CO_2 flooding in this paper: (1) the oil production rate of CO_2 flooding suddenly rises and the curve of oil production rate looks similar to an inverted “V”; (2) the maximum oil production rate is over three times bigger than the average oil production rate.

Chapter 4 - This paper focuses on the polymerization and property assessment of water-soluble AM-co-AMPS polymers for enhanced oil recovery. The AM-co-AMPS polymers are synthesized by acrylamide(AM) and 2-acrylamido-2-methyl-1-propanesulfonic acid(AMPS) in aqueous solution. The synthesis is initiated by (NH4)2S2O8 and NaHSO3 redox system. The effects of initiator concentration, monomers’ concentration, AMPS/AM ratio and the polymerization temperature on the synthetic polymer viscosity are studied based on the single factor experiment. The optimal synthesis conditions of AM-co-AMPS polymer are determined according to golden section method. Property assessment is carried out for AM-co-AMPS polymers under different salinity and temperature, and results are compared with Daqing’s HPAM. The optimal synthesis conditions are initiator concentration of 3mmol%, monomers' concentration of 20%, AMPS/AM ratio of 20%:80%, and the polymerization temperature of 45.5°C. What’s more, The AM-co-AMPS polymer sample employed in the IR characterization was synthesized under the optimal condition discussed above. Then, the property of viscosity of the AM-co-AMPS polymer is carried out in distilled water and a set of formation brines at the temperature of 30°C and 90°C. The results show that under the temperature of 30°C, the viscosity of the target copolymer is 80.5mPa.s and the intrinsic viscosity is 4780mL/g. And the AM-co-AMPS polymer’s somehow ability of long term heat-resistance and salt-tolerance is observed: 1. after 32d of aging at 90°C, it maintained a viscosity higher than 20%; 2. compared with the retention values of the solution in salinity of 1%,1.5% and 2%, after 28d of aging at 90°C, this of the solution in salinity of 0.5% show the maximum retention. The AM-co-AMPS polymer has better thickening properties in the salt water. The salt resistance of AM-co-AMPS polymer has better than DaQing’s HPAM at the temperature of 30°C.

Chapter 5 - In this chapter, the author presents a principally new scientific-technological paradigm of exploitation, preservation, and rehabilitation of petroleum resources.

An oil deposit in the Earth crust is considered as a self-organizing, living fluid-saturated system capable of restoring its balance. Active works during development of an oilfield produce a strong perturbation in the near-equilibrium system and substantially deform its natural parameters. In case of a moderate perturbation, the self-organizing system restores its balance; a prolonged or intensive perturbation considerably exceeding the critical threshold will destroy the system.

Nowadays, the problem of enhanced oil recovery (EOR) and improved oil recovery (IOR) is a key problem in petroleum theory and practice. In world oil-and-gas practice, two different terms are in use: EOR (enhanced oil recovery) meaning intensive, forcible methods; and IOR (improved oil recovery) – advanced and moderate methods.

Chapter 6 - Bottom-water condition has been considered as a challenging issue for implementing Enhanced Oil Recovery (EOR) techniques. In the application of polymer flood which is known as one of effective EOR methods, the existence of strong bottom aquifer leads to severe polymer loss to the aquifer, water fingering, and water production. Because these factors affect economics of the polymer flooding operation, injection design of the polymer flooding should be optimized to maximize profit. This study aims for the evaluation of polymer flooding under bottom water condition considering economic feasibility. Quantitative comparisons considering oil recovery, water production, and economics of polymer injection were made to obtain maximum net present value. The results indicated that oil recovery was increased 5% and water breakthrough was delayed about 40 days to maximize the net present value by 40%. After the optimization of injection designs, impacts of geological uncertainties on net present value were analyzed. According to sensitivity analysis, vertical permeability had most negative impact on the net present value. This study showed that injection of polymers under bottom water condition could be optimized to be an effective EOR technique for maximizing profit.

Chapter 7 - To overcome reduced oil recovery resulting from poor mobility ratio and sweep efficiency by the low viscosity of CO_2, water-alternating-gas (WAG) process has been commonly used as an advantageous alternative. This study examines how to optimize WAG processes in heterogeneous reservoirs by rigorous simulation of WAG injection. Compositional simulations of areally heterogeneous reservoirs were used to quantify the effects of process parameters such as WAG ratio and CO_2 slug size, well parameters including well length and type, and reservoir heterogeneity characteristics such as Dykstra-Parsons' coefficient and correlation lengths on the oil recovery and economic value. Optimization was carried out with net present value (NPV) as an objective function. The main conclusions of this study show that more WAG cycles are required for maximum NPV at a given operation period. Higher water fraction in WAG injection is needed to improve sweep efficiency in more heterogeneous reservoirs. The longer horizontal wells maximize NPV as their orientation and permeability correlation direction is parallel. The proposed methodology has significant improvement on the oil recovery up to 43% and hence yields higher NPV up to 47%. The importance of using the reservoir simulation as a tool for optimum design before implementing WAG injection in practice is evidenced. It also demonstrated that the approach is viable to give the optimistic scenario of oil recovery prediction under WAG injection into heterogeneous reservoirs.

Chapter 8 - In recent years, there has been an increasing interest in modeling miscible flow in petroleum reservoirs. Besides, there are several compositional simulators in order to estimate fluid variables at each time step, however this paper attempts to acquire a simple numerical method based on the compositional simulation. Therefore a two-dimensional, three-phase numerical model reservoirs with formulation details, numerical solution method, and computational results was described for simulation of VAPEX process in heavy oil. The model included the effects of three-phase relative permeability and pressure-dependent fluid properties and also interphase mass transfer of solvent in heavy oil was considered. The three-phase mass balance equations were solved sequentially using finite difference method. Some

VAPEX laboratory data were used for studying the accuracy of this model. Comprehensive and comparative studies together with extensive sensitivity analysis from 0.0178 to 0.00178 cc/s as solvent injection rates were conducted to specify the dependency of the oil recovery, breakthrough time and SOR to injection rate. The simulation results indicated that high injection rates decrease the breakthrough time and lead to more SOR, which is unacceptable from the economical point of view. Therefore, an interesting plane was suggested to control the solvent waste with a high oil recovery.

Chapter 9 - Spontaneous imbibition of water into the matrix blocks is an important mechanism for crude oil recovery from fractured reservoirs. This paper presents a numerical modeling of imbibition phenomenon when water is imbibed by capillary forces into an oil-saturated cylindrical core for studying the effect of boundary condition and shape factor on the spontaneous imbibition. In this modeling work, the authors assume that imbibition is a diffusion process and gravity forces are neglected. They also consider various boundary conditions. Finite difference implicit method is used to solve the spontaneous imbibition equation. Accuracy of the modeling is investigated by comparing the modeling results and the experimental data. Then, imbibition in various directions (r and z) is studied. By using the proposed model, the most effective boundary conditions for achieving the maximum oil recovery can be determined.

Chapter 10 - The reduction of heavy oil viscosity is the most important issue in enhanced heavy oil recovery. The viscosity reduction should be achieved in a quick economical way. In this communication, the effect of different concentrations of various metal oxide including Fe_2O_3, NiO, CuO, ZnO, Al_2O_3, TiO_2 and WO_3 at various temperatures is studied on the heavy oil viscosity. The experimental tests show that some of these nanoparticles decrease the heavy oil viscosity less than 50% in certain concentrations at different temperatures. The reason of this viscosity reduction is that some of the nanoparticles act as a catalyst. The authors' study shows that NiO, Fe_2O_3 are the most effective nanoparticles for viscosity reduction.

Chapter 11 - High potential of enhanced oil recovery (EOR) methods to progressively expand hydrocarbon exploitation makes the decisions on the applicability of certain recovery techniques for Iranian oil reservoirs, which have reached the late stage of their life. Due to high operational costs and high technical sensitivity of these methods particularly in naturally fractured reservoirs because of complexity, natural heterogeneity and non-uniformity, studies are required for selection of the most appropriate EOR method. In this communication, a comprehensive EOR screening for 12 EOR methods by criteria definition based on numerous reported successful EOR projects and presented criteria tables by some previous researchers have been carried out in an Iranian oil reservoir. In addition, for showing the importance of EOR screening approach during selection process of EOR methods, a sensitivity analysis was conducted. The results revealed that polymer, CO_2 miscible and immiscible flooding can be considered technically the most appropriate methods for applying in the Iranian naturally fractured carbonate reservoirs.

Chapter 12 - In this communication, the authors simulate the cyclic steam stimulation (CSS) oil recovery method in the fractured carbonate reservoirs containing heavy-oil and also analyze the influences and contribution of different reservoir properties on ultimate oil recovery and cumulative Steam Oil Ratio (cSOR). Kuh-e Mond field, located in the south western of Iran, is considered as a specific case. The results indicate three different stages of oil production in CSS process in Kuh-e Mond NFR. First, oil production immediately increases due to fast-heating of oil by abrupt injection of steam; second, due to more heating by steam escape through fractures drainage of matrix blocks begins and a reduction in oil rate occurs; third, steam penetrates into upper fractures and therefore a pulse in oil production rate seems. The conclusions indicate that CSS performance within Kuh-e Mond reservoir is impacted by the fracture geometrical properties and how the fluid flow through matrix blocks of Kuh-e Mond reservoir in which permeability, porosity and rock wettability of matrix.

Chapter 13 - Reservoir permeability plays a key role in determining the quality of reservoir for efficient recovery of oil, in particular during enhance oil recovery (EOR) methods. The permeability can be obtained from experimental measurements and or by in-place testing. Both experimental measurement and well test analysis are expensive and time-consuming. In order to find the impact of well log data on permeability, a more powerful method than the traditional ones is necessary. As a result, artificial intelligence technique is one of the most important and efficient soft-computing approaches to approximate any function with finite number of discontinuities by learning the relationships between input and output vectors. In this communication, several nets with various inputs and hidden neurons are developed, and then the best net with optimum input parameters and hidden neuron were selected in order to investigate the effect of input parameters on permeability. Afterward, to confirm the obtained results and study the impact of well log data on permeability, a sensitivity analysis was conducted using sensitivity about the mean (SAM) feature. The results show that permeability has lowest sensitivity to parameter of thermal neutron apparent porosity (NPHI). On the other hand, parameter of corrected log porosity has the highest effect on estimation process of permeability.

In: Enhanced Oil Recovery
Editor: Alicia Knight

ISBN: 978-1-63463-917-0

Chapter 1

ENHANCED OIL RECOVERY: GROWTH, ECONOMIC AND ENVIRONMENTAL BENEFITS AND RISKS

Luis M. Abadie*
Basque Centre for Climate Change, Bilbao, Spain

ABSTRACT

Significant quantities of oil can be extracted using Enhanced Oil Recovery (EOR) methods. One of the main methods of this type is CO_2-EOR, which also has a positive impact on the environment as it results in the practically permanent storage of CO_2, the main greenhouse gas (GHG). The demand brought on by the increase in facilities with CO_2-EOR can also entail additional benefits for CO_2-producing industries, provided that the margins are positive. In recent years oil prices have led to an increase in the number of CO_2-EOR facilities, which means that in the case of reservoirs with favorable characteristics the revenue obtained exceeds the costs incurred, e.g. those of carbon capture and storage (CCS) and transport. The cost of CO_2 is also a basic element in the economics of these facilities: It depends on whether the gas originates naturally or from industry, and on the distance to the point of injection. Moreover, in an economic analysis of investments of this type all risks must be taken into account, including those involving commodity prices. The management options available must also be considered, such as the possibility of temporarily halting oil production if prices fall below a certain threshold level but may recover, or of closing down the operation entirely if it is considered that there is little chance of making a profit in the future or when the maximum amount that can profitably be extracted has been reached.

This chapter examines the extent to which this technology is used and looks at economic and environmental issues, especially the expected profit in line with current and future market oil prices, costs and risks. The options open to facility managers are also examined from an economic viewpoint.

Keywords: EOR, Carbon, uncertainty, investments valuation, real options, stochastic models

---

* Corresponding author: Luis M. Abadie. Basque Centre for Climate Change (BC3), Alameda Urquijo 4, 4°, 48008, Bilbao, Spain. Tel: +34 94-401 46 90. E-mail: lm.abadie@bc3research.org.

1. INTRODUCTION

When its use is economically viable, the technology for extracting oil using CO_2 (CO_2-EOR) enables significant quantities of oil to be obtained and at the same time provides storage for large volumes of CO_2 It thus brings in both financial profits and environmental benefits if the oil extracted is used instead of crude oil obtained by conventional methods and does not entail any increase in consumption.

If the environmental benefits involved are recognized at least in part, for instance by means of an emission allowance scheme or taxation on emissions, or if some kind of state support is provided[1], companies can factor that positive monetary impact into their calculations. An environmental impact that has a positive effect on corporate profits can be seen either as stochastic, as in the case of the European Union Emissions Trading System (EU ETS), or deterministic if the benefit is proportional to the quantity of crude oil extracted or the quantity of CO_2 injected, assuming that the quantities involved can be determined with a high level of certainty.

The viability of CO_2-EOR technology depends strongly on current and future oil prices and on the cost of the CO_2 used. These costs can be broken down into Capture, Transport[2] and Storage, plus possibly a mark-up for the company that sells the CO_2 if it is not the same company that produces oil using CO_2-EOR technology. Capture costs depend basically on whether the gas is from natural or industrial sources[3]; if it is industrial they depend also on the type of industry where it is produced (hydrogen production, natural gas processing, synthetic natural gas, oil refining, power generation, iron and steel production, etc.). Pre-combustion capture is cheaper than post-combustion capture: For instance the Kemper County IGCC Project uses pre-combustion capture, because the type of technology used in an integrated gasification combined cycle (IGCC) is suited to pre-combustion capture. By contrast, the Boundary Dam Integrated Carbon Capture and Sequestration Demonstration Project, which is fired by lignite and brown coal, uses post-combustion capture.

Moreover, for CO_2-EOR technology to be used there must also be pipelines linking the CO_2 sources with the CO_2-EOR facilities, so enough profit needs to be made the oil extracted to cover the cost of constructing and maintaining such pipelines. The type of reservoir (on-shore or off-shore) also impacts the cost.

All these factors mean that margins and risks vary widely from one project to another. Another point to be taken into account is the geological characteristics of the reservoir, which must be suited to the use of this technology. It is geological characteristics that determine, albeit with some degree of uncertainty, what type of technology can be used for extraction (e.g. thermal, CO_2-EOR or other gas injection technologies) and how many barrels of oil can be obtained per tonne of CO_2 injected.

CO_2-EOR technology is more likely to be used when there are mature oil fields and, in the US, when there is cheap, natural-source CO_2 available.

[1] In Texas there are 11,000 permitted CO_2 wells. The state encourages EOR by offering a 50% reduction in production tax.

[2] Transport costs depend strongly on the distance between the CO_2 source and the CO_2-EOR project.

[3] Greater development of this technology in the US would require the use of mainly industrial CO_2 from sources such as industrial facilities, chemical complexes and electric power plants, as only a modest increase in CO_2 from natural sources can be expected.

In the case of on-shore oilfields CO_2-EOR technology has been in use for over 40 years in the US, though it is always being further developed.

The rest of the chapter is structured as follows: Section 2 provides an overview of the current state of the art of this technology in the US and in Large-Scale Integrated Projects (LSIPs) elsewhere in the world. Section 3 reviews the existing literature on the financial aspects of CO_2-EOR. Section 4 analyzes the spot and futures markets for crude oil and carbon, and looks briefly at the stochastic behavior of prices. Section 5 tackles the relevant points for the valuation of a working facility. Section 6 deals with investment risk. Section 7 covers the main points for the valuation of real options in a CO_2-EOR facility. Section 8 concludes.

2. Oil Production from CO_2-EOR

Advances in enhanced oil recovery technology using CO_2 (CO_2-EOR) and high crude oil prices have created new opportunities for CO_2 sequestration. Figure 1 shows the recent trend in crude oil prices and production. There is also more CO_2 available from both natural and industrial sources, e.g. the CO_2 emitted by a fertilizer plant. The development of this technology has also been favored by the fact that the network of pipelines in place has grown over time.

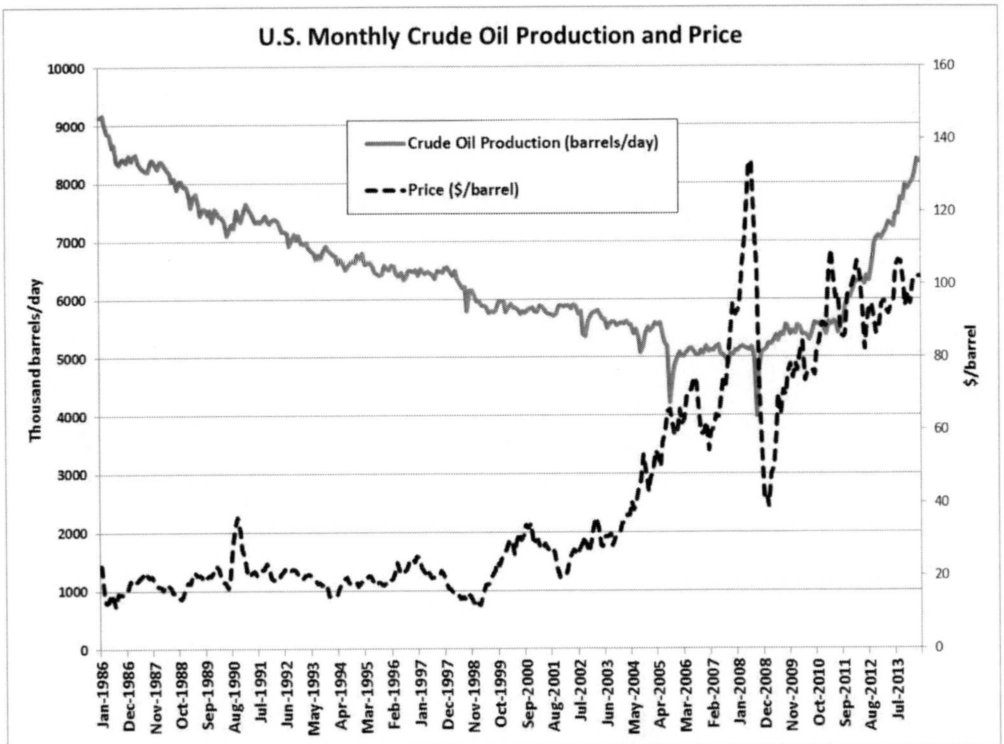

Figure 1. US Monthly Crude Oil Production and Price (graph based on EIA data).

The main obstacle to the implementation of CO_2 sequestration technologies is their high cost.

Table 1 shows that in 2014 a total of 136 CO_2-EOR projects are injecting 68 million metric tons (MMt) per year of CO2 for enhanced oil recovery in the United States, and producing 299,609 barrels of crude oil per day (Kruuskraa and Wallace, 2014) [21]. With an annual figure of 109.36 million barrels, it can be deduced that the average output is 1.61 barrels/tCO_2. Logically, the more barrels per tCO_2 are produced the more financially favorable investment in a CO_2-EOR project is. The 2013 US field production of crude oil is 7,451 thousand barrels/day (EIA) [9], so production from EOR accounts for around 4% of total US crude oil production, which ultimately is a rather low figure. CO_2-EOR currently accounts for 41.63% of total EOR production, and EOR production accounts for 9.6% of total US oil production. As shown in Figure 2, the weight of CO_2-EOR within EOR production as a whole has increased: In 1992 it was 19.07% and it currently stands at 41.63%. This increase has been achieved mainly at the expense of thermal technologies, which have decreased accordingly.

Kruuskraa and Wallace, 2014 [21] estimate that, based on the increased volumes of CO_2 available and the announcement of CO_2-EOR projects, production is likely to double to 638,000 barrels/day by 2020. However these estimates are subject to high levels of uncertainty and could be affected by changes in the price of oil, the cost of CO_2 and new developments in CO_2-EOR technology.

It is these elements that this chapter examines. CO_2-EOR technology development also depends strongly on the completion of major CO_2 pipelines.

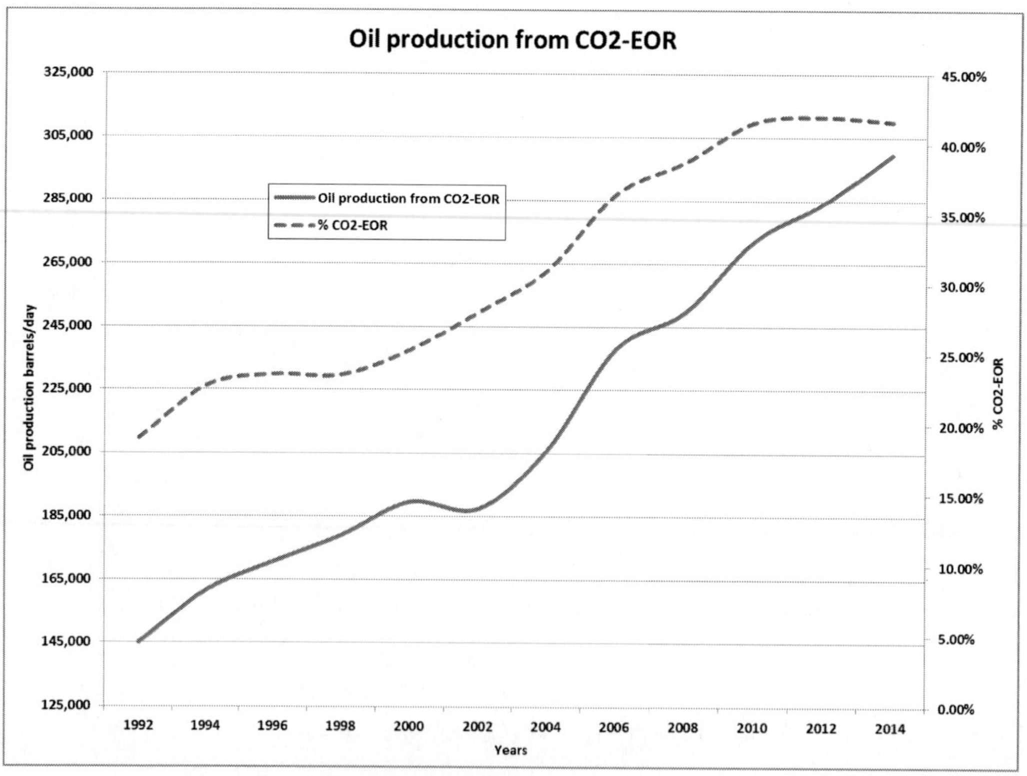

Figure 2. Oil production from CO_2-EOR (based on data from Kuuskraa and Wallace (2014)).

Table 1. US 2014 CO_2-EOR Production

Oil Production		
Active projects	Oil production b/d	
136	299,609	
CO_2 supplies		
Type	Number	CO_2 supply (million tonnes/year)
Natural	5	54.4
Industrial	12	13.6
Total	17	68.0

In the US there are currently 17 sources of CO_2 in use (5 natural and 12 industrial). The total of 68 million tonnes of CO_2 breaks down into 54.4 from natural sources and 13.6 from industrial sources. In 2012, US greenhouse gas emissions totaled 6,526 million metric tonnes of carbon dioxide equivalent (EPA) [10], so although the 68 million used in CO_2-EOR is a large amount it actually accounts for little more than 1% of the total for the country.

According to the Global CCS Institute (Status of CCS Database 07/11/2014 [14]), as of 11 July 2014 there are 16 large-scale integrated projects (LSIPs) using CO_2-EOR around the world that are classed as currently operational or in execution.

Large-scale integrated CCS projects (LSIPs) are defined as those which involve the capture, transport and storage of CO_2 at a scale of:

- at least 800,000 tonnes of CO_2 per annum for coal-based power plants; or
- at least 400,000 tonnes of CO_2 per annum for other emission-intensive industrial facilities (including natural gas-based power generation).

Between them these 16 projects are capable of sequestering around 32.8 million tonnes per annum (mtpa).

As can be seen in Table 2, nine of these projects are located in the US, where the an average volume of CO_2 is 24.7 mtpa, and four in Canada with an average of 5.8 mtpa. There are no CO_2-EOR projects in Europe that can be classed as LSIPs.

The average volume of CO_2 has increased gradually over time as shown in Figure 3.

The electric power industry is capable of producing huge quantities of CO_2 that could be used for CO_2-EOR, but there are only two LSIPs currently at the "in execution" stage. These are the Boundary Dam Integrated Carbon Capture and Sequestration Demonstration Project and the Kemper County IGCC Project, which are due to come on line in 2014 and 2015. Between them they have a CO_2 capture capacity of 4 mtpa.

3. Literature Review

There is a substantial body of literature on EOR methods. Some publications review the status of the methods used while others assess the investments required, using the Net Present Value (NPV) criterion or a valuation under uncertainty using the Real Option Approach. Some of these papers are described briefly below.

Table 2. Large-scale integrated CCS projects (LSIPs)

Project	Average Volume CO_2 (mtpa)	Country
Air Products Steam Methane Reformer EOR Project	1.0	UNITED STATES
Century Plant	8.4	UNITED STATES
Coffeyville Gasification Plant	1.0	UNITED STATES
Enid Fertilizer CO_2-EOR Project	0.7	UNITED STATES
Great Plains Synfuel Plant and Weyburn-Midale Project	3.0	CANADA
Lost Cabin Gas Plant	0.9	UNITED STATES
Petrobras Lula Oil Field CCS Project	0.7	BRAZIL
Shute Creek Gas Processing Facility	7.0	UNITED STATES
Val Verde Natural Gas Plants	1.3	UNITED STATES
Alberta Carbon Trunk Line ("ACTL") with Agrium CO_2 Stream	0.5	CANADA
Alberta Carbon Trunk Line ("ACTL") with North West Sturgeon Refinery CO_2 Stream	1.3	CANADA
Boundary Dam Integrated Carbon Capture and Sequestration Demonstration Project	1	CANADA
Kemper County IGCC Project	3	UNITED STATES
Uthmaniyah CO_2 EOR Demonstration Project	0.8	SAUDI ARABIA
Abu Dhabi CCS Project (formerly ESI CCS Project)	0.8	UNITED ARAB EMIRATES
Petra Nova Carbon Capture Project (formerly NRG Energy Parish CCS Project)	1.4	UNITED STATES

Source: Data gathered from the Status of CCS Project Database (Global CCS Institute).

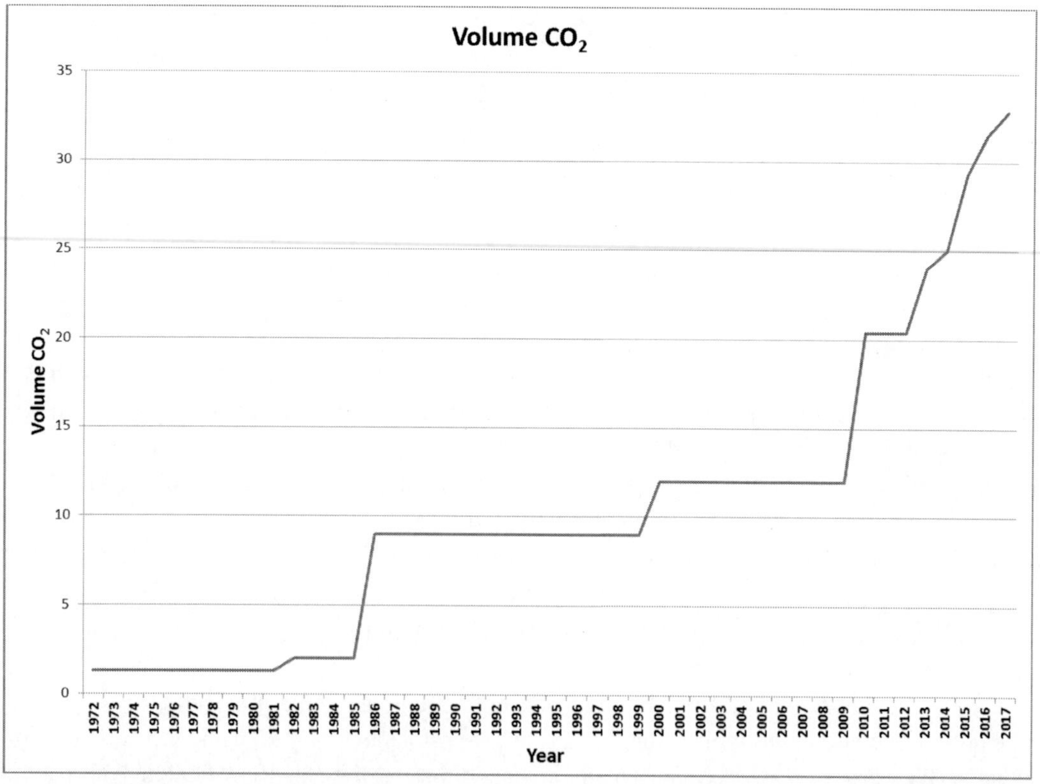

Figure 3. LSIPs and their average CO_2 volume.

Fleten at al. (2010) [11] develop a stochastic dynamic program to decide whether investment in a CO_2 value chain is profitable. In their paper oil and CO_2 prices are modeled as Geometric Brownian Motion (GBM) processes. They use a time frame of 25 years and solve a case study of an investement on the Norwegian continental shelf.

Abadie and Chamorro (2008) [2] examine an option for investing in CCS for a coal-fired power plant, and derive the optimal investment rule. In their paper CO_2 prices are modeled as GBM processes while the electricity price is modeled as a mean-reverting process. They show that uncertainty has a significant effect on investment decisions, and that a high CO_2 price is needed investment in CCS to be worthwhile, so the best option is to wait and see whether prices change favorably.

Heydari et al. (2010) [16] study the option to retrofit a coal-fired power plant with either full CCS (FCCS) or partial CCS (PCCS) retrofits given uncertain electricity, CO_2, and coal prices. The coal and allowance prices are modeled as GBM processes. In this case the authors also conclude that with current market prices the current best option is to wait.

Gaspar et al. (2009) [13] propose a technical and economic model for underground storage of CO_2 emitted by a fertilizer plant in North Eastern Brazil in a hypothetical mature oil reservoir via CO_2-EOR. They use a cash flow model. They find that the feasibility of the project is highly sensitive to oil price, oil production, and CAPEX, and that there would be a contribution from the mitigation of greenhouse gases (GHG) in the form of the sequestration of a significant amount of CO_2 in the reservoir, where it can remain for thousands of years.

Abadie et al. (2014) [3] assess two alternative CCS technologies, one of which uses CO_2 for enhanced oil recovery (EOR) paired with storage in deep saline formations (DSF). They find three sources of risk: the prices of oil, coal and carbon. The numerical method applied is the 3D binomial tree, which is used to assess the optimum time to invest in a technology that operates inflexibly once the investment is decided.

Klokk et al. (2010) [20] present a mathematical model using a net present value criterion for designing a carbon dioxide value chain when CO_2 is deposited in oil reservoirs and it can sometimes be used to achieve additional oil production, i.e. enhanced oil recovery (EOR). The model determines an optimal CO_2 value chain from a fixed set of CO_2 emission points and a set of potential injection sites. It designs a transport network and chooses the best suited oil fields with EOR potential or other geological formations for storage. The model is illustrated by a case study from Norway.

Alvarado and Manrique (2010) [5] review EOR status and opportunities to increase final recovery factors in reservoirs ranging from extra heavy oil to gas condensate. They find that thermal methods, specifically steam injection, still predominate as the preferred EOR methods for heavy oil reservoirs, but that CO_2 injection has become increasingly attractive as an EOR method and potentially as a sequestration strategy in recent years.

Duncan et al. (2009) [7] seek to assess the potential health and safety risks associated with large scale CO_2 sequestration in deep brine. Their paper analyzes the main business risks facing a company engaged in geological sequestration, including the operational risks of capturing, compressing, transporting and injecting CO_2 as well as the risk of accidental CO_2 blowouts from wells. They find that in the case of blowouts, incidents related to CO_2 production wells from natural reservoirs and incidents occurring above production wells resulting from unexpectedly early CO_2 breakthroughs are not directly relevant to understanding the risk of CO_2 sequestration in deep brine reservoirs. These authors assert that the CO_2-EOR industry has a good safety record.

Echevarria, Huaman et al. (2014) [8] review the prospects for development of carbon capture and storage (CCS) technologies to reduce future carbon emissions. Their paper shows the barriers to technology deployment, sets out strategies for accelerating it and lists the stages involved. They analyze the Large-scale integrated CCS projects (LSIPs).

Ferguson et al. (2009) [12] examine domestic oil resources amenable to CO_2-EOR, the size of the related market for CO_2, and the benefits to the power sector from CO_2 sales to the EOR industry. The study finds that, depending on future oil prices and the cost of purchasing CO_2 from power plants and other industrial sources, from 39 to 48 billion barrels of oil could be economically recoverable with CO_2-EOR.

Hammond et al. (2011) [15] analyze six options based on pulverized coal, natural gas combined cycle, and integrated (coal) gasification combined cycle power stations for carbon capture and storage (CCS) facilities coupled to power plants.

According to Herzog (2011) [17] there are significant challenges to be addressed in scaling CCS up from today's megaton level to the gigaton level that it needs to reach if it is to help mitigate global climate change. In this author's opinion, the two areas of biggest concern are storage capacity and costs. It has not yet been demonstrated that enough storage capacity exists to support CCS at the gigaton scale, and the cost of CCS mitigation may be greater than is politically acceptable for the next couple of decades.

Jaramillo et al. (2009) [19] analyze the net life cycle of CO_2 emissions in an EOR system.

Muggeridge et al. (2014) [22] summarize the breadth of EOR processes, the history of their application, and their current status. They discuss some examples of existing EOR projects in the mature oil province of the North Sea.

Thomas (2008) [23] examines thermal and non-thermal oil recovery methods. He finds that nearly $2.0 \cdot 10^{12}$ barrels of conventional oil and $5.0 \cdot 10^{12}$ barrels of heavy oil will remain in reservoirs worldwide after conventional recovery methods have been exhausted. Much of this oil could be recovered by Enhanced Oil Recovery (EOR) methods.

Finally, a method for valuation under uncertainty is described in Dixit and Pindyck (1994) [6], and application to energy assets using the Real Option Approach is examined in Abadie (2009) [4] and Abadie and Chamorro (2013) [1].

4. Commodity Futures and Spot Markets

Quotes on futures markets provide highly important information: given that they are certain prices with no counterparty risk, they can be used to obtain the Net Present Value of certain quantities of commodities to be delivered in the future merely by discounting the prices traded at the riskless rate. The price of government debt with a given maturity is usually taken for calculating this riskless rate. For example, for the US case the market yield on US Treasury bonds with 20-year constant maturity can be used, the value of which was 0.029 as of 08/22/14[4].

These markets can also provide a great deal of information useful for assessing whether prices are performing according to certain stochastic processes and deciding, when relevant, between geometric Brownian motion (GBM) and mean-reverting processes. Quotes on these markets enable the parameters of the deterministic part of the differential equation to be

[4] www.federalreserve.org/releases/h15/data.htm accessed on 08/30/2014.

calculated (in the risk-neutral world), while volatility can be calculated by using the quotes for commodities on the spot market.

They also help to estimate the real future behavior of prices, though such estimates are usually not very reliable. The risk premium at time t for a contract with maturity T, $RP(t,T)$ is defined as the difference between the corresponding future price and the expected value in the real world, i.e. $RP(t,T)=F(t,T)-E_t(S_T)$.

The most complicated part is usually determining the risk premium, without which there is no reliable way of determining $E_t(S_T)= F(t,T)- RP(t,T)$. As mentioned, however, for purposes of assessment $E_t(S_T)$ does not need to be known: it suffices to use $F(t,T)e^{-(T-t)r}$ for a single flow, where r is the riskless rate. Considering that the riskless rate has a time structure that is significant for valuation purposes, it is possible to discount an interest rate as a function of time $r=f(T-t)$, and even if the volatility is believed to be significant the riskless rate can be modeled as stochastic, though this complicates the valuation process.

In the case of CO_2-EOR there are markets that provide a great deal of information on crude oil and carbon commodities.

Specifically, they provide a lot of information on the value of the parameters of the deterministic side of the differential equation, always in the risk-neutral world, which is the world applicable to the futures market and the one which is relevant for valuation.

For information on the financial aspects of futures markets see Hull (2011) [18].

In the case of futures markets with GBM-type stochastic processes the value of the drift $\alpha-\lambda$ [5] can be easily and accurately estimated.

The same goes for $\frac{kS_m}{k+\lambda}$ and $k+\lambda$ [6] in the case of the IGBM model. In the mean-reverting case these markets provide information on the long-term equilibrium value and the rate at which the expected price approaches that value in the riskless world (futures market).

4.1. T-West Texas Intermediate Light Sweet Crude Future

Figure 4 shows the prices for T-West Texas Intermediate Light Sweet Crude Futures and ICE Brent Futures on 10/22/2014.

From an examination of Figure 4 it is easy to deduce that a mean reverting process is better suited to representing the stochastic behavior of crude oil prices[7]. These prices and their volatility are among the most significant elements for valuing investment under uncertainty.

In the case of an Inhomogeneous Geometric Brownian Motion (IGBM) mean-reverting model such as the one described in Abadie and Chamorro (2013) [1] the parameters of the model can be estimated using equation (1):

$$F(t,T_2)=\frac{kS_m}{k+\lambda}+[F(t,T_1)-\frac{kS_m}{k+\lambda}]e^{-(k+\lambda)(T_2-T_1)} \tag{1}$$

[5] The components are the drift in the real world α and the market price of risk λ. But only α-λ needs to be known.
[6] Where k is the reversion rate in the real world, S_m is the long-term equilibrium value also in the real world.
[7] The statistical demonstration of this lies outside the scope of this chapter.

where it is really only necessary to estimate two parameters: the long-term reversion value of the futures market $\frac{kS_m}{k+\lambda}$ and the reversion rate $k+\lambda$ such that the sum of the squares of the differences between the real and estimated furure is minimised for each date, and the following values are obtained for that day: $\frac{kS_m}{k+\lambda} = 87.32$ and $k+\lambda = 0.55$[8].

This calculation is based on the future with the shortest maturity T_1 (Oct. 13), which has a price of $93.65/barrel on that date. As can be seen in Figure 5, the fit using the values of the two parameters is very good.

In accordance with the model used, taking $t=0$ as the starting point the current value per barrel extracted per annum between τ_1 and τ_2 is:

$$V(\tau_1,\tau_2) = \int_{\tau_1}^{\tau_2} F(0,t)e^{-rt}dt = \frac{kS_m}{r(\lambda+k)}[e^{-r\tau_1} - e^{-r\tau_2}] + \frac{S_0 - \frac{kS_m}{\lambda+k}}{\lambda+k+r}[e^{-(\lambda+k+r)\tau_1} - e^{-(\lambda+k+r)\tau_2}] \qquad (2)$$

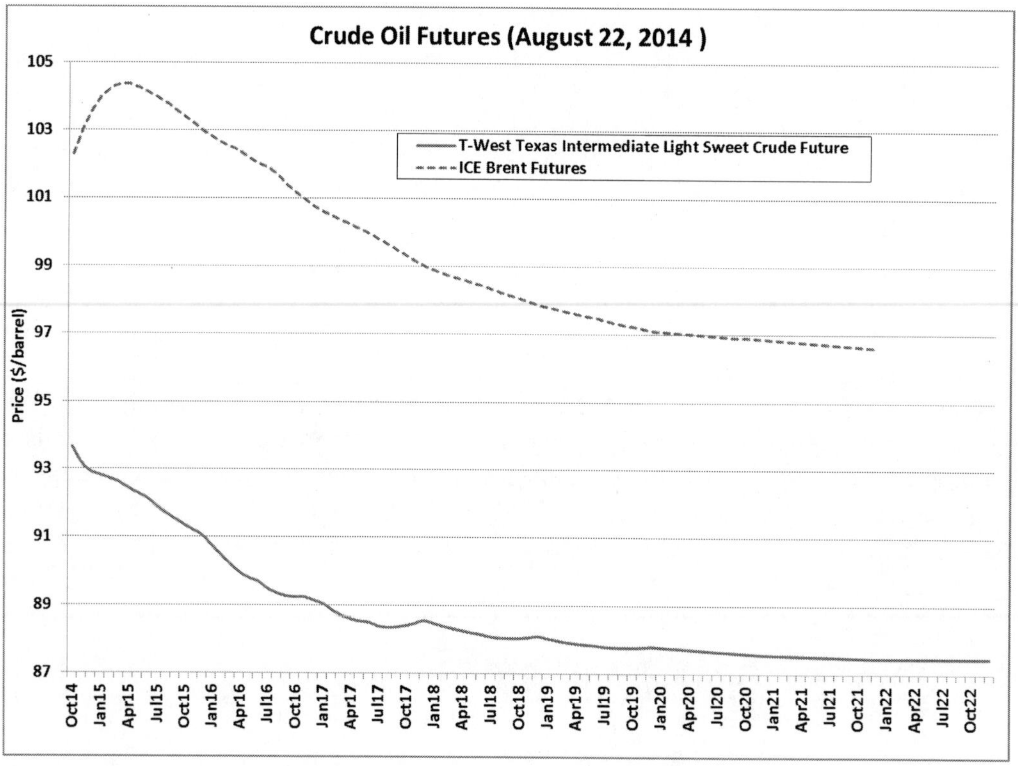

Figure 4. Crude Oil Futures Quotes 08/22/2014.

[8] Quotes for several days could also have been used.

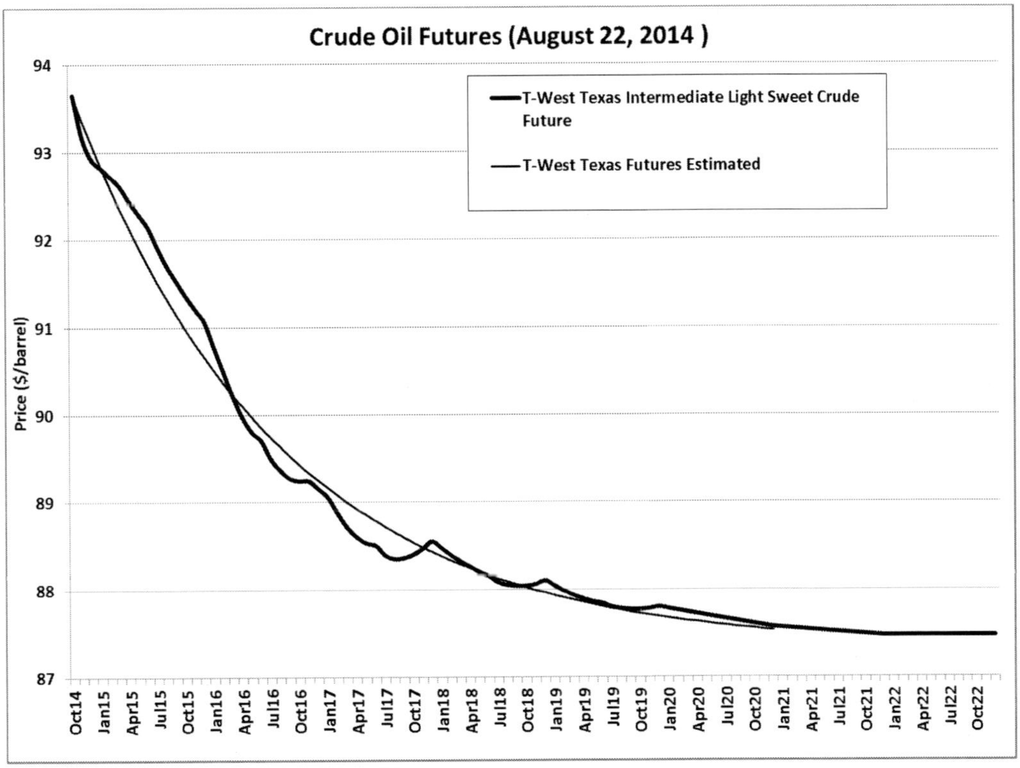

Figure 5. Crude Oil Futures Quotes 08/22/2014 and ajusted IGBM Model.

Constant extraction of a barrels of oil per annum between τ_1 and τ_2 gives a current income of $aV(\tau_1, \tau_2)$. This valuation is consistent with market quotes.

To analyze risks and use the real option approach, volatility also has to be estimated. Using spot prices for EIA from the beginning of 2010 to 08/25/2014, a volatility figure of 0.2653 is obtained for crude oil.

4.2. ICE EUA Futures

Figure 6 shows the quotes for ICE EUA futures on 08/22/2014. It is immediately apparent that the prices quoted are compatible with a GBM type model, given that the price increases exponentially as the maturity period *(T-t)* of the futures contract increases. These quotes are significant for investment in CO_2-EOR only if the quantities of CO_2 stored give rise to emission allowance profits.

The curve of the futures market for the GBM model is given by equation (3)

$$F(t,T) = A_t e^{(\alpha-\lambda)(T-t)} \tag{3}$$

using a procedure similar to the case of IGBM, a value of $\alpha - \lambda = 0.03$ can be obtained. This figure is very close to the riskless rate r=0.029.

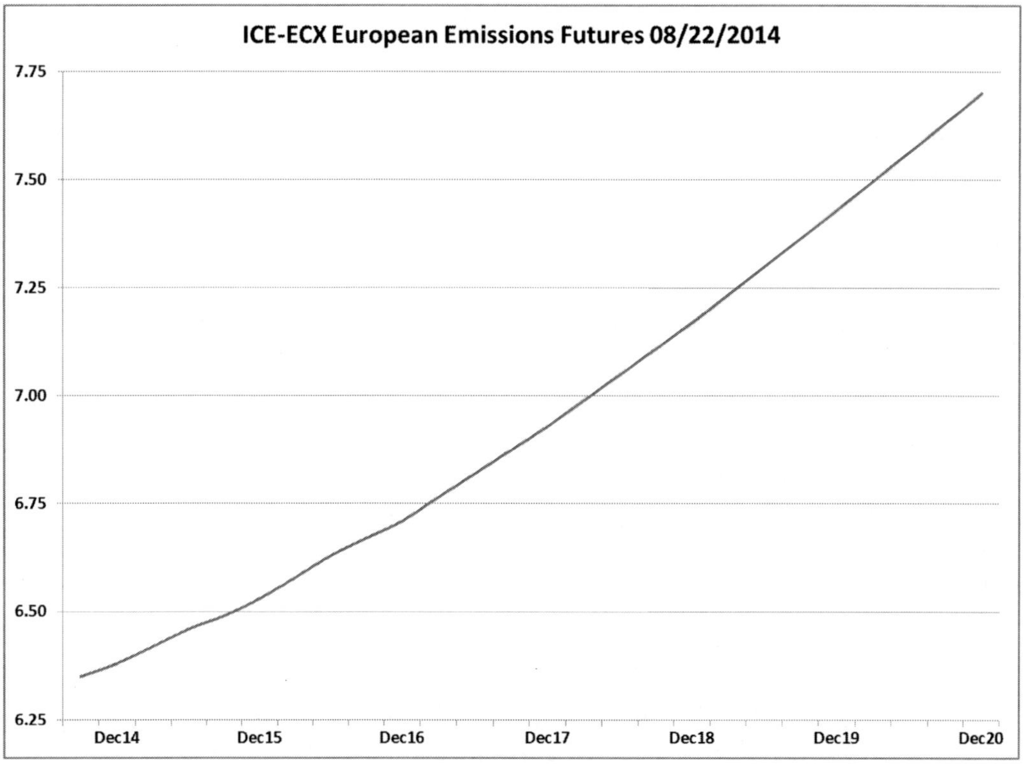

Figure 6. ICE-ECX European Emission Futures 08/22/2014.

As can be seen, the price of emission allowances is currently very low and provides scant incentive for investment to reduce GHG emissions.

Volatility rates can be calculated using spot prices. For 2014 (up to 08/25/2014) the figure obtained is 0.8662, which is very high. Prices are very low but very volatile. Figure 7 shows the behavior of emission allowance prices in the first part of 2014.

5. Net Present Value of Investment

Using risk-neutral valuation, the method used for futures market quotes, it is possible to obtain a Net Present Value (NPV) even under condition of uncertainty, provided that the uncertainty lies in commodities that are quoted on futures markets and that those markets have sufficient liquidity and enough quotes with long maturity periods.

The Net Present Value (NPV) of investment in a CO_2-EOR facility can be calculated using the components of equation (4).

$$NPV = NPV_O + NPV_A - NPV_C - I \tag{4}$$

where NPV_O is the profit from the sale of the crude oil extracted, NPV_A is the NPV of the potential financial profit for the company from the sequestration of CO_2, NPV_C is the NPV for the costs of CO_2 and I is the current amount of the remaining costs, both fixed and variable.

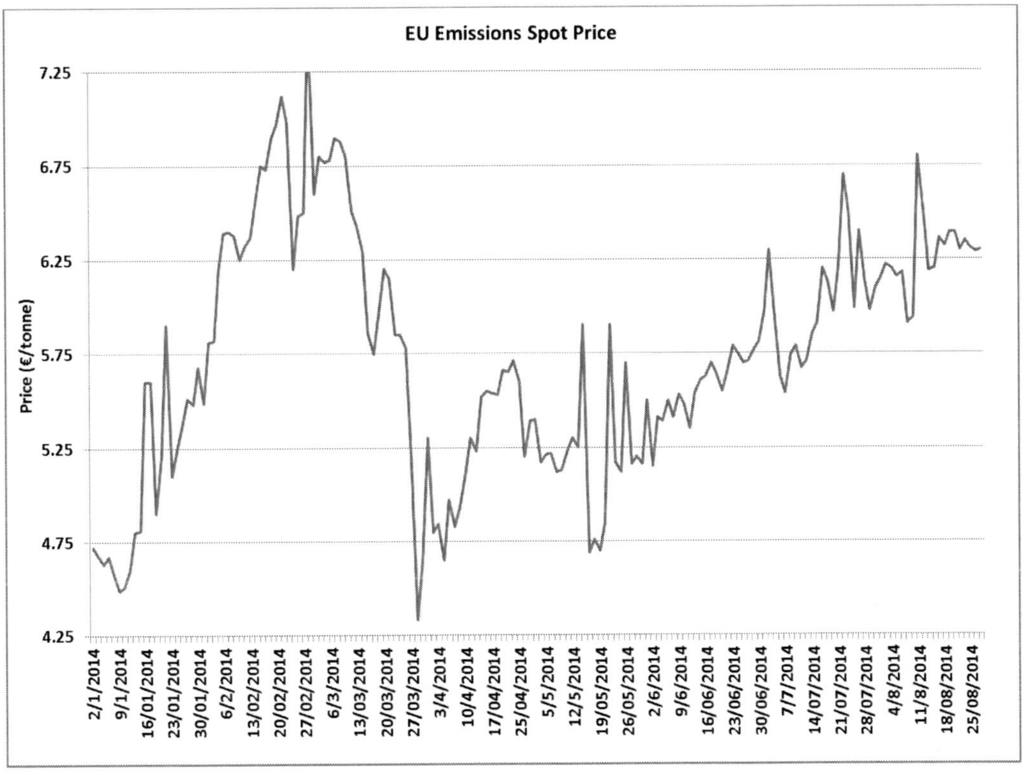

Figure 7. European Emission Spot Prices.

Based on the futures equation (5) used for crude oil:

$$F(0,t) = \frac{kS_m}{k+\lambda} + [S_0 - \frac{kS_m}{k+\lambda}]e^{-(k+\lambda)t} \quad (5)$$

If the quantity of crude oil extracted is deterministic $q(t)$, the NPV of income from crude oil can be calculated by using equation (6).

$$NPV_O = \int_{\tau_1}^{\tau_2} q(t)F(0,t)e^{-rt}dt \quad (6)$$

Another relevant element for evaluating investment is how much CO_2 needs to be injected per barrel of crude oil extracted. Normally a substantial amount of CO_2 has to be injected before the first barrel is extracted. CO_2 costs (NPV_C) also depend on whether the CO_2 is from natural or industrial sources.

If it is industrial they also depend on what type of industry produces the gas. For the capture, transport and storage of CO_2 generated by power plants (given that the amount of electricity generated decreases due to the energy penalty of capture), any potential future increase in electricity prices could result in a higher NPV_C.

Transporting the gas through pipelines also has its costs, which increase with the distance between the CO_2 source and the oil field.

There may also be profit associated with storing CO_2 (NPV_A). For instance in the case of the EU ETS Scheme, if *g(t)* tonnes of CO2 are stored their current value can be calculated based on futures equation (7):

$$F(0,t) = A_0 e^{(\alpha-\lambda)t} \tag{7}$$

where A_0 is the carbon spot price.

The NPV is given by equation (8)

$$V(\tau_1,\tau_2) = \int_{\tau_1}^{\tau_2} g(t) A_0 e^{(\alpha-\lambda)t} e^{-rt} dt \tag{8}$$

There may also be other forms of subsidies that need to be incorporated into the NPV, depending on their characteristics.

There are also other fixed and variable costs that must be incorporated when calculating the NPV. These costs are represented here by *I*.

The NPV can be calculated for current commodity prices $NPV(S_0,A_0,0)$ or for certain future conditions $NPV(S_T,A_T,T)$ when it is assumed that only two stochastic processes are relevant and the rest of the results are deterministic, and may in any case depend on the time of investment *T*.

6. Investment Risk

When there is uncertainty there is risk in investments if they are not hedged on organized or OTC (over-the-counter) markets. Even if the NPV is calculated as per Section 5, if the operation is not hedged, e.g. by selling crude oil output on a deferred basis, income may vary upwards or downwards. The distribution of the NPV can be obtained by means of a Monte Carlo simulation if the parameters of the stochastic process have been calculated.

If only the risk in crude oil prices is considered those prices can be simulated by using equation (9) from Abadie and Chamorro 2013 [1].

$$S_{t+\Delta t} = \frac{kS_m}{k+\lambda} + [S_t - \frac{kS_m}{k+\lambda}]e^{-(k+\lambda)\Delta t} + \sigma S_t \sqrt{\Delta t}\varepsilon_t \tag{9}$$

where σ is the instantaneous crude oil volatility and ε_t stands for a random sample from a *N(0,1)* distribution. In this case steps of size Δt are used and the value obtained, ε_t, can be used to calculate $S_{t+\Delta t}$ from S_t. For each step the NPV is obtained as $q_{\Delta t} S_t e^{-rt}$, where it is assumed that there is constant production $q_{\Delta t}$ at each step. The NPV for each path is the sum of those for the steps discounted.

Table 3. Real Option Valuation

t=	S	V	V-I	$(V\text{-}I)e^{-rt}$	$p(V\text{-}I)e^{-rt}$
1	S^+=122.10	V^+=1,385.23	385.23	374.22	136.07
1	S^-=71.83	V^-=1,298.40	298.40	289.87	184.47
0	S = 93.65	V =1,336.09	336.09	336.09	336.09

A number of paths *N* is obtained, each with an NPV, which enables a frequency distribution to be worked out. The average NPV across all paths should almost coincide with the figure obtained using equation (6).

If the stochastic model envisages to sources of risk then one must consider correlation in the generation of random samples.

7. THE REAL OPTION APPROACH

There are a number of real options for an investment project in CO_2-EOR.

The first is the possibility of postponing investment until more favorable conditions arise. At first sight carbon prices as per the quotes in Figure 6 and potential technological development stands as an argument in favor, while crude oil futures prices as per the quotes in Figure 4 stands as an argument in favor. However, volatility is a highly significant factor that generally tends to affect investment negatively, thus encouraging postponement.

This option can be assessed with the help of a binomial tree in which it is considered that uncertainty exists only in crude oil prices, and that currently S_0=\$93.65/barrel, $\frac{kS_m}{k+\lambda}=87.32$, σ=0.2653, r=0.029 and $k+\lambda=0.55$. For the sake of illustration, consider Δt= 1 year, though in real life it is better to divide time into smaller steps. After one period, Δt= 1 year, oil prices according to Table 3 may rise to $S^+=S_0e^{\sigma\sqrt{\Delta t}}=122.10$ or fall to $S^-=S_0e^{-\sigma\sqrt{\Delta t}}=71.83$. Denote $\Delta X=\sigma\sqrt{\Delta t}$. This gives $\mu=\frac{(k+\lambda)(S_m-S_0)}{S_0}-\frac{o^2}{2}=-0.07237$ and the probabilities in the risk-neutral world are $p_u=\frac{1}{2}+\frac{\mu\Delta t}{2\Delta X}=0.3636$ $p_d=1-p_u=0.6364$.

If one barrel per annum is to be obtained from the outset and for 20 years, and the current value of the other deterministic costs is *I*=1,000, then a market valuation can be obtained according to equation (6), with the calculation process reflected in Table 3.

Therefore the initial investment value is 336.09 and the continuation value is 136.07+184.47=320.54. Thus, at current prices, it would be preferable to invest immediately given that the investment value is higher than the continuation value.

There is a value S^* that triggers immediate investment for values $S \geq S^*$. In the example given here that value would be S^*=\$77.84/barrel.

It can be checked that three conditions are met in the example:

a) $p_u+p_d=1$.
b) The first moment: $\mu\Delta t=p_u\Delta X-p_d\Delta X$.

c) The second moment[9]: $\Delta X^2 = \sigma^2 \Delta t$.

Apart from the "wait" option there are other real options that could be assessed in a more complex model, e.g.:

a) the option to abandon the CO_2-EOR project, with the abandonment costs and residual value, if conditions make operating it unprofitable and there expectations for the future are not favorable;
b) the option of operating the facility flexibly, and halting oil extraction when prices are unfavorable; and
c) the option of changing the size of the facility.

The model could also be made more complex by including more stochastic processes to represent the behavior of other risks, correlated or otherwise. Cases in point include the costs of emission allowances, investment costs and the cost of geological uncertainty.

CONCLUSION

Technology for extracting crude oil using CO_2 (CO_2-EOR) has been around for more than 40 years. It has enabled increasing amounts of crude oil to be extracted and at the same time enabled significant amounts of CO_2 to be sequestered, thus benefiting the environment, especially when it is considered that demand for crude oil has not increased as a result of the use of this technology. Moreover, the use of CO_2-EOR has enabled Carbon Capture and Storage (CCS) techniques to be tried out. However, the technique still accounts for only a small proportion of crude oil production in the US. It has so far been based on the use of CO_2 from cheaper, natural sources and no large quantities of additional CO_2 from such sources are expected to become available in the future.

The electric power industry can produce huge quantities of CO_2, which could be used for CO_2-EOR, but there are currently only two LSIPs at the "in execution" stage, scheduled to come on line in 2014 and 2015. In Europe there are no projects at all that could be classed as LSIPs for CO_2-EOR.

The recent expansion of this technology has been favored by crude oil prices, which are higher than in earlier times.

The fact that there are both spot and futures markets for crude oil makes it possible to carry out a valuation of such facilities consistent with the prices on those markets. This chapter shows the financial characteristics of the commodities markets relevant to investment in CO_2-EOR and analyses when it is preferable to opt for a mean-reverting stochastic process and when a Geometric Brownian Motion (GBM) process is the best option.

The chapter also shows how to conduct a valuation on an facility of this kind that is up and running, and how to assess the risks derived from crude oil prices. Finally, it explains how to conduct a valuation considering the option of postponing investment, and how to determine the trigger price S*.

[9] This last equation is an approximation. It is more accurate for lower values of Δt.

It is suggested that in a valuation of real life case all financially relevant options should be considered, including the "abandon" option and the option of operating the facility flexibly.

ACKNOWLEDGMENTS

Luis M. Abadie is grateful for financial support received from Fundación Repsol under the Low Carbon Programme (www.lowcarbonprogramme.org). The author also thanks the Spanish Ministry of Science and Innovation for support via research project ECO2011-25064, and the Basque Government for support via project GIC12/177-IT-399-13.

REFERENCES

[1] Abadie, L. M. and Chamorro, J. M. (2013). *Investment in Energy Assets Under Uncertainty*. London: Springer-Verlag.

[2] Abadie, L. M. and Chamorro, J. M. (2008). European CO_2 prices and carbon capture investments. *Energy Economics*, 30, 2992-3015.

[3] Abadie, L. M., Galarraga, I. and Rübbelke, D. (2014). Evaluation of Two Alternative Carbon Capture and Storage Technologies: A Stochastic Model. *Environmental Modelling and Software,* 54, 182-195.

[4] Abadie, L. M. 2009. Valuation of Long-Term Investments in Energy Assets under Uncertainty. *Energies* 2009. 2(3). 738-768.

[5] Alvarado, V. and Manrique, E. (2010). Enhanced Oil Recovery: An Update Review. *Energies*, 3, 1529-1575.

[6] Dixit, A. K., Pindyck, R. S. (1994). *Investment Under Uncertainty*. Princeton University Press.

[7] Duncan, I. J., Nicot, J.-P. and Choi, J.-W. (2009). Risk Assessment for future CO_2 Sequestration Projects Based CO_2 Enhanced Oil Recovery in the US, *Energy Procedia*, 1, 2037-2042.

[8] Echevarria Huaman, R. N. and Jun, T. X. (2014). Energy related CO_2 emissions and the progress on CCS projects. *Renewable and Sustainable Energy Reviews*, 31, 368-385.

[9] EIA (2014). US Field Production of Crude Oil. http://www.eia.gov/dnav/pet/hist/LeafHandler.ashx?n=petands=mcrfpus2andf=m.

[10] EPA (2014). National Greenhouse Gas Emissions Data. http://www.epa.gov/climatechange/ghgemissions/usinventoryreport.html.

[11] Fleten, S.-E., Lien, K. and Ljø nes, K. (2010) Value chains for carbon storage and enhanced oil recovery: optimal investment under uncertainty. *Energy Systems*, 1, 457-470.

[12] Ferguson, R. C., Nichols, C. and Leeuwen, T. V. (2009). Storing CO_2 with Enhanced Oil Recovery, *Energy Procedia*, 1, 1989-1996.

[13] Gaspar Ravagnani, A. T. F. S., Ligero, E. L. and Suslick, S. B. (2009) CO_2 sequestration through enhanced oil recovery in mature oil field. *Journal of Petroleum Science and Engineering*, 65, 129-138.

[14] Global CCS Institute. *Status of CCS Project Database*. http://www.globalccsinstitute.com/projects/status-ccs-project-database.

[15] Hammond, G. P., Ondo Akwe, S. S. and Williams, S. (2011). Techno-economic appraisal of fossil-fueled power generation systems with carbon dioxide capture and storage. *Energy*, 975-984.

[16] Heydari, S., Ovenden, N. and Siddiqui, A. (2010) Real options analysis of investment in carbon capture and sequestration technology. *Computational Management Science*, 9, 109-138.

[17] Herzog, H. J. (2011). Scaling up carbon dioxide capture and storage: From megatons to gigatons. *Energy Economics*, 33, 597-604.

[18] Hull, J. C. (2011) *Options, Futures and Other Derivatives* (8th Edition), Prentice Hall.

[19] Jaramillo, P., Griffin, W. M. and Mccoy, S. T. (2009). Life Cycle Inventory of CO_2 in an Enhanced Oil Recocery System. *Environmental Science and Technology*, 43, 8027-8032.

[20] Klokk, Ø., Schreiner, P. F., Pages-Bernaus, A., and Tomasgard, A. (2010) Optimizing a CO_2 value chain for the Norwegian Continental Shelf. *Energy Policy*, 38, 6604-6614.

[21] Kuuskraa, V. and Wallace, M. (2014). CO_2-EOR set for growth as new CO_2 supplies emerge. *Oil and Gas Journal*, 112, 66-70.

[22] Muggeridge, A., Cocking, A., Webb, K., Frampton, H., Collins, I., Moulds, T., and Salino, P. (2014). Recovery rates, enhanced oil recovery and technological limits. *The Philosophical Transaction of the Royal Society A*, 372, 1-25.

[23] Thomas, S. (2008). Enhanced Oil Recovery - An Overview. *Oil and Gas Science and Technology*, 63, 9-19.

In: Enhanced Oil Recovery
Editor: Alicia Knight

ISBN: 978-1-63463-917-0

Chapter 2

CONTRIBUTIONS OF THE CHEMICAL INDUSTRY TO ENHANCED OIL RECOVERY

Christian Bittner*
BASF SE, GMM/S – J550, Ludwigshafen, Germany

ABSTRACT

This commentary deals with the notable contributions to enhanced oil recovery (EOR), particularly of large chemical companies. In general, EOR encompasses a bundle of different methods for the recovery of additional amounts of original oil in place – e.g. by injection of gas, steam, microbes or chemicals. Especially the latter method is focused on the application of polymers and surfactants and is an attractive option to improve the economics of petroleum production (chemical EOR is good for brown fields and for even green fields). As the development of a reservoir can be very challenging, an interdisciplinary team should be formed and work together to achieve the successful design and implementation of an EOR program. Such an approach allows the combination of the different kinds of needed expertise and avoids problems on the long run (e.g. lack of supply of material or undesired performance of material in the field).

The expertise, in particular, of large chemical companies experienced in polymer and surfactant production is reflected by following points:

- Selection of relevant raw materials, which are available in the desired amounts
- Secure a reliable feedstock supply by either being backward integrated or by having long term supply contracts in place
- Capabilities to scale-up manufacturing from lab to industrial scale production without changing product quality (from couple of 100 kg for single-well test to 100000 to for full field implementation)
- Utilize state of the art production processes to ensure cost-effectiveness
- Leverage global production footprint to guarantee supply reliability and cost effectiveness
- Benefit from a global supply chain network and expertise to supply large quantities of chemicals even to remote locations

* E-mail: christian.bittner@basf.com.

- Manage regulation needs (e.g. REACH)
- Contribute own R&D resources to develop new compounds with an improved performance profile or to identify more cost-effective alternative solutions.
- Another important factor, which should not be neglected, is the financial strength of a global chemical company. As a full field implementation can last several years and requires significant amounts of chemicals, a strong chemical partner can cope with investment needs to fulfill even a strong growth in chemical demands. This fact guaranties long-term supply of the needed volume of chemicals in desired quality.

INTRODUCTION

In general, oil production can be divided into three types of recovery. [1] Primary oil recovery relies on the pressure in the reservoir (e.g. due to dissolved gas), but only 5 – 10 % of original oil in place (OOIP) are accessible with this method. Once pressure is decreased, one tries to re-increase the pressure in the reservoir by injection of e.g. water (or gas). This type of production is called traditionally secondary oil recovery and one has to drill additional wells for the injection of the water or gas. On average 30% OOIP can be recovered from the reservoir. At a certain point of time the injected water breaks through and forms ‘by-passes’ of the oil bearing zones (the so-called ‘fingering’). As a matter of consequence, the oil cut in the produced liquid decreases significantly and more and more water will be produced (water will be re-injected). To produce additional amounts of OOIP several other methods for tertiary oil recovery are known - such as injection of gas, steam, microbes or chemicals. Nowadays, one could also directly start with enhanced oil recovery after the primary recovery.

CHEMICAL ENHANCED OIL RECOVERY

The chemical enhanced oil recovery (CEOR) [2] is one of the most attractive method among the above described methods because of the significant amount of incremental oil, which could be produced with this approach. CEOR uses polymers and surfactants. Polymeric thickeners are used to increase the viscosity of the water for improving the macroscopic displacement (avoids fingering). Surfactants are used to lower the interfacial tension between oil and water to ultralow values. Consequently, capillary forces will be overcome and the microscopic displacement will be improved (capillary trapped oil will be mobilized). By the combination of surfactants and polymers up to another 30% of OOIP can be produced. Another advantage of CEOR is, that invest into the infrastructure is manageable (chemicals are injected through the same wells as water is injected; investment into tanks and dissolution units is needed). A matured field can be re-vitalized and a risky (and costly) exploration can be avoided. The *International Energy Agency* has stated in their world energy outlook 2013 [3], that CEOR is even more profitable than primary recovery from tight oil reservoirs (return on investment: 197% versus 100%). But one has to keep into mind, that implementation of a CEOR project will take much longer compared to primary recovery from tight oil reservoir using hydraulic fracturing and that production profiles are different (payback time: 4 years versus 2 years) [3]. Over this period of time and after, the oil price should be at least stable to ensure the promising economics. The reason for the significant lead time of a CEOR project is, that several stages have to be passed. After reservoir is found to be in principle suitable for CEOR, the chemicals have to be tested in the lab including core flood tests. Then, a single well test will be done and several 100 kg of chemicals will be needed. If single well test is successful, a pilot test will be done. Modelling of the reservoir and the injection strategy is very important. At this stage several 100 to 1000 to of chemicals will be injected and it will last ~ 2 years. After the pilot, the full-field implementation will follow. This will require investments in the field but also on the side of the chemical vendor as significant amounts of chemicals will be needed (depending on the field size more than 20000 to of chemicals per year).

To cope with the challenging conditions in the reservoir, even new chemicals have to be developed. They have to withstand salinities up to 200000 ppm total dissolved salt or temperatures up to 130°C for several years. Solutions for 'easy' fields (sandstone, low salinity, and low temperature) already exist, but there is still room for improvement. However, beside the chemistry several other factors have a huge impact on the later oil production (e.g. geology of the reservoir, flow in the reservoir, robust formulation of the chemicals, etc.). Therefore, it is necessary to form an interdisciplinary team. The collaborative work is a key to achieve the successful design and implementation of an EOR program. Another highly important factor for the success of a CEOR project has not been discussed in the community over the past years: it is the early involvement of particularly chemical companies experienced in the production of polymers and surfactants at large scale. Expertise of such companies is e.g. to select the right raw materials and to scale-up and produce the chemicals on large scale in same quality as in case of developmental samples. This will avoid lack of supply of chemicals or change of their performance during the full-field implementation. These and further aspects of the expertise will be explained below in detail.

KEY CONTRIBUTIONS OF CHEMICAL INDUSTRY

In order to flood a reservoir with aqueous solutions of polymers and surfactants, significant amounts of chemicals are needed. This can be more than 20000 to of chemicals per year for only one field. Therefore, it is necessary to choose even at the beginning of a CEOR project the relevant raw materials, which are available in the desired amounts. Production of coupling products, which will not be used, should be avoided. To give an example, feedstock for a surfactant such as an alkyl ether sulfate will be discussed. Fatty alcohols are made out of fats and oils from plants, which are available on million to base. One obtains alcohols with a distribution in alkyl chain length ranging from 6 to 22 carbon atoms. [4] They are separated by distillation and one obtains certain cuts. Short chain alcohols are used e.g. for plasticizers, while alcohols with medium chain length (e.g. $C_{12}C_{14}$) are used as building blocks for detergents. Pure alcohol (no alcohol mixture) is isolated out of the former alcohol cut by a further distillation. However, it is a specialty for certain niche applications. Researchers have looked for long-chain alcohols to make EOR surfactants out of it. Such surfactants can give ultralow interfacial tension with crude oils following the simplified rule, that interfacial tension becomes lower as longer the alkyl moiety of the surfactant is. From rapeseed oil a $C_{18}C_{20}C_{22}$ alcohol cut is isolated and further distilled. If one would aim only for pure C_{22} alcohol as feedstock for an EOR surfactant, one would produce also the same amount of C_{18} alcohol and C_{20} alcohol. Latter alcohols cannot be used in many other applications. Therefore, the pure C_{22} alcohol becomes much more expensive - not only due to the additional distillation step but also due to the coupled products (C_{18} alcohol and C_{20} alcohol) and their disposal. It is important to balance the performance (of the surfactant) with the financial impact of the raw materials and the production costs. As a matter of consequence, a much better strategy is to build surfactants on the existing alcohol cuts.

Due to long term scope of CEOR projects a reliable feedstock supply needs to be considered from the beginning, even if the chemicals are based on building blocks, which are globally available in large amounts. Two options for ensuring supply are either by backward integration or long term supply contracts. For example propylene oxide is available in large amounts and gives beneficial properties to EOR surfactants, if it is used as their feedstock. But it is also used for large scale and further growing production of polyurethanes (suitable as isolation foams beside many other applications). Therefore, long-term supply contracts in the industry exist to guarantee the supply for such applications. One has to address need for additional significant amounts of propylene oxide in an early stage, in order to trigger investment into additional chemical plants capacity.

After the right building blocks have been chosen and an EOR chemical with suitable performance in lab tests has been identified, the right production process and the scale-up without performance lost comes into play. A state-of-the art production process should be chosen to ensure cost-effectiveness also with regard to the manufacturing costs. Chemical companies, which are experienced in production of large quantities of surfactants or polymers, usually provide samples of such a quality, which is equal to the quality of samples from large scale production. But beside the experienced chemical companies also other provider of new chemicals exist (e.g. research institutes or universities) and these providers might not be aware of difficulties, which could occur within the scale-up. In the lab stage, samples in kg scale are used. However, their production process should be done in such a way, that one could scale it up. 'Academic' processes, which might be more convenient in the lab, should be avoided, because they are most times not economic for production on large scale. For example sulfation of an alkyl alkoxylate might be done with sulfamic acid in the laboratory. Sulfamic acid is a solid and can be easily handled by a technician, if only several grams of sulfamic acid are added to some hundred grams of alkyl alkoxylate. In case of the production of 10 to of alkyl ether sulfate the handling of sulfamic acid is very difficult. One has to add several 100 kg of a solid to the reaction. Addition of a solid requires a special (and expensive) dosage unit or additional workforce, which leads to higher manufacturing costs. Instead, sulfation with gaseous sulfur trioxide is much less expensive on large scale. As different kinds of reagents are used the product composition and its performance can differ. Even if the same process is used in the lab as later in the production, one has to take into account, that conditions are still not the same. In a 2 liter vessel the volume to surface ratio is totally different compared to the ratio in a 16 m^3 vessel. This will have an effect on the heat transfer and the transport of substances from the gas phase to the liquid phase. As a matter of consequence, these factors might affect the formation of side products and thereby the performance of the final product. Companies with world-scale production units are skilled to transfer products from the laboratory to large scale production. Factors from above are known and measures are taken to avoid performance change.

The next issue is the transport of the EOR chemicals to the oil reservoir. This might be sometimes very challenging as fields are in remote locations. A global production footprint and a skilled global supply chain are the two key factors to guarantee supply reliability and cost effectiveness. One has to compare the cost savings due to large scale production with the cost savings due to production close to the oil field. World scale production lowers the manufacturing costs, while the other factor decreases the overall freight costs. Large chemical companies have a global production with world scale units in the region. Thereby, one benefits from world-scale and close distance. The supply chain should be skilled to handle transport of large quantities of chemicals. A global supply chain is aware of the regional differences and local transportation rules or issues. Large companies have experience in transportation of large quantities of chemicals – e.g. supply of the big soapers such as P&G or Unilever with laundry surfactants and polymers for 20 years and more. Even if there is a surprising problem in a production plant of the chemical vendor and chemicals cannot produce on this site for a certain point of time, a skilled supply chain will manage the production and transportation of the desired chemical from another production site to the customer. This will guarantee a continuous supply of the customer and his own production has not to be stopped.

In the last decade research groups at universities and institutes started the development of EOR chemicals. This has led to interesting new structures and tests with surprising results have been done. However, regulation needs (such as e.g. REACH) have sometimes not been taken into account. Consequently, registration of such structures can require more (and expensive) tests. A delay in production can be the result and of cause higher costs. This can be avoided in the upfront, if one is aware of regulation needs and one selects chemical structure carefully.

Many chemical companies initiated programs for the development of EOR chemicals, too. These own R&D resources contributed to identification of new compounds with improved performance profile or of more cost-effective alternative solutions. In particular large chemical companies reflected the important factors from above. Close cooperation with oil companies would enable the chemical company to improve performance of the EOR chemical further. The oil company would benefit too as they would be aware of alternative solutions and no surprising problems with respect to the chemical performance and supply during the implementation of the EOR project.

Another important factor, which should not be neglected, is the financial strength of a global chemical company. As a full field implementation can last several years and requires significant amounts of chemicals, a strong chemical partner can cope with investment needs to fulfill even a strong growth in chemical demands. This fact guaranties long-term supply of the needed volume of chemicals in desired quality. Compared to a small(er) chemical company with less financial strength there is no risk, that the company will become insolvent [5] during a crisis of the economy and cannot supply the customer with material anymore.

Conclusion

Chemical enhanced oil recovery is focused on the application of polymers and surfactants and is an attractive option to improve the economics of petroleum production. For a successful implementation of a CEOR project, a team out of different disciplines should be formed. The chemical companies should be involved in an early stage to avoid problems in a later stage of the project. The expertise, in particular, of large chemical companies, which are experienced in polymer and surfactant production, comprises the selection of the suitable raw materials and of the most fitting chemical structures, the capability to scale-up the chemicals without performance loss, and the security of supply (even on the long run) due to a strong supply chain.

References

[1] Lake, L.W., 'Enhanced Oil Recovery', Prentice Hall, Englewood Cliffs, 1989.

[2] Bittner, C., 'Impact of Innovations on Future Energy Supply – Chemical Enhanced Oil Recovery (CEOR)', *Chimia* 2013, *67*, 724-728.

[3] International Energy Agency, 'World Energy Outlook 2013', 2013.

[4] Noweck, K., Grafahrend, W., 'Fatty alcohols', *Ullmann's Encyclopedia of Industrial Chemistry*, 2012, *14*, 117-141.

[5] Statista, 'Number of insolvencies in the chemical industry in the U.S. from 2000 to 2009', 2014. Available from: http://www.statista.com/statistics/272764/number-of-insolvencies-in-the-chemical-industry-in-the-us/

In: Enhanced Oil Recovery
Editor: Alicia Knight
ISBN: 978-1-63463-917-0

Chapter 3

RESEARCH OF SC-CO_2 IN IMPROVING THE RECOVERY OF HEAVY OIL

Li Zhaomin and Lu Teng
China University of Petroleum, China

ABSTRACT

Enhanced oil recovery (EOR) becomes increasingly important to the petroleum industry. SC-CO_2 has been successful applied to a large extent under some favorable reservoir conditions. In this chapter, mutual interactions between each of three different crude oils and SC-CO_2 are experimentally studied. First, the solubility of CO_2 in ultra-heavy oil is conducted to determine the viscosity reducing behavior of SC-CO_2. Second, the interfacial tension between CO_2 and crude oil is measured as a function of the temperature and pressure. Third, CO_2 core displacement experiments are conducted to investigate the production characteristics of SC-CO2 flooding. It is found that the CO_2 solubility in the ultra-heavy oil increases as the pressure increases, but the increasing amplitude becomes increasingly smaller. CO_2 has an obvious effect on reducing the viscosity of ultra-heavy oil; CO2 is able to reduce the viscosity by 99.64%. As the pressure increases, the equilibrium interfacial tension of the CO_2-crude oil system decreases. When the pressure is greater than 3.0MPa, the influence of temperature on the interfacial tension becomes increasingly obvious; as the temperature increases, the interfacial tension of CO2-crude oil becomes greater. As the pressure increases, the extent of CO_2 flooding enhancing oil recovery increases. Compared to the asphaltene content of crude oil, the content of the oil produced is little different when the injection pressure is smaller. When the injection pressure is large, the asphaltene content of the oil produced significantly decreases because precipitation of asphaltene occurs, and the precipitation adsorbs and deposits into the porous media. During the CO_2 flooding, the oil production rate will suddenly rise before gas breakthrough occurs, which is followed by a phase of rapid decline in the oil production rate. The curve of oil production rate looks similar to an inverted "V". The maximum oil production rate can be 3.09~9.18 times that of the average oil production rate and can be 1.52~3.39 times of the gas breakthrough oil production rate. There are two conditions that can be used to predict the gas breakthrough in CO_2 flooding in this paper: (1) the oil production rate of CO_2 flooding suddenly rises and the curve of oil production rate looks similar to an inverted "V"; (2) the maximum oil production rate is over three times bigger than the average oil production rate.

1. Study on the Solubility of CO_2 in Ultra-Heavy Oil

The solubility of CO_2 in crude oil is the volume of CO_2 able to be dissolved per unit volume of the stock tank oil at a given pressure and temperature (at standard conditions). The measurement of the solubility of CO_2 in crude oil is as follows: CO_2 and the crude oil at a given percentage are charged into a PVT cylinder; the gas-oil mixture volume of which can be altered by the movement of the piston (the pressure can also be altered), thereby putting the corresponding volume and pressure on the paper. The pressure under the breakpoint and volume of CO_2 on the curve is known as the bubble point pressure. Next, the volume of CO_2 is increased and the pressure is measured to obtain the bubble point pressure curves under the different volume of CO_2. This method is useful in measuring the bubble point pressure and the volume factor under the different crude oil types, temperatures, and volumes of CO_2. The density of the oil-gas mixture can be calculated by the pressure and volume factor [1-3].

1.1. Experiment Apparatus

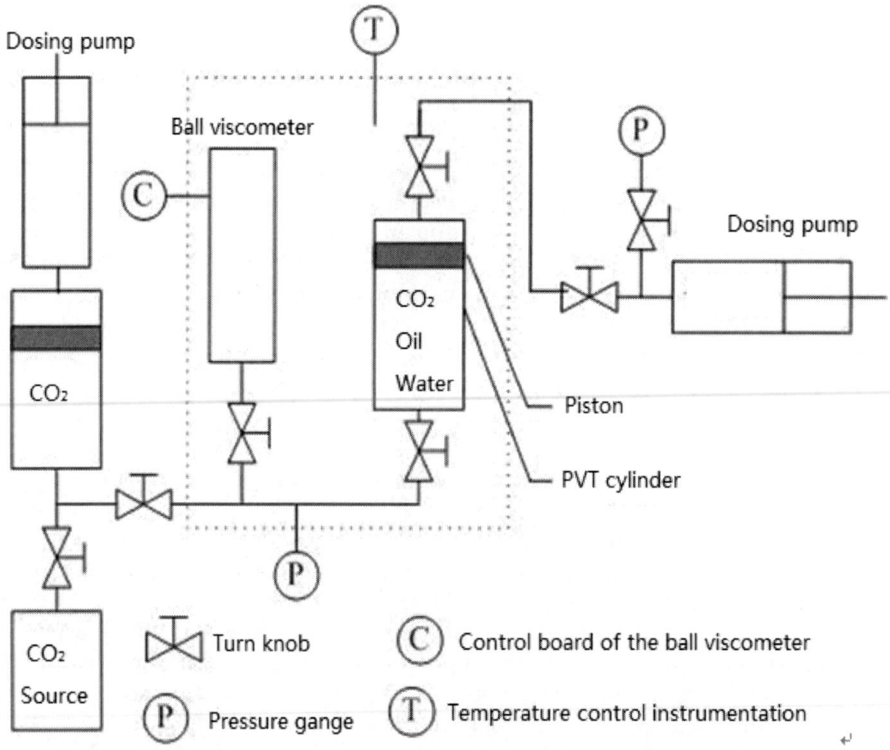

Figure 1.1. PVT apparatus for high temperature and pressure.

CO_2 can be dissolved fully into the crude oil or water-in-oil emulsion using a high pressure PVT cylinder. The PVT cylinder utilizes the heating control system and stirring apparatus to reduce the dissolving equilibrium time. In the heating control system, the active zone is 0-200℃, the control accuracy is ±0.1℃, the highest pressure of the gauge is 40 MPa and the accuracy figure is 0.4. The measurement unit for the oil-gas mixture viscosity under high temperature and pressure is the ball viscometer. For the ball viscometer, the temperature is 0-200°C, the operate miss is ±0.01°C, the highest pressure is 60 MPa and the viscosity is 0.1-100000 mPa·s.

Table 1.1. Oil properties of Z411-P2 well

Number	Density (g/cm^3)	Viscosity (mPa·s)	Sulfur (%)	Colloid (%)	Asphalt (%)	Wax (%)	Freeze Point (%)
Z411-P2	1.0095	11290/80°C	0.94	47.85	10.69	1.59	27

1.2. Test Sample

The oil sample used in testing is from the Z411-P2 well, and the water sample is also from the Z411-P2 well. The oil and water properties are shown in Tables 1.1 and 1.2.

1.3. Test Conditions

The conditions for the water-in-oil emulsion are as follows: water saturation is from 0 to 50%, the test temperature ranges from 60 to 200℃, and the gas-oil ratio ranges from 1 to 126. Before testing, the water-in-oil emulsion and CO_2 are mixed evenly in a high pressure PVT cylinder under a constant temperature for 4h. Then, the bubble point pressure, the volume factor, the density, and the viscosity are measured.

Table 1.2. Water properties of Z411-P2 well

Number	Chlorion(mg/L)	Total Solids (mg/L)	Water Type
Z411-P2	8479	14287	$CaCl_2$

1.4. Results and Discussion

1.4.1. The Properties of CO_2 Dissolved in the Degassing and Dehydration Oil

For the different CO_2 solubilities in the oil, the saturation pressure, volume factor, density, and viscosity were measured under the temperatures of 60°C, 80°C, 100°C, and 120°C. The results and discussion are as follows:

1) Confirm the highest solution of CO_2 in the ultra-heavy oil in the Z411-P2 well

Figure 1.2 shows that the solubility of CO_2 in the ultra-heavy oil is able to reach 120sm^3/m^3 under 30MPa; however, in the clean oil, the solubility is able to reach 400sm^3/m^3 [4], which is much higher than the solubility in the Z411-P2 well. The CO_2 solubility in the crude oil of Z411-P2 increases as the pressure increases, but the increasing amplitude becomes increasingly smaller. The solubility will not change when the saturation pressure is higher than 30 MPa. The mechanism of CO_2 dissolved in the crude oil is to extract the light hydrocarbons, so the solubility is mainly related to the properties of the crude oil. The ultra-heavy oil has a heavy API, a less light hydrocarbon, and more colloid and asphalt, which leads it to have a very limited ability to dissolve CO_2.

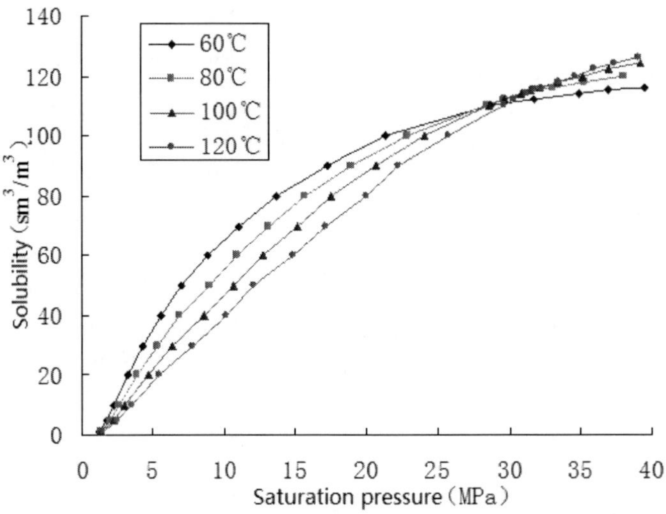

Figure 1.2. Curve of CO_2 solubility-saturation pressure under different temperatures.

Figure 1.2 shows that when the pressure is lower than 28MPa, the crude oil under the same saturation pressure dissolves a smaller volume of CO_2 while the environmental temperature increases. When the pressure surpasses 28MPa, it can dissolve more CO_2 when the environmental temperature increases. This phenomenon is under the joint influence of the change in both the CO_2 molecule's locomotivity and the oil molecule interval. First, crude oil has a larger molecule interval under low pressure. Then, the molecular motion of CO_2 has a greater influence on the solution of CO_2. As the temperature increases, the motion of CO_2 molecule becomes more intense, thereby making it easier for CO_2 to vaporize instead of dissolving in the crude oil. Meanwhile, the increasing saturation pressure compresses the crude oil, which causing its density to increase. The molecular gap of the oil is also an important influencing factor. A lower temperature leads to a higher oil density and a more compact molecule arrangement, causing more difficulty in the dissolution of CO_2 into crude oil.

Reservoir Engineering provides plenty of ways to determine a certain solution of gas in crude oil. Methods such as the equations presented by standing, nomograms, Lasater, and Vasquez and Beggs are the most frequently applied ones [5]. However, considering the ultra-heavy oil has a high density and viscosity, applying these equations to calculate the amount of CO_2 dissolved in ultra-heavy oil results in great error compared to the actual solution of CO_2. Chung and his partners put forward a new expression to calculate the solution of CO_2 in ultra-heavy oil on the basis of many experimental results with a calculating error within 5%.

Table 1.3. Comparison of calculated data and experimental data for solubility-saturation pressure

Gas-oil	Dissolving capacity (MPa)											
ratio	60℃(C)	60℃(E)	Error	80℃(C)	80℃(E)	Error	100℃(C)	100℃(E)	Error	120℃(C)	120℃(E)	Error
1	1.16	1.21	4.31	1.27	1.28	0.79	1.37	1.32	3.65	1.45	1.39	4.14
5	1.76	1.66	5.68	1.99	1.89	5.03	2.23	2.03	8.97	2.48	2.28	8.06
10	2.28	2.38	4.39	2.67	2.76	3.37	3.06	3.26	6.54	3.55	3.81	7.32
20	3.22	3.02	6.21	3.91	3.71	5.12	4.67	4.37	6.42	5.45	5.25	3.67
30	4.28	4.48	4.67	5.36	5.56	3.73	6.41	6.71	4.68	7.89	8.24	4.44
40	5.58	5.64	1.08	6.97	6.92	0.72	8.62	8.69	0.81	10.18	10.28	0.98
50	7.11	7.38	3.80	9.01	9.31	3.33	10.77	11.15	3.53	12.15	12.81	5.43
60	8.85	8.45	4.52	11.02	10.53	4.45	12.81	12.31	3.90	14.95	14.35	4.01
70	11.12	11.58	4.14	13.16	13.56	3.04	15.15	15.65	3.30	17.12	17.82	4.09
80	13.71	13.83	0.88	15.68	15.57	0.70	17.56	17.68	0.68	19.95	19.78	0.85
90	17.25	17.88	3.65	18.92	18.79	0.69	20.61	20.76	0.73	22.25	22.38	0.58
100	21.34	22.25	4.26	22.87	23.87	4.37	24.01	25.21	5.00	25.72	26.82	4.28
110	29.72	29.98	0.87	28.48	28.64	0.56	28.61	28.93	1.12	29.67	28.82	2.86

The CO_2 solution expression [6] is as follows:

$$R_s = 0.178*[a_1\gamma^{a_2}T^{a_7} + a_3T^{a_4}\exp(-145a_5P + a_6/145P)]^{-1} \tag{1.1}$$

$$T=1.8t+32$$

where R_s—Solution, m^3/m^3;

γ - Oil relative density;

t – Temperature, °C;

P - Pressure, MPa;

$a_1=0.4934\times10^{-2}$; ;

$a_2=4.0928$;

$a_3=0.571\times10^{-6}$;

$a_4=1.6428$;

$a_5=0.6763\times10^{-3}$;

$a_6=-781.334$;

$a_7=-0.2499$.

2) Confirm the volume factor of the CO_2 dissolved in the ultra-heavy oil

After CO_2 was dissolved into the ultra-heavy oil, the volume factor of the ultra-heavy oil is small; thus, increasing the amount of dissolved CO_2 will increase the volume linearly. As figure 1.3 shows, when the experiment reaches the highest dissolved pressure of CO_2, the bulk coefficient of the mixture is approximately 1.32, but it can reach 1.83 in clean oil [7]. The difference between the factors is caused by the different solutions of CO_2 [8]. The volume of CO_2 dissolved in the oil can increase the initial oil volume expansion, the elastic energy, and the production.

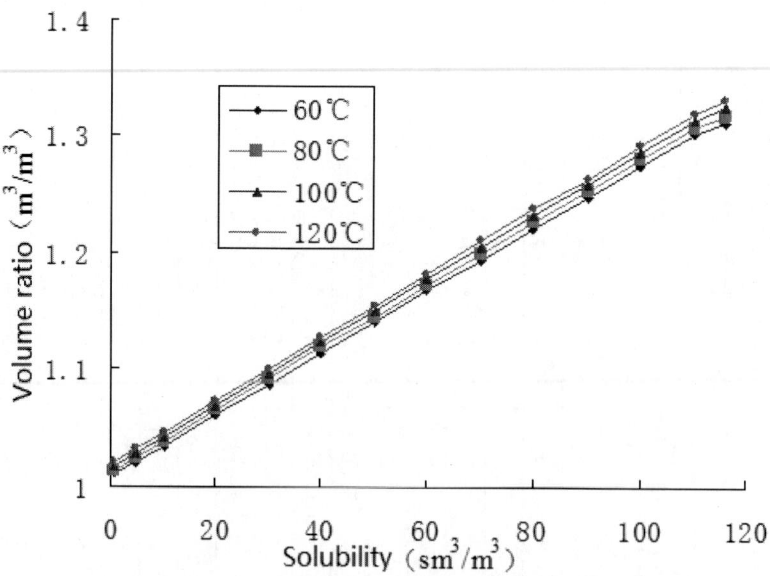

Figure 1.3. Curve of CO_2 solubility-volume ratio under different temperatures.

The volume expansion not only can increase the elastic energy for the layer, but it also can transform the immobile oil into an active oil. In addition, it can reduce the IOS (irreducible oil saturation) and enhance the oil recovery. As figure 1.3 shows, temperature has little effect on the volume factor of the ultra-heavy oil with the same solution and under the same pressure. Increasing the temperature is able to increase the volume of gas-oil factor slightly.

Vasquez and Beggs [5] combined the bubble point pressure with the DGOR (dissolved gas-oil ratio), resulting in an equation on the formation volume factor of the saturated oil:

$$B_0 = 1 + C_1 R_s + C_2(1.8t - 28)(\gamma_{API} / \gamma_{gc}) + C_3 R_s (1.8t - 28)(\gamma_{API} / \gamma_{gc}) \quad (1.2)$$

where B_0 —Volume factor under the pressure P and temperature T, m^3/m^3;

R_s - Solution under the pressure P and temperature T, sm^3/m^3;

T - Temperature of the crude oil, ℃;

γ_{API} - Gravity of the crude oil, °API;

γ_{gc} - Relative density of the gas, Air = 1.

C_1, C_2 and C_3 are parameters related to the density of the crude oil; they are 4.677×10^{-4}, 1.751×10^{-5} and -1.811×10^{-8}, respectively. For the experiment, we used experimental data to perform the regression, and suitable C_1, C_2 and C_3 values for the ultra-heavy oil on Shengli Oil field were determined: 4.6078×10^{-4}, 2.4329×10^{-5} and 1.4162×10^{-8}, respectively.

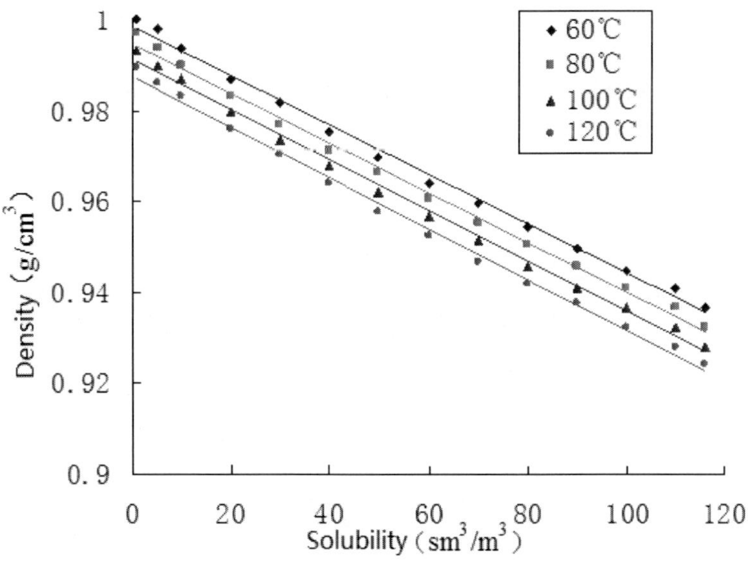

Figure 1.4. Curve of CO_2 solubility-density ratio under different temperatures.

3) Determine the density of the ultra-heavy oil after dissolving CO_2

As shown in Figure1.4, the density of the ultra-heavy oil that dissolved CO_2 decreases linearly. The density of the ultra-heavy oil decreases by 7 percent when dissolving CO_2

(solution is 110sm^3/m^3). Previous research showed that the density of crude oil (absorbing CO_2) is firmly related to its own characteristics. Ultra-heavy oil contains a low percent of intermediate hydrocarbon. The density of the ultra-heavy oil keeps declining while absorbing CO_2. The reduction in the density of the oil results from the compressibility of CO_2 and the oil itself [9]. In Figure 1.4, under the same saturation pressure, the increasing temperature could result in a decrease in the density of an oil-air mixture of the same weight. CO_2 molecules and light hydrocarbons in the oil-air mixture become easier to vaporize as the environmental temperature increases. High temperatures accelerate the motion of the gas molecules, increasing molecular collisions and expanding the intervals between molecules as well. Thus, the density of the oil-air mixture decreases as solubility increases.

4) Determine the viscosity of the extra-heavy oil after dissolving CO_2

(1) Research on the viscosity of the ultra-heavy oil under the saturation pressure

Figure 1.5 showed that the viscosity of the extra-heavy oil sample extracted from the western part of the Z411-P2 well declines rapidly as the CO_2 solubility increases. However, the decreasing rate lessens as the solubility increases. The viscosity of the ultra-heavy oil saturated with CO_2 decreases when the temperature rises, but low temperature promotes a remarkable reduction in viscosity. CO_2 has an obvious effect on reducing the viscosity of ultra-heavy oil; in the test range, CO_2 was able to reduce the viscosity by 99.64%. As shown in Table 1.4, the dropping rate is 277.78. However, generally, crude oil can only reduce its viscosity by approximately 90% at most [10], which is due to the relative change of the intermolecular forces of oil molecules compared to the original system after dissolving CO_2. The former molecular forces between liquid molecules change into molecular forces between liquid molecules and gas molecules. The laminated structure of macromolecules in the ultra-heavy oil colloid and asphalt is undermined by CO_2 molecules dissolved in the sample as well as the large decrease in the intermolecular forces and frictional resistance between molecules. As a result, the viscosity of the oil is significantly reduced.

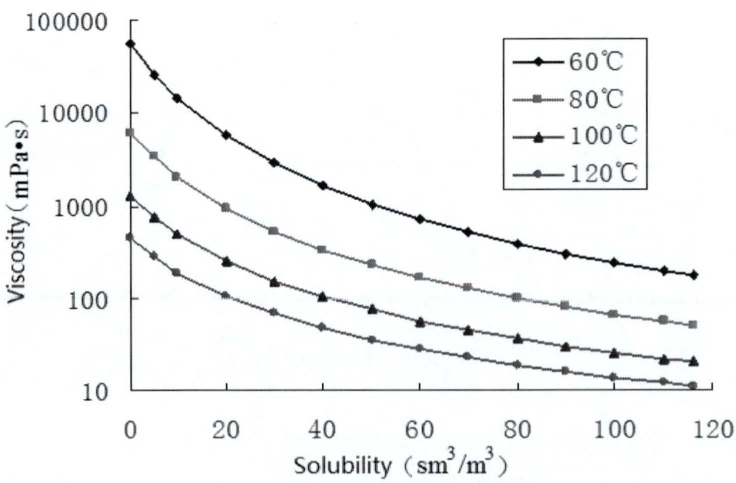

Figure 1.5. Curve of CO_2 solubility-viscosity under different temperatures.

Table 1.4. Effect of CO_2 solubility on viscosity of ultra-heavy oil

Solution	Viscosity reducing rate (%)				Viscosity reducing ratio			
(sm^3/m^3)	60°C	80°C	100°C	120°C	60°C	80°C	100°C	120°C
5	53.9	42.86	35.82	33.4	2.17	1.75	1.56	1.50
10	75.29	65.2	58.47	55.67	4.05	2.87	2.41	2.26
20	89.53	86.04	79.38	75.8	9.55	7.16	4.85	4.13
30	94.56	91.63	87.34	83.94	18.38	11.95	7.90	6.23
40	97.19	94.23	92.4	89.94	35.59	17.33	13.16	9.94
50	98.16	96.14	94.57	92.08	54.35	25.91	18.42	12.63
60	98.63	97.16	95.44	93.79	72.99	35.21	21.93	16.10
70	99.13	98.05	96.67	94.65	114.94	51.28	30.03	18.69
80	99.25	98.46	97.25	95.93	133.33	64.94	36.36	24.57
90	99.44	98.7	97.9	96.57	178.57	76.92	47.62	29.15
100	99.58	98.99	98.05	97.22	238.10	99.01	51.28	35.97
110	99.64	99.14	98.48	97.86	277.78	116.28	65.79	46.73

A significant amount of funds and time are spent in measuring the viscosity break of CO_2 for the ultra-heavy oil, and the steps are complex. CO_2 is dissolved in the ultra-heavy oil by using a high pressure PVT cylinder. Then, the mixture is transferred to a high temperature and pressure ball viscometer, and the viscosity is measured. It is important to select or deduce the formula to forecast the viscosity break of CO_2 for the ultra-heavy oil by using experimental data. This study has demonstrated that the formula suggested by Beggs and Robinson [11] can be used for the computational solution and experimental data to calculate the viscosity. An accuracy formula can be obtained by modifying the experimental data, as follows:

$$\mu_{os} = A\mu_{od}^{B} \tag{1.3}$$

and:

$$A = 10.7(5.6184R_s + 100)^{-0.515} \tag{1.4}$$

$$B = 5.446(5.6184R_s + 150)^{-0.338} \tag{1.5}$$

where μ_{os}—Viscosity of the saturated crude oil, mPa·s;

A—A function on DGOR;

μ_{od}—Viscosity of the dead crude oil, mPa·s;

B—A function on DGOR.

The viscosity of the oil-gas mixture can then be calculated under the different DOGs and temperatures. Under 60°C, 80°C, 100°C, 120°C, we obtained average relative errors of 4.52%, 4.64%, 5.34% and 5.66%, respectively, and the average error was 5%. This result can satisfy the project needs. It shows that the formula can accurately calculate the viscosity of the mixture after the dissolution of the CO_2.

(2) Research on the viscosity of ultra-heavy oil containing dissolved CO_2 under supersaturated pressure

The viscosity-temperature characteristics of ultra-heavy oil dissolved CO_2.

The viscosity of degassed crude oil was measured under the conditions of: a dissolved gas-oil ratio of $50sm^3/m^3$ and a pressure of 13MPa with the temperature of the gas-oil mixture at 50°C-60°C.

As shown in Figures 1.5 and 1.6, the viscosity of the gas-oil mixture containing dissolved CO_2 decreases rapidly as the temperature increases; it forms a linear relationship with temperature on the semilog coordinate. The viscosity reduction ratio gently decreases; it is still above 90% under 120°C. We can see that CO2 and the steam huff and puff collaborate in exploiting the ultra-heavy oil reservoir. This will superimpose their advantage in heavy oil exploitation, bringing out the best of each other and having an obvious effect.

Table 1.5. Viscosity-temperature characteristics of CO_2 and ultra-heavy oil system

Temperature(℃)	Viscosity of degassed oil (mPa·s)	Viscosity of gas-oil mixture (mPa·s)	Viscosity reducing ratio (%)	Viscosity reducing ratio
50	252000	3945	98.43	63.88
60	55279	1279	97.69	43.22
70	15889	504	96.83	31.53
80	6167	242	96.08	25.48
90	2470	118	95.22	20.93
100	1259	78	93.8	16.14
110	713	51	92.85	13.98
120	439	37	91.57	11.86

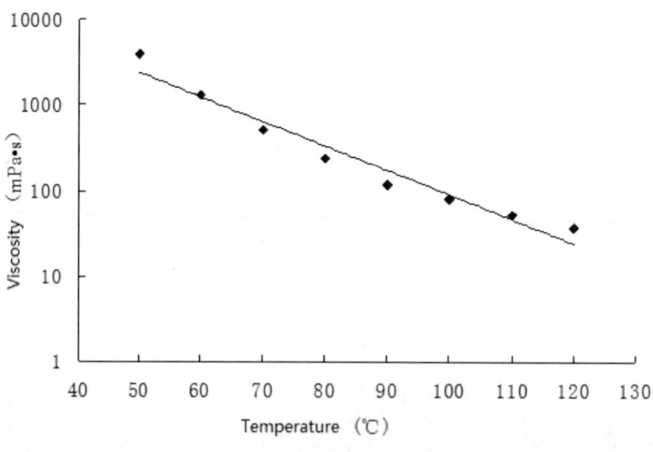

Figure 1.6. Viscosity-temperature curve of CO_2 and ultra-heavy oil system.

The relationship of viscosity-pressure after CO_2 dissolved in the ultra-heavy oil.

The viscosity of gas-oil mixture was measured under the following conditions: the DGOR (dissolved gas-oil ratio) was 50 sm^3/m^3, the temperature was 70°C, and the pressure was variable. Figure 1.6 shows that when the pressure of the gas-oil mixture is higher than the bubble point pressure, the viscosity will slightly increase after the CO_2 dissolves. This phenomenon is very beneficial in the process of injecting steam. For injecting steam, the pressure of affected radius is increased, but the mobility contrast of the mixture and hot water remains at a low level. On one hand, this can prevent the cusping from the high mobility contrast of injecting steam; on the other hand, the CO_2 continue to be dissolved in the ultra-heavy oil, which not only increases the swept volume but also increases the cleaning efficiency.

Table 1.6. Viscosity-pressure characteristics of CO_2 in ultra-heavy oil system

Pressure (MPa)	Viscosity of the gas-oil mixture (mPa·s)	Viscosity reducing rate (%)	Viscosity reducing ratio
10	464	97.08	34.24
11	473	97.02	33.61
12	481	96.97	33.05
13	485	96.95	32.78
14	498	96.87	31.93
15	510	96.79	31.17
16	525	96.7	30.28
17	533	96.65	29.83
18	547	96.56	29.07
19	558	96.49	28.49
20	572	96.4	27.8

1.4.2. The Dissolved CO_2 Characteristics of Different Water Degassed Oils

At 80°C, the corresponding saturation pressure, bulk coefficient, density and viscosity of different CO_2 solubilities of water-in-ultra-heavy-oil emulsions with water cuts of 10%, 30%, and 50%, respectively, were tested. The following section analyzes the influence of ultra-heavy oil emulsions with different water cuts on the characteristics of CO_2 dissolution:

(1) As shown in Figure 1.7, as the water cut value increases, the amount of dissolved CO_2 decreases under the same saturation pressure. With an increase in saturation pressure, the differences caused by the water cut is increasingly obvious. The biggest dissolved CO_2 volume in the ultra-heavy oil of the Zheng 411 reservoir is approximately 110 sm^3/m^3. The saturation pressure needed to achieve the dissolved amount is 30MPa, but under this pressure, the solubility of CO_2 in water is less than 40sm^3/m^3. In addition, after the saturation pressure reaches higher values than 10MPa, solubility will change slowly when the pressure changes. Thus, as the water cut of the crude oil increases, the capacity to dissolve CO_2 decreases. In

addition, as the pressure increases, the more obvious the difference when compared to pure oil.

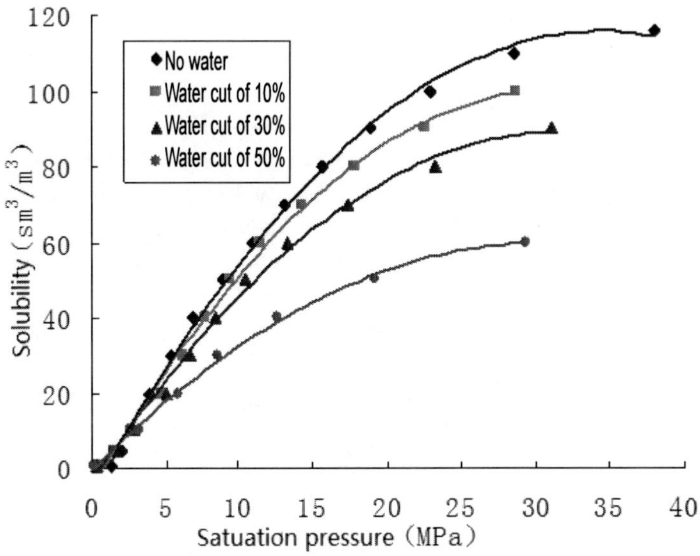

Figure 1.7. Curve of CO_2 solubility-saturation pressure under different water cuts.

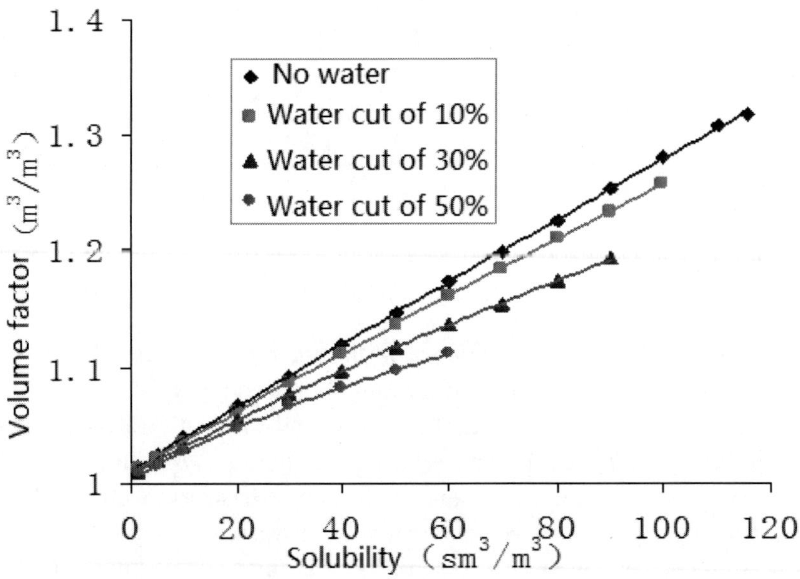

Figure 1.8. Curve of CO_2 solubility-volume ratio under different water cuts.

(2) As shown in Figure 1.8, under the same saturation pressure, as the water cut increases, the bulk coefficient of the ultra-heavy oil able to dissolve the same amount of CO_2 decreases [12]. This is because the volume expansion of ultra-heavy oil dissolved CO_2 is bigger than that of the water. When there is little water contained in the crude oil, the largest bulk coefficient of ultra-heavy oil dissolved CO_2 is 1.32. In comparison, the largest bulk

coefficient of water dissolved CO_2 is less than 1.05[11], which is far less than the expansion ability of ultra-heavy oil.

(3) As shown in Figure 1.9, as the water cut of the ultra-heavy oil with the same amount of dissolved CO_2 increases, the density increases. When the water cut is more than 30%, the density of the ultra-heavy oil containing a certain amount of dissolved CO_2 will increase with an increase in the amount of dissolved CO_2. The entirely different change in density between the ultra-heavy oil and water after dissolving CO_2 is the cause of this phenomenon. When there is little water, the density of ultra-heavy oil decreases upon an increase in the amount of dissolved CO_2. However, for water, the density increases upon an increase in the amount of dissolved CO_2 [13].

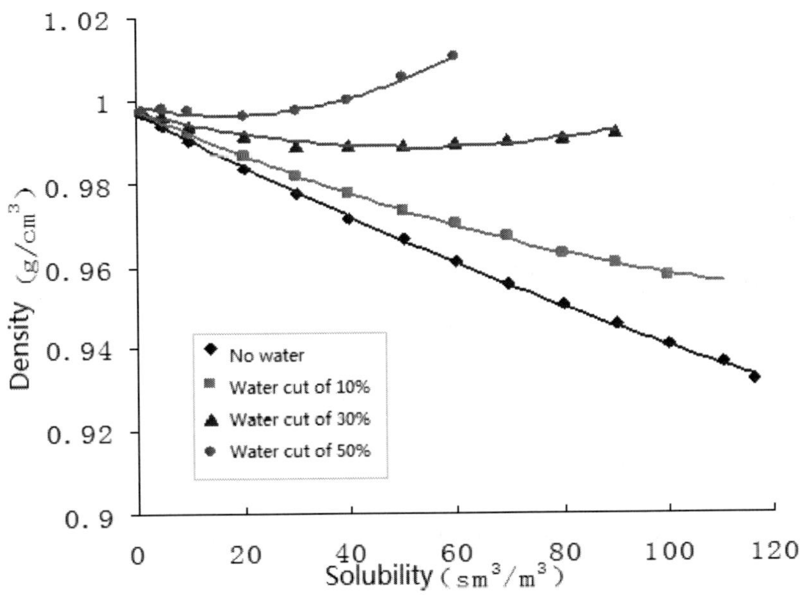

Figure 1.9. Curve of CO_2 solubility- density under different water cut.

(4) As shown in Figure 1.10, the viscosity of the oil-in-water emulsion increases dramatically upon an increase in the water cut. Under 80°C, the viscosity of water-in-ultra-heavy oil emulsion with a water cut of 50% is four times higher than the viscosity of ultra-heavy oil without water. As the water cut amount increases, the viscosity of the emulsion will decrease more rapidly upon an increase in CO_2 dissolution. When the CO_2 solubility in the water-in-ultra-heavy oil reaches $50sm^3/m^3$, the water-in-oil emulsion with different water cut amounts has nearly the same viscosity.

The experimental results show that as the water cut of crude oil increases, the more obvious the effect of reducing its viscosity by CO_2 becomes. This shows that CO_2 has a positive effect in breaking the water-in-ultra-heavy oil emulsion. After a long period of steam stimulation, the water cut of crude oil around the pity bottom, which is seriously emulsified, increased and the viscosity of crude oil increased greatly during the late steam stimulation. However, after injection of CO_2, viscosity can be significantly reduced, which will increase oil production.

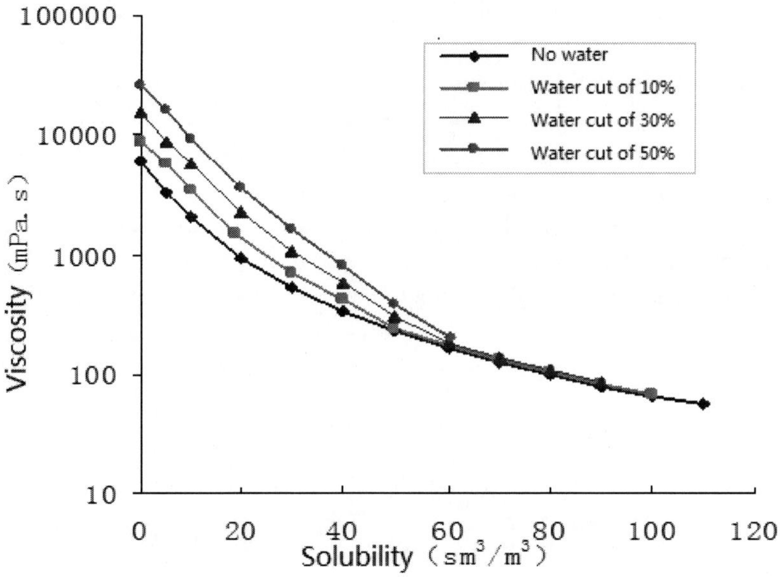

Figure 1.10. Curve of CO_2 solubility-viscosity under different water cuts.

2. CO_2-Crude Oil Interfacial Tension Experiment

2.1. Experimental Medicine

Crude oil: degassed oil of the Xinjiang Karamay oilfield; crude oil viscosity of 760000 m·Ps when at 20°C.

CO_2: Qingdao Tianyuan gas manufacturing company, purity 99.9%.

Table 2.1. SARA test of oil used in this work

Sample	Saturation (w %)	Aromatic (w %)	Colloid (w %)	Pitch (w %)	Total amount (w %)
Crude oil	40.20	21.57	36.94	1.29	100

2.2. Experimental Apparatus

The experimental apparatus used in this experiment is the Tracker interfacial tensions/rheometer. This device is shown in Figure 2.1 and mainly consists of a high temperature and high pressure vessel with a window, a pressure and temperature control system, a CO_2 injection system, a motor driving system and a computer data acquisition system. The high temperature and high pressure unit can reach maximums of 200°C and 20MPa. The electric heating mode, full-automatic acquisition and analysis of droplet shape are used to measure the gas-liquid and liquid-liquid interfacial tension.

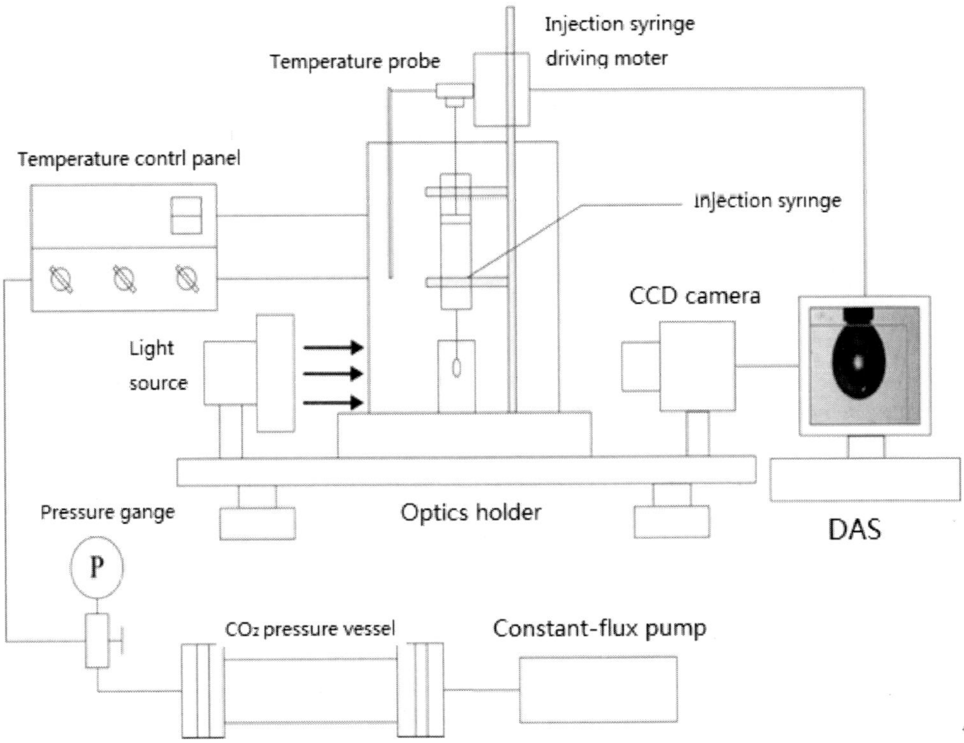

Figure 2.1. Schematic diagram of the experimental set-up for IFT measurement.

Pressure vessel: Haian oil technology instrument co. LTD, volume 1 L, maximum pressure: 50MPa, temperature range: normal~120°C.

Pressure gage: Shanghai automation instrument co., LTD, range of measurement: 0~25MPa.

2.3. Experimental Procedure

- The system leakproofness test under high temperature and high pressure uses nitrogen to test the experimental facility gas tightness.
- Before each test, acetone and alcohol are used to rinse the high temperature and high pressure container and injection syringes. They are then cleaned with distilled water.
- The injection syringe is installed, the window position adjusted, and low-pressure CO_2 is used to empty the high temperature and high pressure container.
- The pressure and temperature of the high temperature and high pressure container are increased until they become stable.
- The drive motor is used to make the liquid form a pendant drop on the syringe needle point.
- The density of CO_2 and droplets are input, using the measurement software to measure droplet size and interfacial tension.
- The interfacial tension of the droplet is measured 3~5 times under each condition to ensure the experiment has good repeatability.

2.4. Experimental Results and Analysis

(1) The influence of temperature and pressure

Figure 2.2 is a diagram that shows the relationship between the interfacial tension and pressure of the CO_2-crude oil system under different temperatures. The figure shows that as the pressure increases, the equilibrium interfacial tension of the CO_2-crude oil system decreases.

In addition, upon an increase in pressure, the reduction amplitude of the equilibrium interfacial tension decreases.

The equilibrium interfacial tension of the CO_2-crude oil system gradually tends to be approximately 5.5mN/m.

The influence of the law of temperature on interfacial tension is similar to that of the CO_2-water system.

When the pressure is less than 3.0MPa, the interfacial tension values measured under different temperatures are basically similar, and the influence of temperature on the equilibrium interfacial tension is not obvious.

When the pressure is greater than 3.0MPa, the influence of temperature on the interfacial tension becomes increasingly obvious; as the temperature increases, the interfacial tension of CO_2-crude oil becomes greater [14-16].

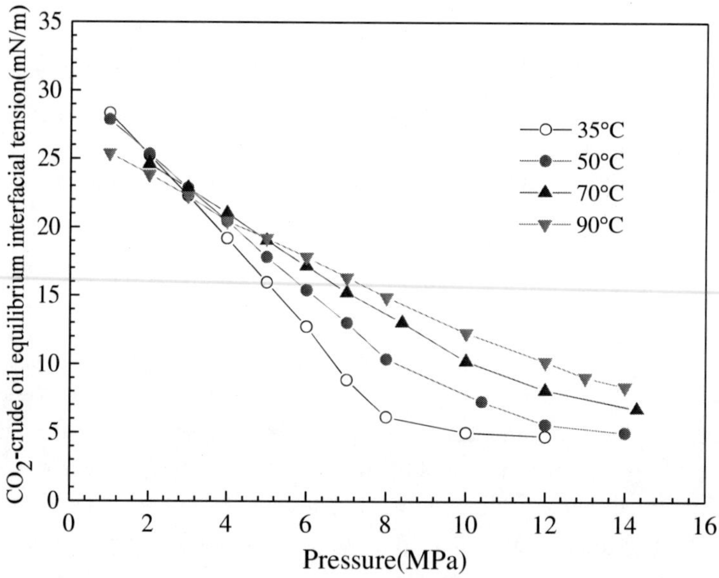

Figure 2.2. IFT of CO_2-oil system as a function of pressure at different temperatures.

(2) The effects of CO_2 solubility

The CO_2 solubility calculation formula suggested by Emera and Sarma [17] was used to calculate the CO_2 solubility in the crude oil under different temperatures and pressures:

$$s = 2.238 - 0.33y + 3.235y^{0.6474} - 4.8y^{0.25656} \tag{2.1}$$

$$y = \gamma \left[0.006897 \times \frac{(1.8T + 32)^{0.8}}{P_s} \right]^{\exp\left(\frac{1}{MW}\right)} \quad (2.2)$$

S represents the CO_2 solubility in the crude oil mole fraction. T represents the experimental temperature in °C. P_s represents the saturation pressure in MPa. MW represents the average molecular weight of crude oil in g/mol. ′Y represents the relative density of crude oil under 15.56 ℃.

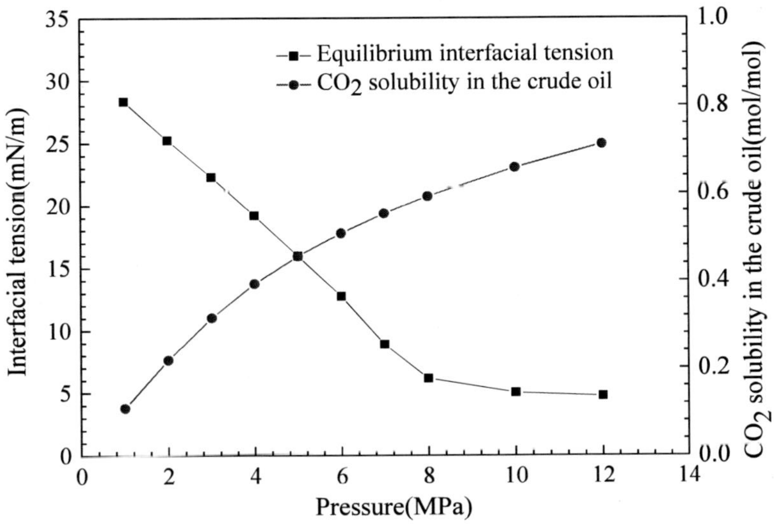

(a) 35℃

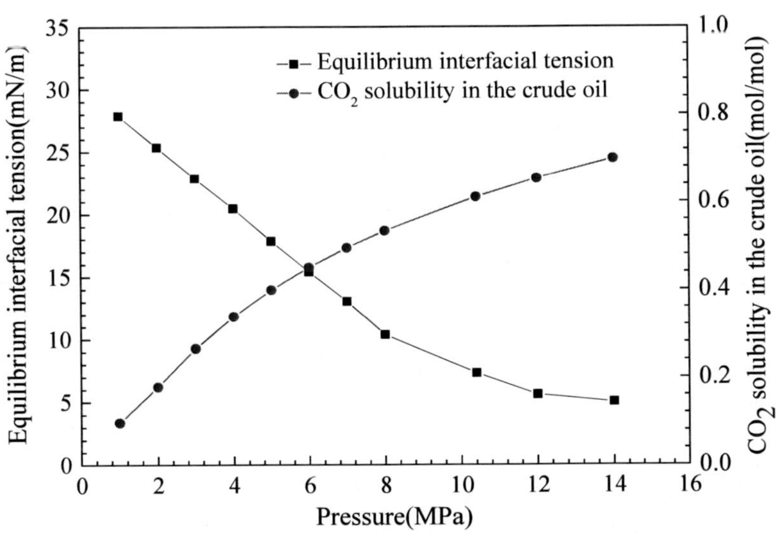

(b) 50℃

Figure 2.3. (Continued)

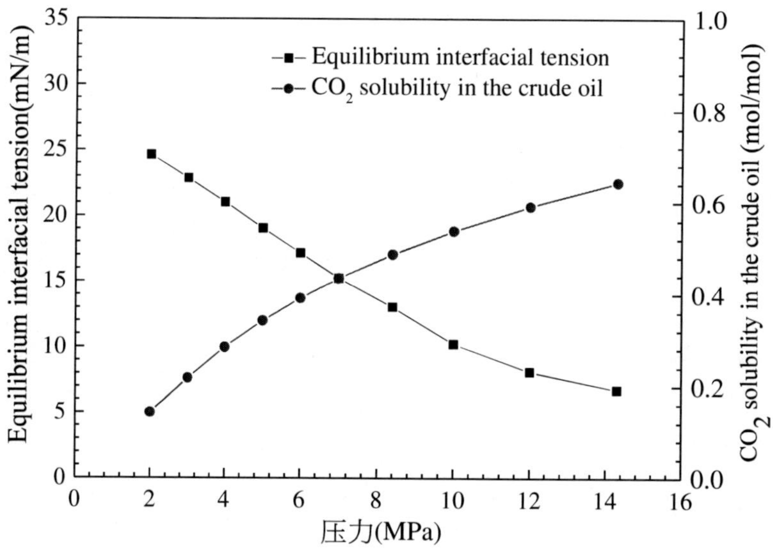

(c) 70°C

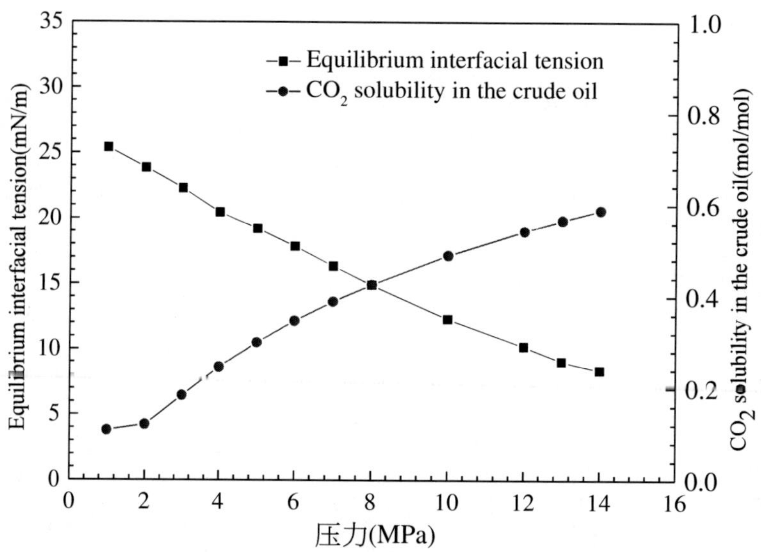

(d) 90°C

Figure 2.3. IFT of CO_2-oil system as a function of CO_2 solubility in oil at different temperatures.

Figure 2.3 is a diagram that shows the relationship between the solubility of CO_2 in crude oil and the change of the CO_2-crude oil interfacial tension. Additionally, the figure shows that upon an increase in pressure, the CO_2 solubility in the crude oil increases and the interfacial tension of CO_2-crude oil system decreases, which is similar to the change observed in the CO_2-water system. This suggests the dissolution of CO_2 in the crude oil contributes to the reduction of the CO_2-crude oil system interfacial tension, thereby enhancing the displacement efficiency of CO_2. Correlation analysis of the CO_2 solubility and interfacial tension was performed, and the spearman correlation coefficient was calculated.

The data shown in Figure 2.4 suggests that the CO_2 solubility in the crude oil and the equilibrium interfacial tension of CO_2-crude oil system show nearly the biggest negative correlation, in which the spearman correlation coefficient is -0.9863 [18-20].

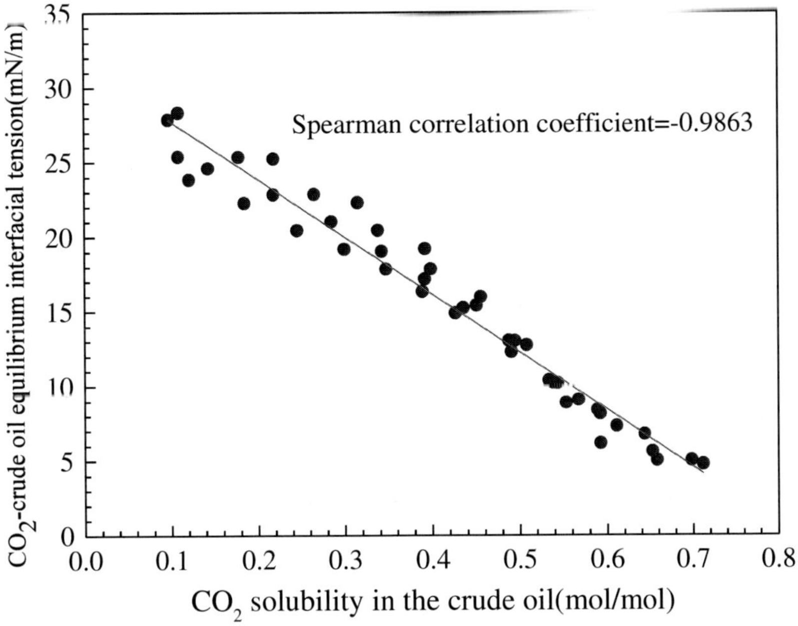

Figure 2.4. IFT of CO_2-oil system as a function of CO_2 solubility in oil.

3. CO_2 Core Displacement Experiment

3.1. Experimental Objective

Through the long core displacement experiment, the law of CO_2 enhancing oil recovery and the influence of the law of the depressurizing exploitation after shut-in diffusion to enhance oil recovery under different temperatures and pressures is researched. Second, the basic properties of the oil produced in different stages of displacement and the changing law of four components as well as the influence law of temperature and pressure on CO_2 flooding is researched. Finally, the production characteristics of CO_2 flooding is researched.

3.2. Experimental Drug

Crude oil: the simulated oil consisting of crude oil and kerosene.
CO_2: produced by Qingdao Tianyuan gas manufacturing co., LTD.
Distilled water: water that underwent distillation.

3.3. Experimental Apparatus and Flow Chart

(1) Experimental apparatus

Double-plunger metering pump, middle pressure vessel, pressure transmitter, sand-filling module tube, gas flow meter, balance, back-pressure valve, manual pump, etc.

(2) Flow chart of experimental apparatus

The flow chart is shown in Figure 3.1.

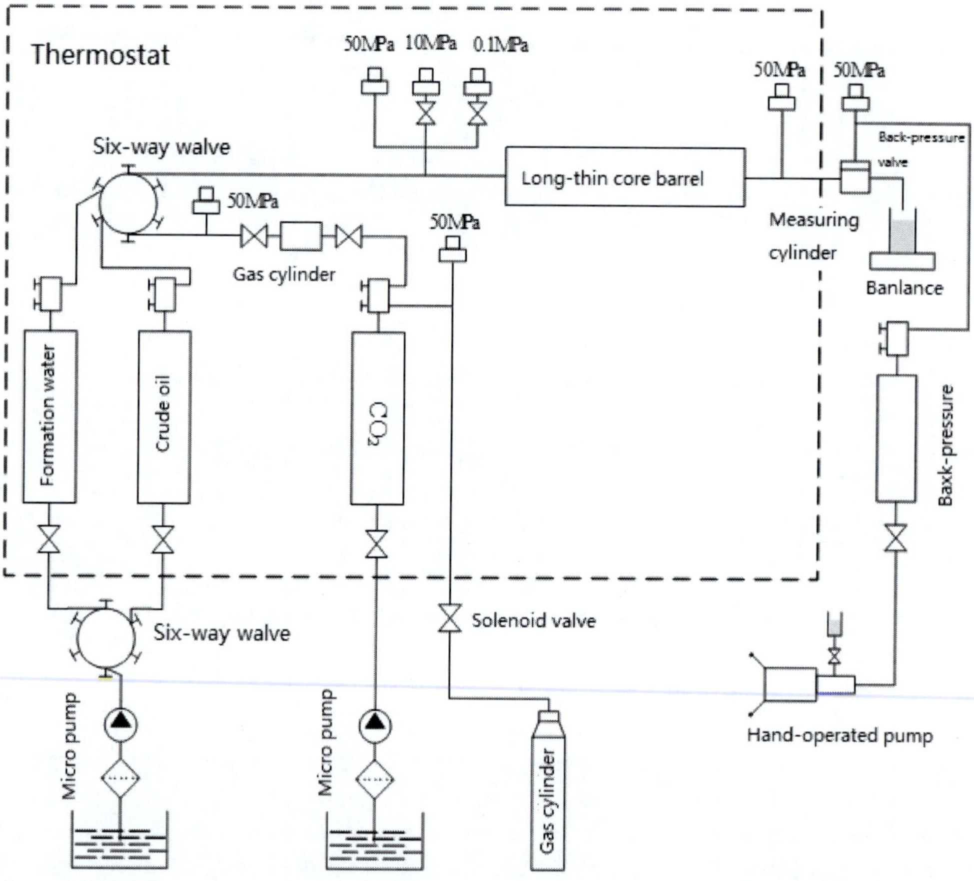

Figure 3.1. Experimental set-up for CO_2 flooding.

3.4. Experimental Core Data

The CO_2 long thin tube core displacement experiment uses the manual-work sand-filling core model in which the length of the core is 60 centimeters and the core diameter is 2.5 centimeters. The basic data of the experimental core is shown in table 3.1.

Table 3.1. Core parameters for CO_2 flooding experiments

Core number	Diameter (cm)	Length (cm)	Pore volume (mL)	Porosity (%)	Water phase permeability (10^{-3} μm^2)	Initial oil saturation (%)
#1	2.5	60	91.00	29.93	219.28	80.22
#2	2.5	60	104.00	34.21	254.34	80.77
# 3	2.5	60	107.00	35.19	320.98	72.90
#4	2.5	60	102.00	33.55	222.77	67.65
#5	2.5	60	88.00	28.95	205.08	78.41
#6	2.5	60	96.00	31.58	228.86	71.88
#7	2.5	60	88.00	28.95	217.48	84.09
#8	2.5	60	96.00	31.58	226.85	73.96
#9	2.5	60	92.00	30.26	175.51	88.04

3.5. Experimental Procedure

1. The core tube is filled according to the experimental condition, vacuumized for 3 hours, and saturated with water. Its volume is then measured.
2. Core displacement is conducted with distilled water at set constant flows of 1.0 mL/min and 2.0 mL/min. The core permeability under different flow rates is measured twice; their average is the water phase permeability of the core.
3. The experimental core is allowed to equilibrate to the constant experimental temperature for 5 hours. It is then saturated with oil at a constant flow velocity of 0.1 mL/min.
4. A gas flow meter is used to control the displacement of CO_2 at a rate of 1.5 mL/min until the injective volume reaches 2.0PV. At this point, the oil recovery no longer changes, thus stopping the displacement.
5. The entry and exit sides of the core barrel are closed, causing CO_2 to diffuse in the core barrel for 30 min. The exit entry is then opened to conduct the depressurizing production at the rate of 0.1MPa/min until the pressure drops to 1.0 MPa.
6. A densimeter and rotary viscosimeter are used to measure the density and viscosity of the crude oil, respectively, before and after CO_2. A 4-component analysis on the crude oil is produced.
7. The oil sand left in the core barrel is removed along the direction of displacement, which is equal to three in this procedure. The oil content of the oil sand and the four components of the crude oil in the three sections are then measured, respectively.

3.6. Experimental Results and Analysis

(1) Analysis of CO_2 flooding oil recovery

The experiment records data during the CO_2 displacement process, the CO_2 gas breakthrough oil recovery, the oil recovery at the end of the gas flooding, and the oil recovery of depressurizing production after the CO_2 diffusion in the core for 30 min when the pressure drops to 1 MPa. The oil recovery for CO_2 flooding under different experimental conditions is shown in Figure 3.2.

Table 3.2. The oil recovery for CO_2 flooding

Core number	Experimental temperature (°C)	Gas injection pressure (MPa)	Gas injection volume until gas breakthrough (PV)	Oil recovery until gas breakthrough (%)	Oil recovery of CO_2 flooding (%)
#1	60.0	2.2	0.21	14.68	37.47
#2	60.0	5.2	0.10	19.13	37.53
#3	60.0	9.3	0.32	19.50	60.91
#4	60.0	14.5	0.15	12.47	74.71
#5	60.0	20.3	0.14	10.16	79.34
#7	22.0	2.2	0.15	11.84	35.85
#8	22.0	5.2	0.81	17.78	35.94
#9	22.0	9.2	0.46	15.47	38.21

Figures 3.2 and 3.3 show the change in CO_2 flooding oil recovery with CO_2 injection PV number at 60 and 22°C, respectively. As shown in the figures, upon an increase in the CO_2 injection PV number, the oil recovery increases gradually and reaches its largest value at approximately 1.5 PV.

The biggest CO_2 injection volume for each group of experiments is 2.0PV, and when the injection volume reaches 2.0 PV, the core does not produce oil anymore. When the CO_2 injection pressure is low, P=2.0 MPa, the CO_2 flooding initial oil recovery increases rapidly. However, under a similar CO_2 injection PV number, the initial oil recovery increases slowly when the pressure is 9.3 MPa because the CO_2 solubility in the crude oil is high under high pressure and most of the supercritical CO_2 injected at the beginning dissolved in the crude oil; they were not significantly affected by the displacement [21-23]. At 60°C, the oil recovery increases as the CO_2 injection pressure increases. However, at 22°C, the injection pressure does not have an obvious effect on the oil recovery; when the CO_2 injection pressure increases, the oil recovery remains unchanged.

Figures 3.4 and 3.5 show the change in CO_2 flooding oil recovery, gas injection volume and oil recovery during the CO_2 gas breakthrough with the CO_2 injection pressure at 60°C and 22°C, respectively. As shown in Figure 3.2, when the temperature is 60°C and the CO_2 injection pressure increases, the CO_2 will transform from a gas state to a supercritical state. When the CO_2 is in a gas state and the injection pressure increases, the oil recovery remains basically unchanged.

The oil recovery is 37.47% at a pressure of 2.2MPa and is 37.53% at a pressure of 5.2 MPa. When the CO_2 undergoes a supercritical state and the injection pressure increases, the oil recovery increases rapidly. When the injection pressure is higher than the MMP=16.3 MPa, the extent of the increase will become smaller, and the oil recovery is able to reach 79.34% at this point because as the system pressure increases, the CO_2 solubility in the crude oil increases and the interfacial tension of the CO_2-crude oil system decreases. This effect will lead to a higher CO_2 displacement efficiency and greater oil recovery. When the pressure is higher than the MMP, the interfacial tension of the CO_2-crude oil system will not change anymore, so the extent of the increase in the oil recovery becomes smaller [24-27].

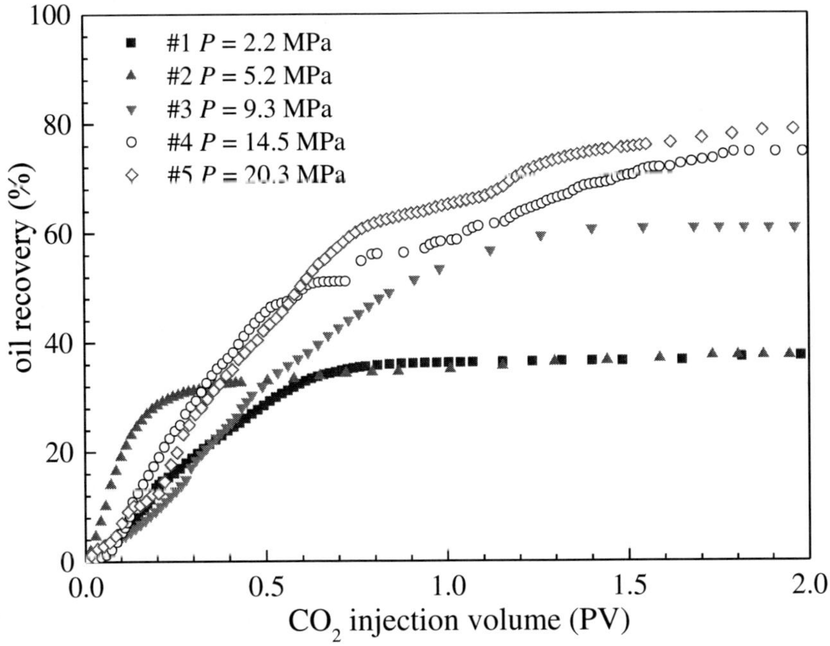

Figure 3.2. CO_2 flooding oil recovery for different pressures at 60°C.

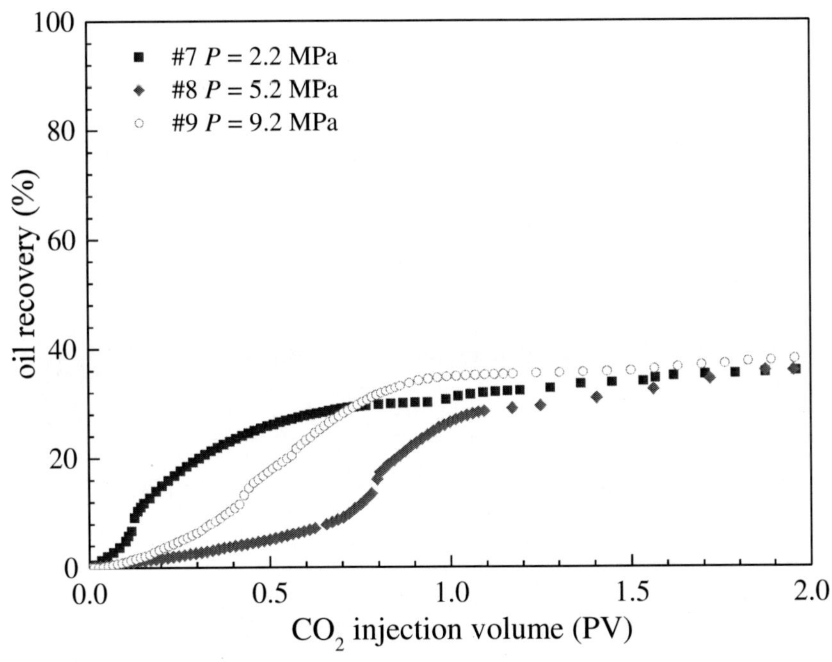

Figure 3.3. CO_2 flooding oil recovery for different pressures at 22°C.

By comparing the oil recovery of the gas breakthrough with the CO_2-flooding oil recovery, we can see that as the pressure increases, the extent of CO_2 flooding enhancing oil recovery increases. Because the CO_2 injection volume of gas breakthrough represents the speed of CO_2 gas channeling, as the injection volume during the gas breakthrough decreases, the CO_2 flooding gas channeling increases in speed. As shown in Figure 3.4, the injection volume of CO_2 flooding during gas breakthrough decreases gradually upon increases in pressure in the gaseous and supercritical CO_2 areas, respectively. Its sudden rise near the critical point suggests that as the injection pressure increases, the easier the CO_2 gas channeling becomes. In addition, gaseous CO_2 flooding is easier for gas channeling than supercritical CO_2 flooding [28, 29].

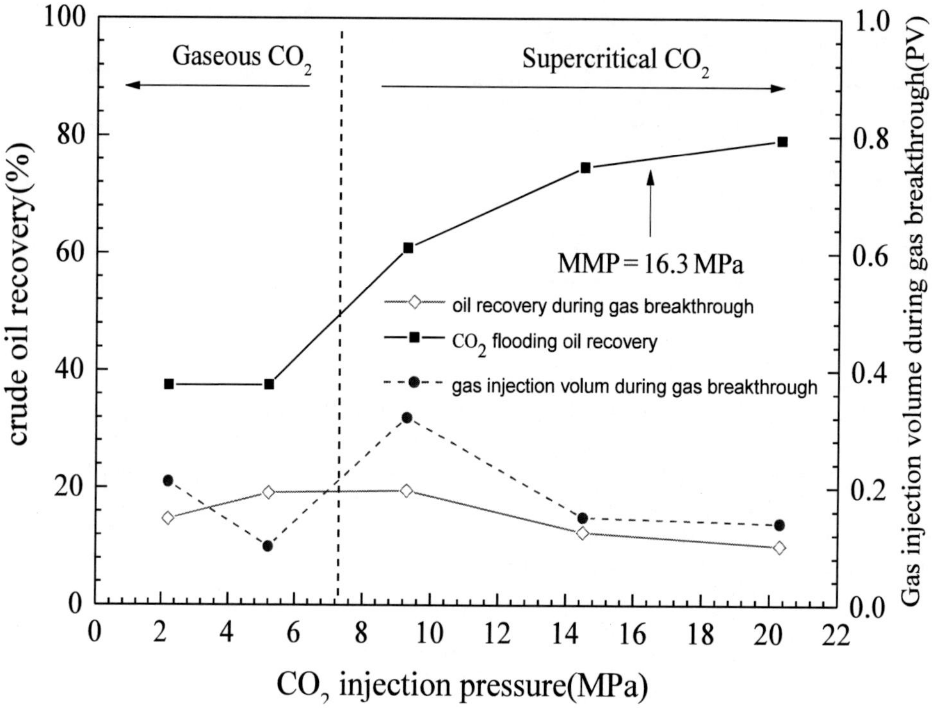

Figure 3.4. CO_2 flooding oil recovery for different pressures at 60℃.

Table 3.3. The results for CO_2 blowout

Core number	Experimental temperature (°C)	Gas injection pressure (MPa)	CO_2 flooding oil recovery (%)	Residual oil saturation (%)	Depressurizing production oil recovery (%)	Overall oil recovery (%)
#1	60.0	2.2	37.47	50.16	6.06	43.53
#2	60.0	5.2	37.53	50.46	7.01	44.54
#3	60.0	9.3	60.91	28.50	5.23	61.60
#4	60.0	14.5	74.71	17.11	4.11	78.82
#5	60.0	20.3	79.34	16.20	2.78	82.12
#7	22.0	2.2	35.85	53.95	1.81	37.66
#8	22.0	5.2	35.94	47.37	3.44	39.38
#9	22.0	9.2	38.21	54.40	11.72	49.93

Figure 3.5 shows that as CO_2 transforms from the gaseous state to the liquid state, CO_2 flooding oil recovery remains basically unchanged and the oil recovery of gas breakthrough remains basically the same. While the CO_2 injection volume of the liquid CO_2 flooding during gas breakthrough is obviously greater than that of the gaseous CO_2 flooding, the gas channeling speed of gaseous CO_2 flooding is faster than that of the liquid CO_2 flooding.

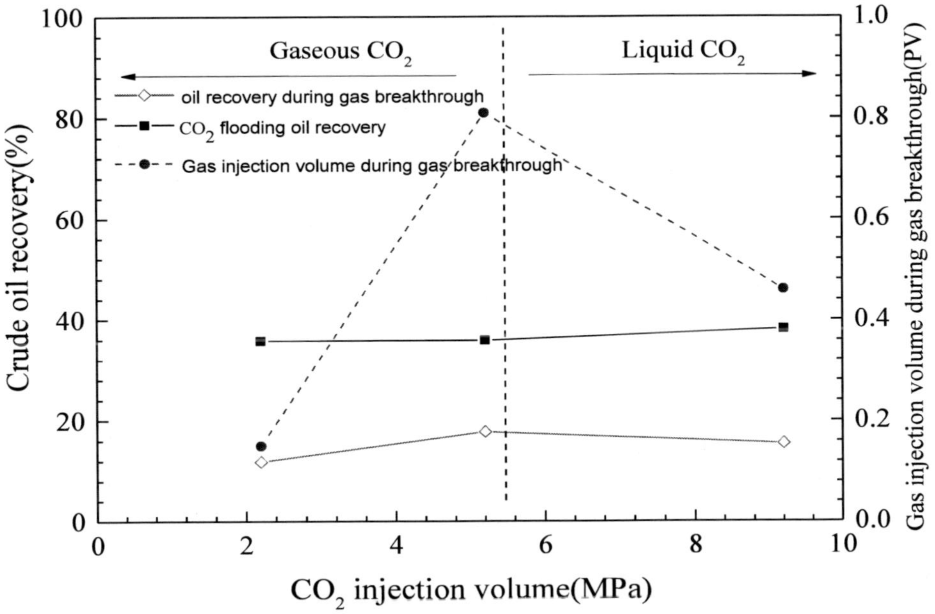

Figure 3.5. CO_2 flooding oil recovery for different pressures at 22 °C.

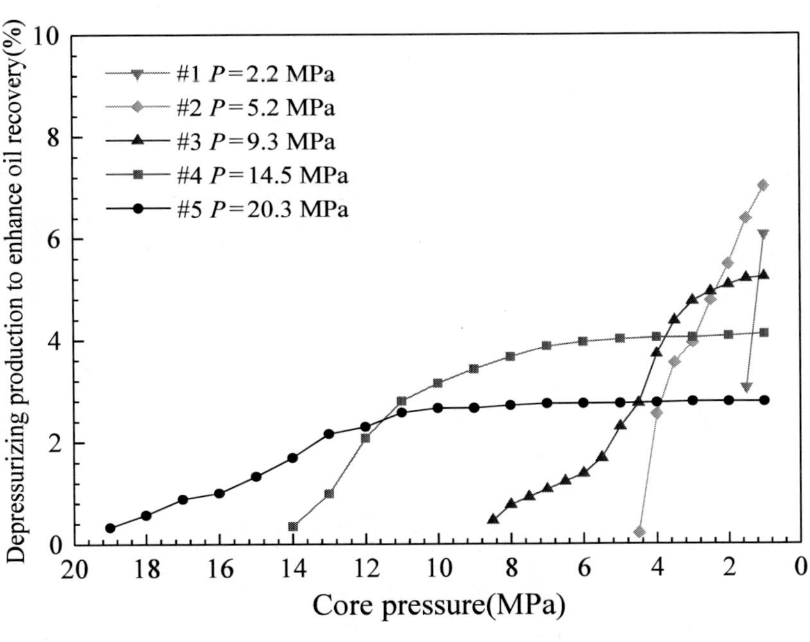

Figure 3.6. Blowout oil recovery for different pressures at 60°C.

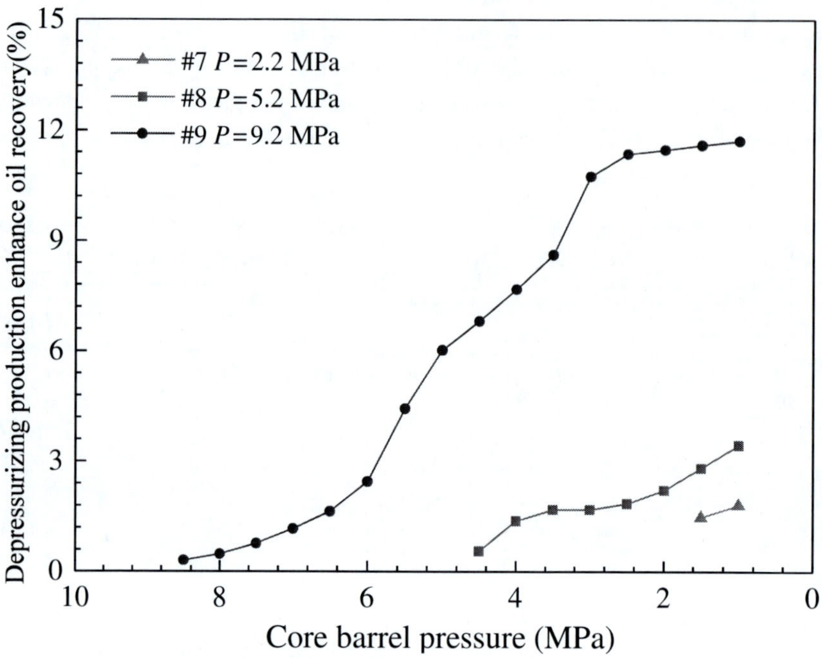

Figure 3.7. Blowout oil recovery for different pressures at 22°C.

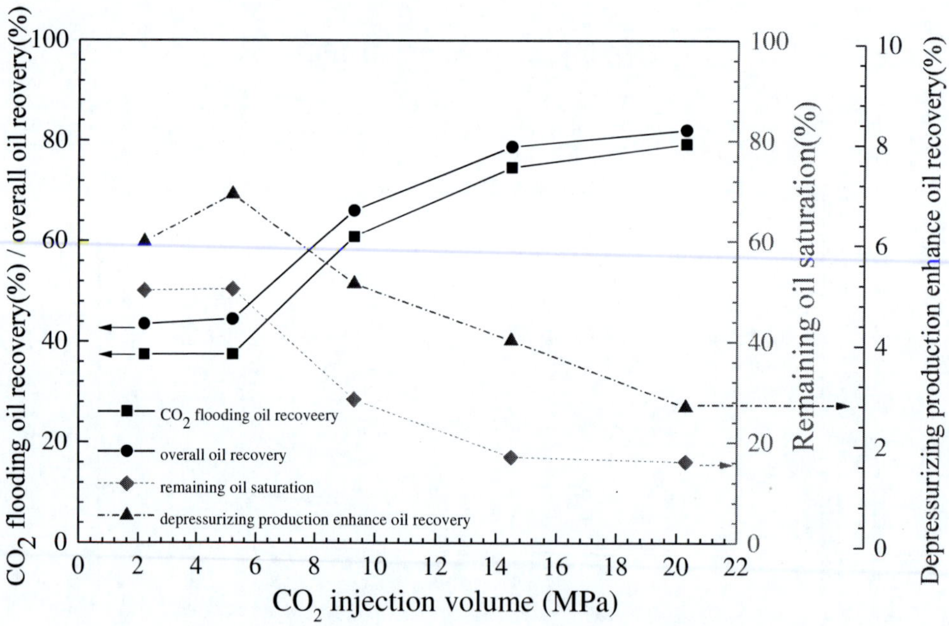

Figure 3.8. Oil Recovery and oil saturation as a function of pressure at 60°C.

(2) Depressurizing production oil recovery analysis

The exploration result of conducting depressurizing production after 30 min of CO_2 diffusion is shown in Figure 3.3. Figures 3.6 and 3.7 show that conducting depressurizing production after closing the exit and entry ends for 30 min will result in a gradual increase in the oil recovery accompanied by a decrease in formation pressure. The depressurizing production under different pressures at 60°C can improve the oil recovery from 2.78% to 7.01%. The depressurizing production at 22°C can enhance oil recovery from 1.81% to 11.72%. The depressurizing production enhances oil recovery by approximately 5.27% in average.

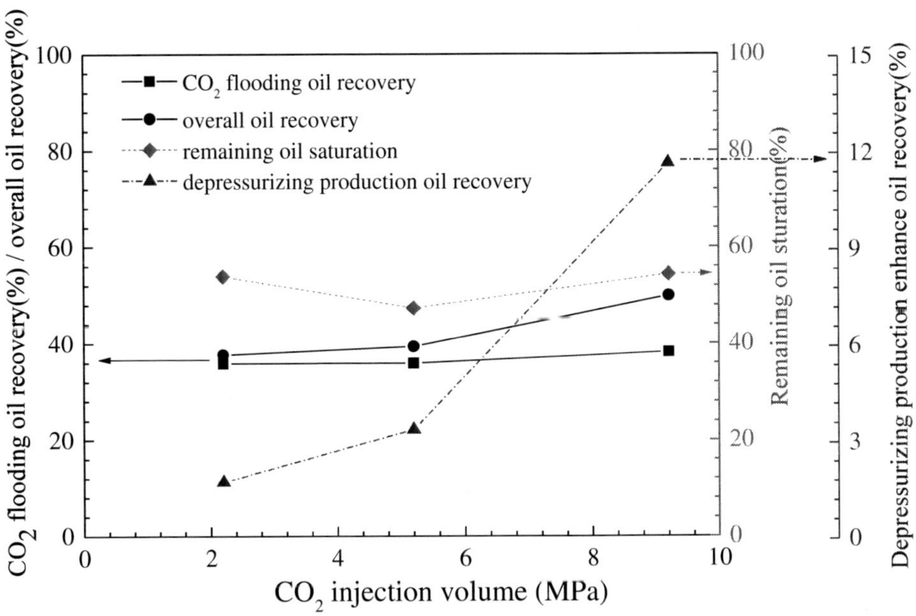

Figure 3.9. Oil Recovery and oil saturation as a function of pressure at 22°C.

As shown in Figures 3.8 and 3.9, through the analysis of the CO_2 flooding oil recovery, the depressurizing production oil recovery, the overall oil recovery and the residual oil saturation of core barrel after CO_2 flooding, the depressurizing production enhanced oil recovery at 60°C decreases upon an increase in the CO_2 injection pressure. However, at 22°C, it increases as the CO_2 injection pressure increases. One reason for this phenomenon is that the depressurizing production EOR is mainly affected by 2 factors: first, the residual oil saturation of core barrel after CO_2 flooding, and second, the interaction of diffusion and dissolution between the CO_2 and oil recovery during the shut-in diffusion process. Between these factors, the first is most dominant. Figure 3.8 shows that upon an increase in the CO_2 injection pressure, the CO_2 flooding oil recovery increases obviously, so the residual oil saturation of the core barrel decreases. As the residual oil saturation decreases, the effect of depressurizing production becomes less obvious and the EOR decreases. When the temperature is 22°C, as shown in Figure 3.9, the CO_2 flooding oil recovery under different injection pressures is basically the same, so the residual oil recovery remains fundamentally unchanged. At this point, the dominant factor that affects the depressurizing production effect is the diffusion and dissolution between CO_2 and crude oil. As the pressure increases, the CO_2 dissolution in the crude oil increases, resulting in a more obvious depressurizing production effect. Thus, when the temperature is 22°C, as the pressure increases, the extent of EOR increases [30-36].

(3) Analysis on gas to oil exchange ratio of CO_2 flooding

The average gas to oil exchange ratio of CO_2 flooding is defined as the ratio of the amount of CO_2 injected into the core and oil production by CO_2 flooding. The calculation results for the average gas to oil exchange ratio of CO_2 flooding are shown in Table 3.4. Figure 3-10 and Figure 3.11 show that the average gas to oil exchange ratio before gas breakthrough is significantly higher than the total average gas to oil exchange ratio when the pressure is less than the critical pressure of CO_2 (7.38 MPa). This is because after gas breakthrough, the gas production substantially increases, the oil production decreases and the gas to oil exchange ratio of CO_2 decreases. In addition, as the injection pressure increases, the average gas to oil exchange ratio decreases. When the pressure increases, the oil production increases; however, at higher pressures, a larger amount of substance of CO_2 exists, which is why the gas to oil exchange ratio decreases [37-39].

Table 3.4. Results for gas to oil exchange ratio of CO_2 flooding

Core	Test temperature (°C)	Gas injection pressure (MPa)	Average gas to oil exchange ratio before gas breakthrough (g/g)	Total average gas to oil exchange ratio (g/g)
# 1	60.0	2.20	12.41	3.36
# 2	60.0	5.20	12.54	1.26
# 3	60.0	9.30	1.49	0.76
# 4	60.0	14.50	0.81	0.36
# 5	60.0	20.30	0.66	0.35
# 7	22.0	2.20	12.74	2.82
# 8	22.0	5.20	0.95	0.80
# 9	22.0	9.20	0.31	0.17

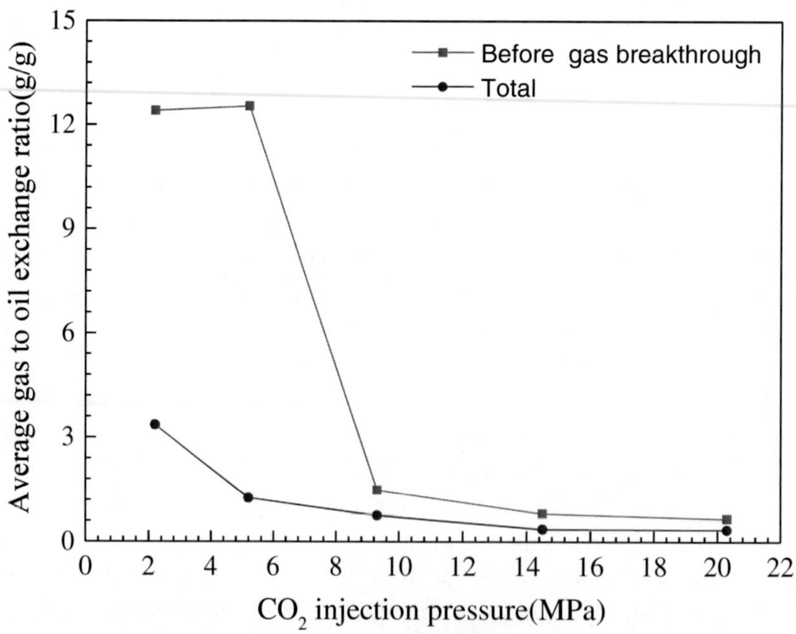

Figure 3.10. Gas to oil exchange ratio as a function of pressure at 60°C.

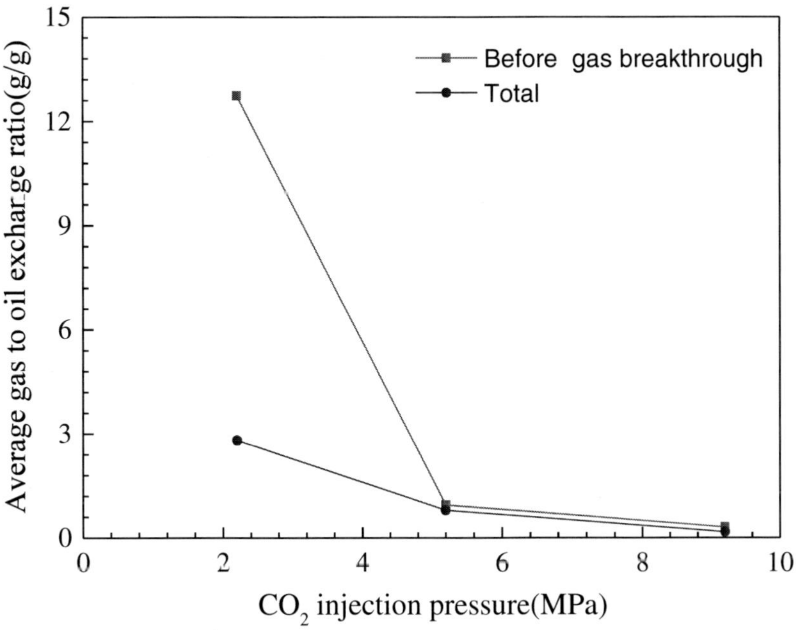

Figure 3.11. Gas to oil exchange ratio as a function of pressure at 22°C.

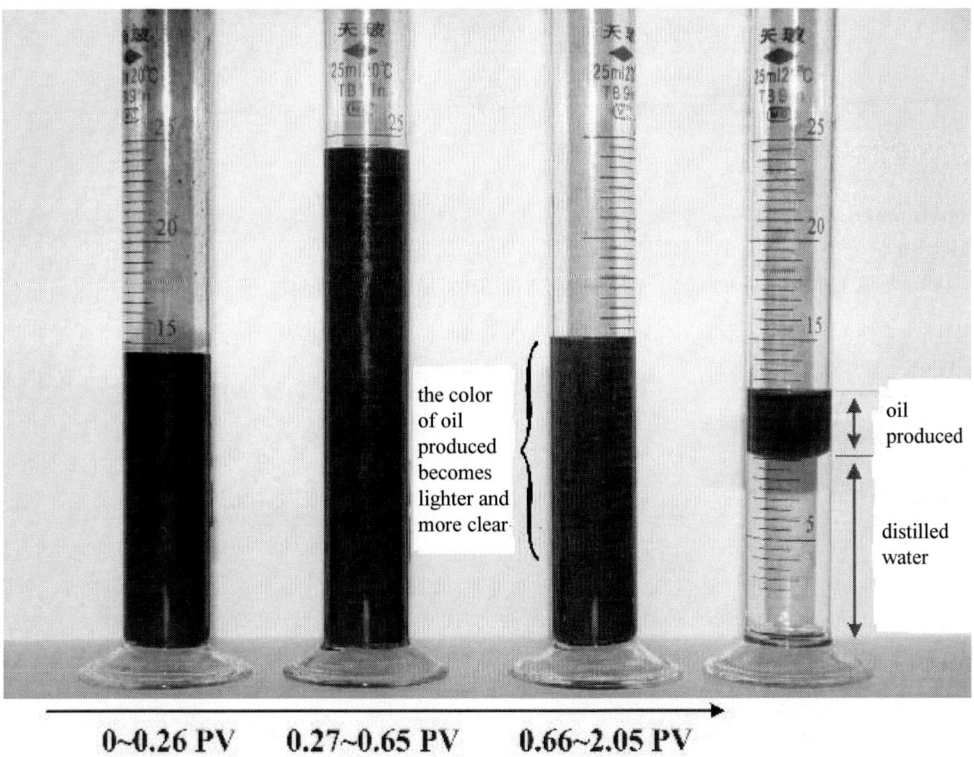

Figure 3.12. CO_2-produced oil as a function of time for # 5 (60°C, 20.3 MPa).

(4) Analysis on the basic properties of oil produced by CO_2 flooding

The oil produced in different production stages in experiment #5 is shown in Figure 3.12. Figure 3.12 shows that as CO_2 flooding proceeds, the color of oil produced becomes lighter and clearer. This phenomenon occurs because during CO_2 flooding, the light components of the oil produced are extracted by CO_2, and the amount of the substance of oil produced is decreasing. The experiment #5 pressure is greater than minimum miscibility pressure for the CO_2-crude oil, indicating that the experiment has reached the CO_2 miscible flooding.

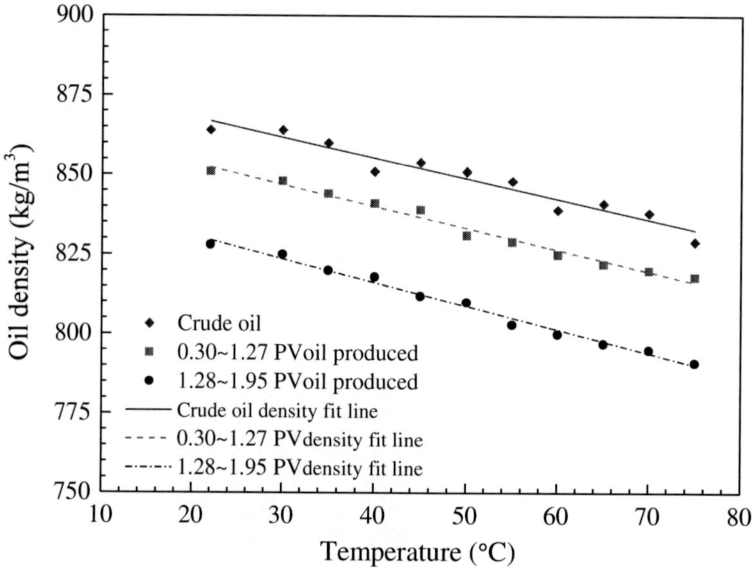

Figure 3.13. CO_2-produced oil density as a function of temperature for # 6 (60 ℃, 20.5 MPa).

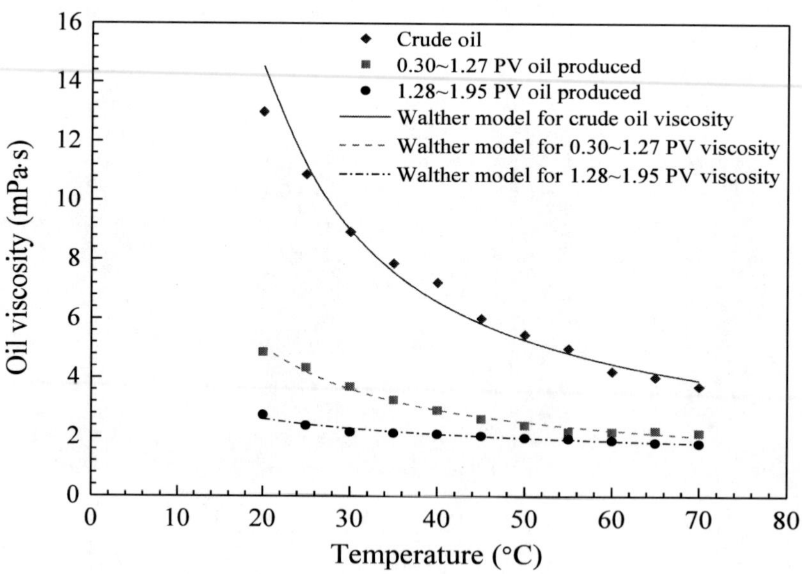

Figure 3.14. CO_2-produced oil viscosity as a function of temperature for # 6 (60 ℃, 20.5 MPa).

To further analyze the properties of the oil produced in different production stages of CO_2 flooding, a detailed experimental study on the basic property of oil produced in different stages of experiment #6 was performed. The results of measuring the density and viscosity of the oil produced in different production stages are shown in Figures 3.13 and 3.14.

Figure 3.13 and 3.14 show that the properties of oil produced in different production stages of CO_2 flooding change. Upon an increase in the production time, the viscosity and density of the oil produced decrease. The viscosity and density of crude oil are 864 kg/m^3 and 12.99 mP s at room temperature, respectively.

The viscosity and density values of oil produced when the injected CO_2 volume is in the range of 0.30 to 1.27 PV are 4.86 mP·s and 851 kg/m^3, respectively, and they are 2.75 mP·s and 828 kg/m^3 when the injected CO_2 volume is in the range of 0.30 to 1.27 PV at room temperature, respectively. The density fit lines and the viscosity Walther models of oil produced in different production stages are as follows:

$$\rho_{\mathrm{oil}} = -0.64T + 881.06 \quad R^2 = 0.9301 \tag{3.1}$$

$$\rho_{\mathrm{oil,1}} = -0.68T + 867.27 \quad R^2 = 0.9822 \tag{3.2}$$

$$\rho_{\mathrm{oil,2}} = -0.74T + 845.80 \quad R^2 = 0.9879 \tag{3.3}$$

$$\log[\log(\mu_{\mathrm{oil}} + 0.7)] = 0.6725 - 0.4609\log T \quad R^2 = 0.9639 \tag{3.4}$$

$$\log[\log(\mu_{\mathrm{oil,1}} + 0.7)] = 0.4539 - 0.4418\log T \quad R^2 = 0.9839 \tag{3.5}$$

$$\log[\log(\mu_{\mathrm{oil,2}} + 0.7)] = -0.0022 - 0.2165\log T \quad R^2 = 0.9517 \tag{3.6}$$

The density and viscosity values of the oil produced after gas breakthrough under different CO_2 injection pressures at normal temperature and pressure (22 °C, 0.1 MPa) were measured; these results are shown in Figure 3.15. Figure 3.15 shows that upon an increase in injection pressure, the density and viscosity of the oil produced after gas breakthrough decline.

As the injection pressure increases, the amount of dissolved CO_2 in the oil increases and the interaction between the CO_2 and the oil is stronger. As a result, the CO_2 extracts the light components of oil inside core tubes more easily, and the oil produced after gas breakthrough fully contacts CO_2. These phenomena are why the components of the oil produced are lighter and the density and viscosity of the oil produced are smaller [40, 41].

(5) Rules of the asphaltene of oil produced

Wang et al. [42] defined the threshold pressure of precipitation of asphaltene (P_{asp}) as the lowest possible pressure at which the interaction between CO_2 and oil causes precipitation of asphaltene.

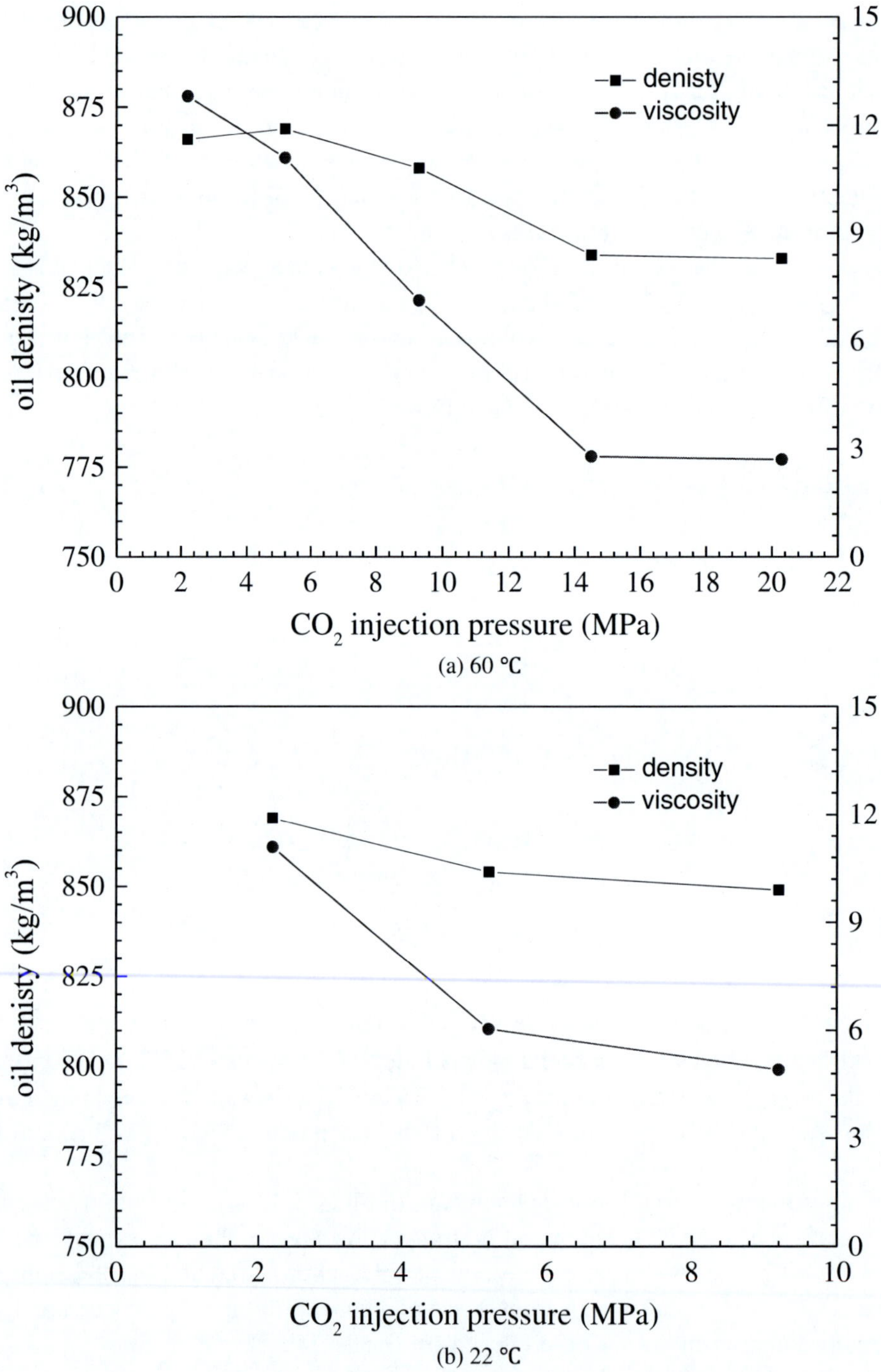

Figure 3.15. CO_2-produced oil density and viscosity as a function of injection pressure.

At pressures less than $P_{asp,}$ the interaction between CO_2 and oil would not cause the precipitation of asphaltene. On the contrary, at pressures above P_{asp}, precipitation would occur. The four components of the oil produced before and after gas breakthrough under different experimental conditions were measured, and the results are shown in Figure 3.15 and Figure 3.17. Compared to the asphaltene content of crude oil, the content of the oil produced is little different when the injection pressure is smaller (when the temperature is 60°C, the pressure is smaller than 9.3 MPa, and when the temperature is 22°C, the pressure is smaller than 2.2 MPa). This phenomenon occurs because the injection pressure does not reach P_{asp}. However, the saturates of the oil produced decrease and aromatics increase.

When the injection pressure is larger (when the temperature is 60°C, the pressure is larger than 14.5 MPa, and when temperature is 22°C, the pressure is larger than 5.2 MPa), the asphaltene content of the oil produced significantly decreases because precipitation of asphaltene occurs, and the precipitation adsorbs and deposits into the porous media. Thus, P_{asp} can be predicted in the range of 9.3~14.5 MPa when the temperature is 60 °C and in the range of 2.2 MPa~5.2MPa when the temperature is 22°C [43].

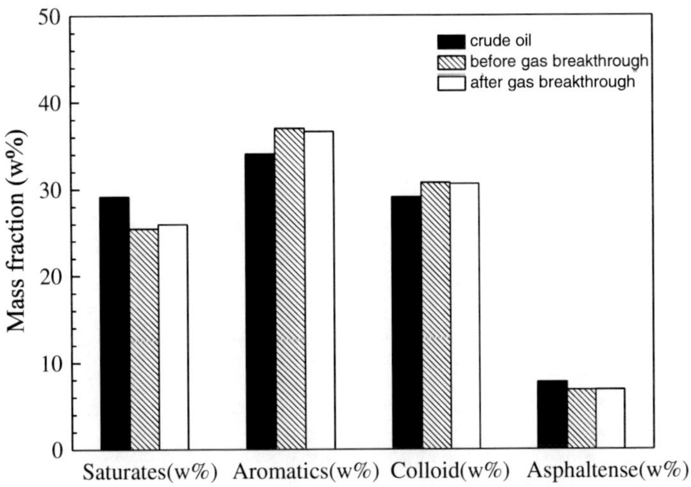

(a) Experiment # 1 (60°C, 2.2 MPa)

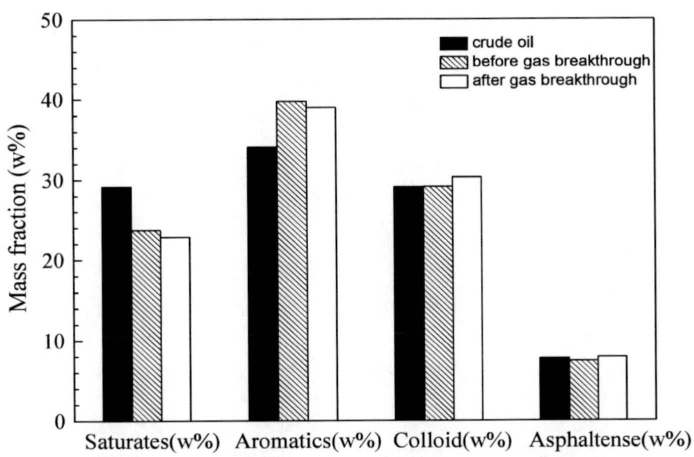

(b) Experiment # 2 (60°C, 5.2 MPa)

Figure 3.16. (Continued)

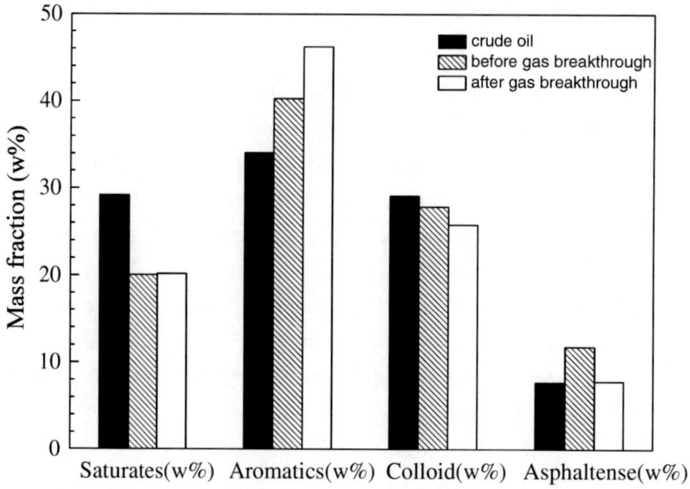

(c) Experiment # 3 (60°C, 9.3 MPa)

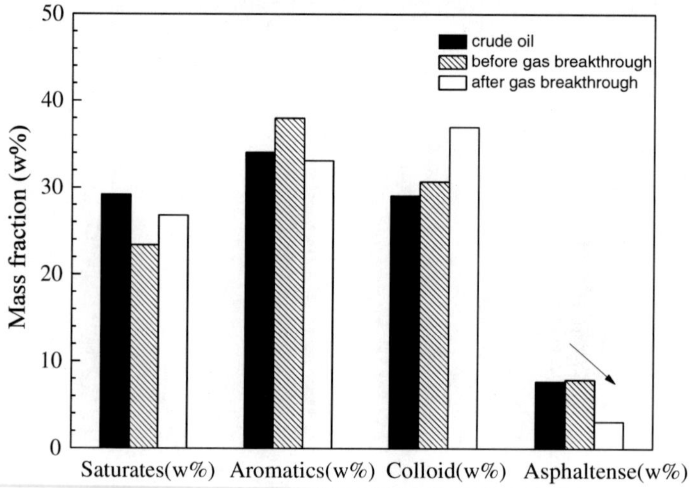

(d) Experiment # 4 (60°C, 14.5 MPa)

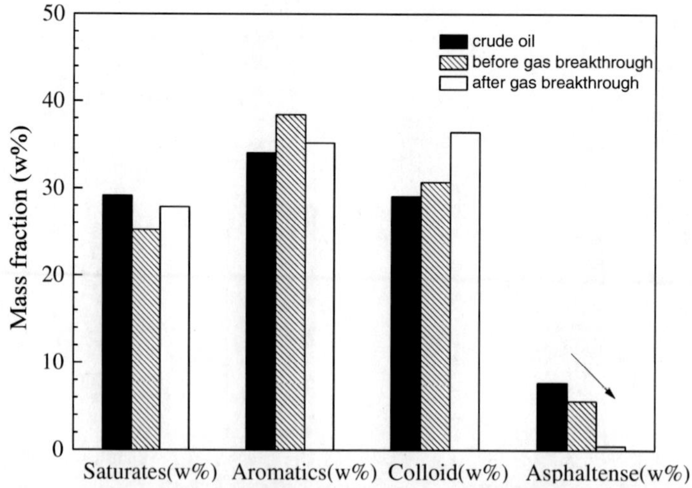

(e) Experiment # 5 (60°C, 20.3 MPa)

Figure 3.16. SARA test of CO_2-produced oil before and after gas breakthrough at 60°C.

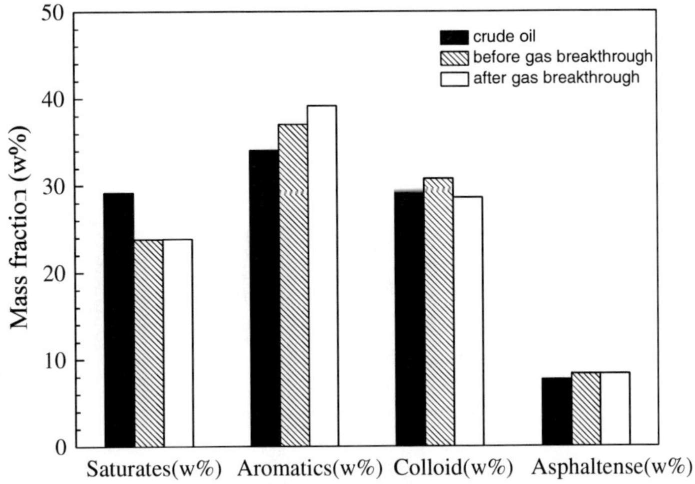

(a) Experiment # 7 (22°C, 2.2 MPa)

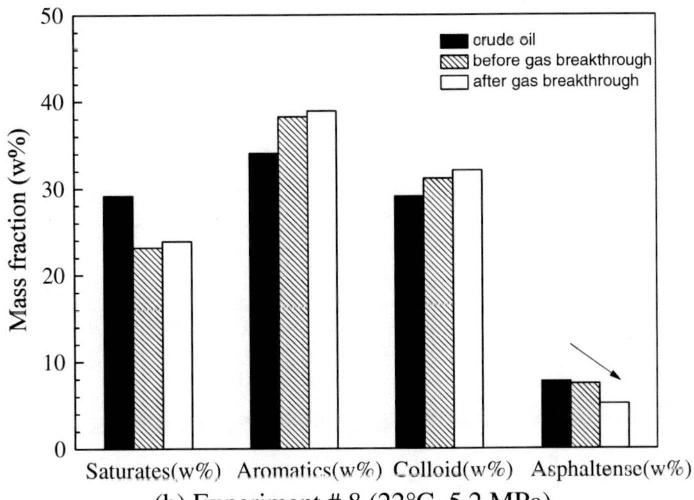

(b) Experiment # 8 (22°C, 5.2 MPa)

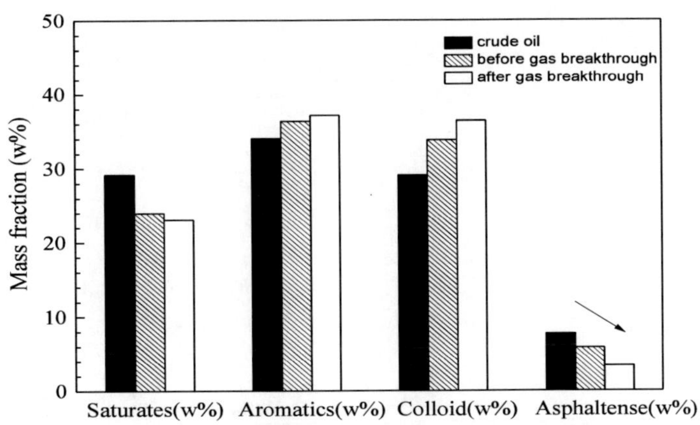

(c) Experiment # 9 (22°C, 9.2 MPa)

Figure 3.17. SARA test of CO_2-produced oil before and after gas breakthrough at 22°C.

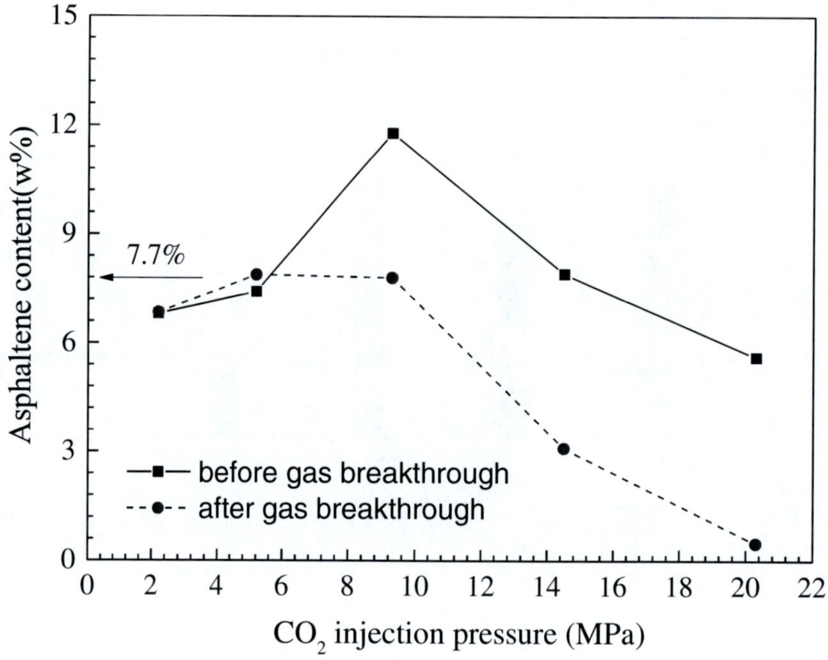

Figure 3.18. Asphaltene content of CO_2-produced oil as a function of pressure at 60°C.

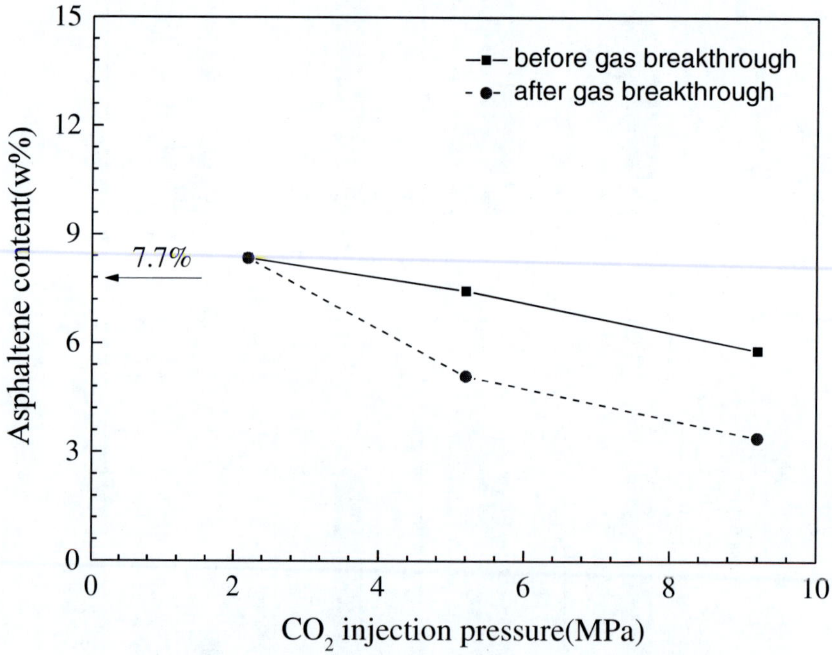

Figure 3.19. Asphaltene content of CO_2-produced oil as a function of pressure at 22°C.

The relationship between the asphaltene content of the oil produced and CO_2 injection pressure are shown in Figure 3.18 (the temperature is 60°C) and Figure 3.19 (the temperature is 22°C). In these figures, the initial asphaltene content of the crude oil is 7.7%. Upon an increase in the injection pressure, the asphaltene content decreases. On one hand, the increase in the system pressure causes more asphaltene precipitation and adsorption on the rock surface or depositing at the pore throat. On the other hand, as the pressure increases, the CO_2 extraction capability strengthens. CO_2 extracts and outputs light components of oil, which leads to a significant reduction in the asphaltene content of the oil produced. For example, when the CO_2 injection pressure is 20.3 MPa at 60°C, the asphaltene content of the oil produced is just 0.48% after gas breakthrough [44, 45].

In addition, when the pressure is larger than P_{asp}, the asphaltene content of oil produced before gas breakthrough is significantly higher than that after gas breakthrough. This is mainly due to the full contact between the CO_2 and oil produced after gas breakthrough. When there is a large amount of CO_2 dissolved in the oil, precipitation of asphaltene in the core tube increases, thus resulting in the reduction of the asphaltene content of the oil produced.

(6) Change rules of asphaltene of residual oil in the core tube

Figure 3.20 is the schematic diagram of oil sand collection after the completion of the displacement experiment. The oil sand in the core tube was removed after completing the experiment and divided into three equal portions along the direction of CO_2 flooding: upper, middle and lower portions. The oil content of the oil sands and the asphaltene of residual oil were measured for each of these portions.

The analysis results of the oil content are shown in Figure 3.21. Along the direction of CO_2 flooding, the oil content of oil sands from the outlet end of the core tube is highest, while that from the inlet end is lowest. Residual oil is mainly distributed in the back of core tube after CO_2 flooding. For instance, when the pressure is 2.2 MPa and the temperature is 60°C, the oil content of oil sands from the inlet end of the core tube is just 7.32% while that from outlet end is 12.39%.

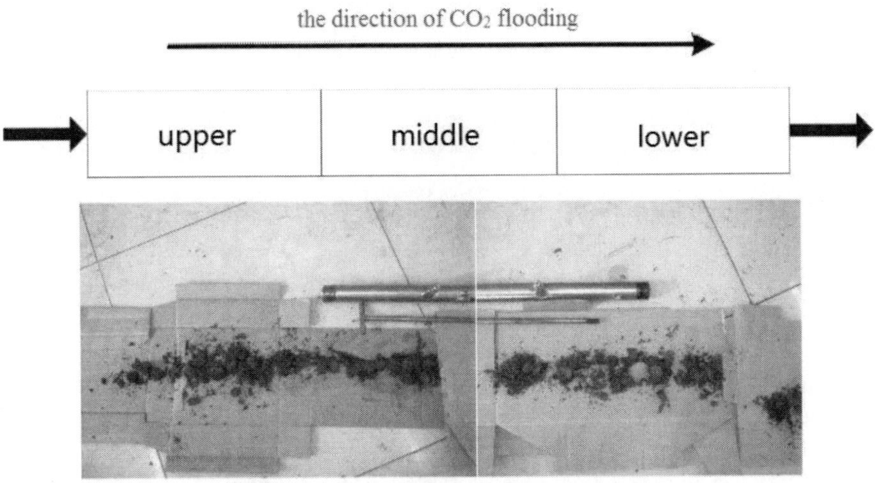

Figure 3.20. Schematic diagram of oil sand collection.

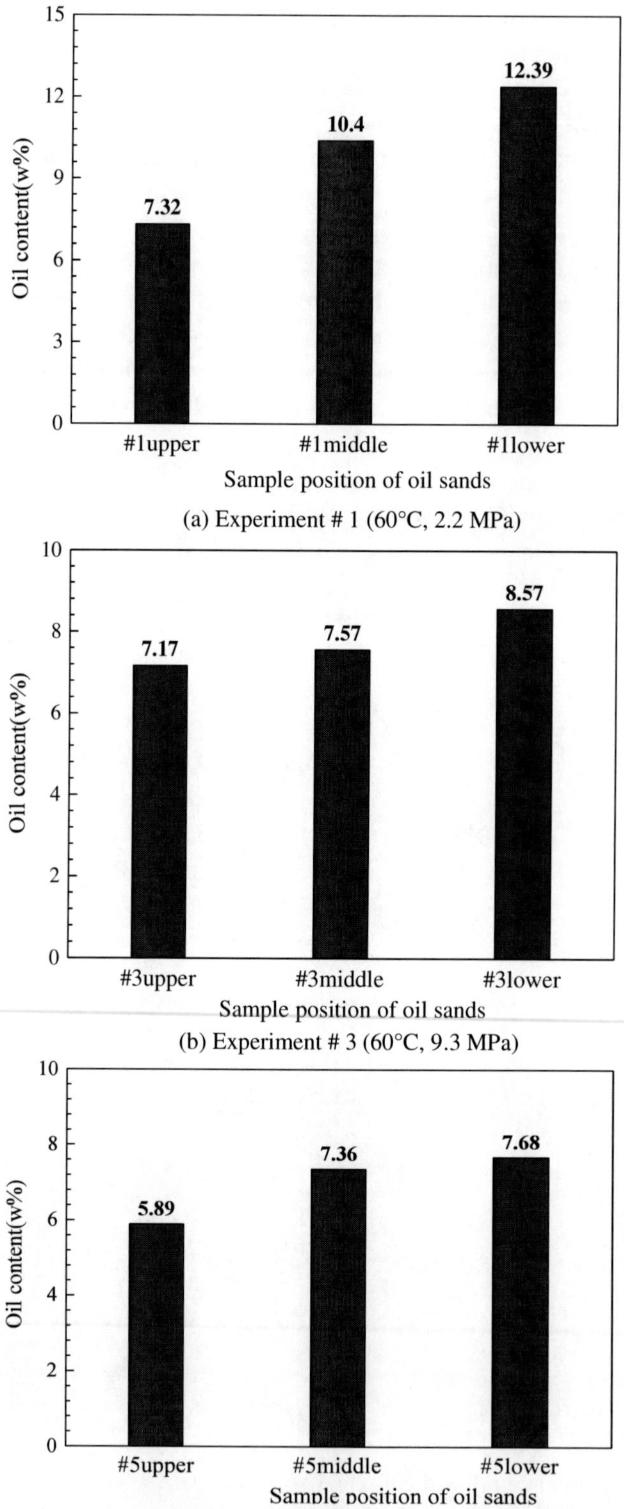

(c) Experiment # 5 (60°C, 20.3 MPa)

Figure 3.21. Oil content of the residual oil in oil sand collected in Tests # 1, # 3 and # 5.

Figure 3.22 shows the contrast diagram for the asphaltene content of crude oil, oil produced and residual oil. The asphaltene content of the residual oil is significantly higher than that of the crude oil, while that of the oil produced is lower than that of crude oil. The ranges of the asphaltene content of the residual oil and the oil produced are 11.76%~19.58% and 0.48%~6.84%, respectively, while that of crude oil is 7.70%.

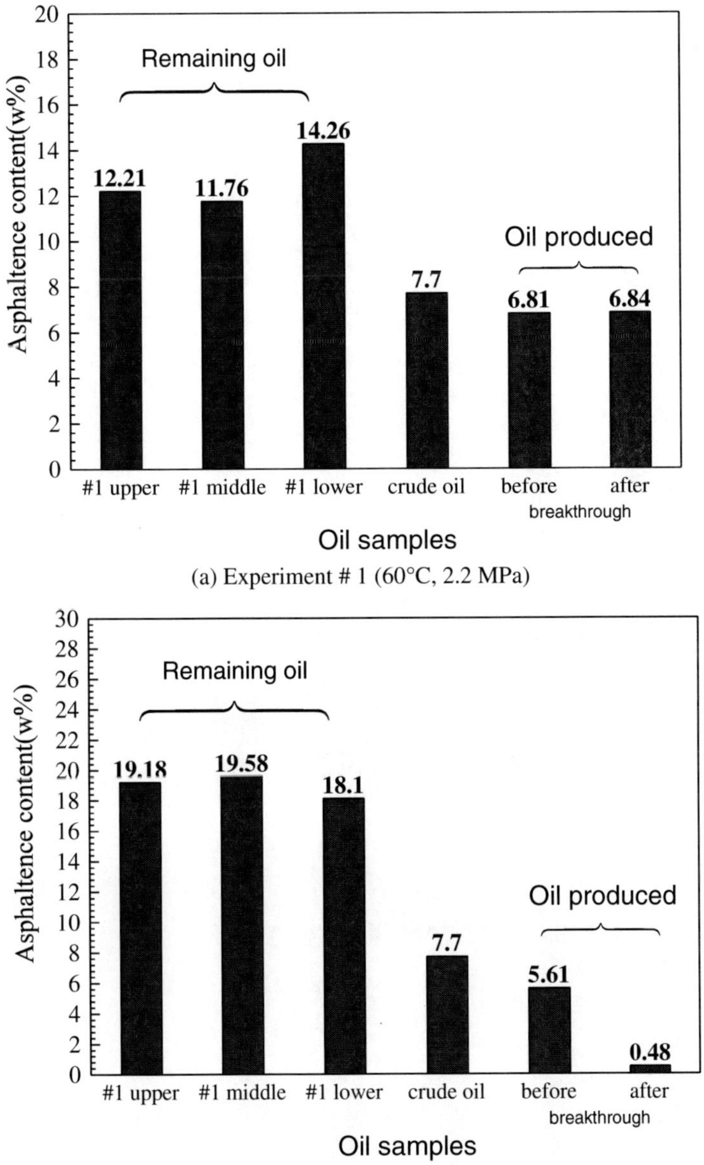

(a) Experiment # 1 (60°C, 2.2 MPa)

(b) Experiment # 5 (60°C, 20.3 MPa)

Figure 3.22. Asphaltene content of the CO_2-produced oil and residual oil.

Thus, the asphaltene content of the residual oil is higher (4.06%~11.88%) than that of crude oil, while that of the oil produced is lower 0.86%~7.22% than that of the crude oil. The experimental results directly demonstrated that the dissolution of CO_2 in oil leads to the precipitation of asphaltene and precipitation deposits in the core tube, which is why the asphaltene content of the oil produced is significantly reduced. However, a large amount of asphaltene depositing or adsorbing during formation may lead to a decrease in the permeability of formation and even formation plugging, which would seriously affect CO_2 flooding [46-48].

(7) Analysis of the oil producing characteristics

Figure 3.24 and Figure 3.25 are variation diagrams showing the relationship between the oil production rate and the CO_2 injection volume under different temperature and pressure conditions.

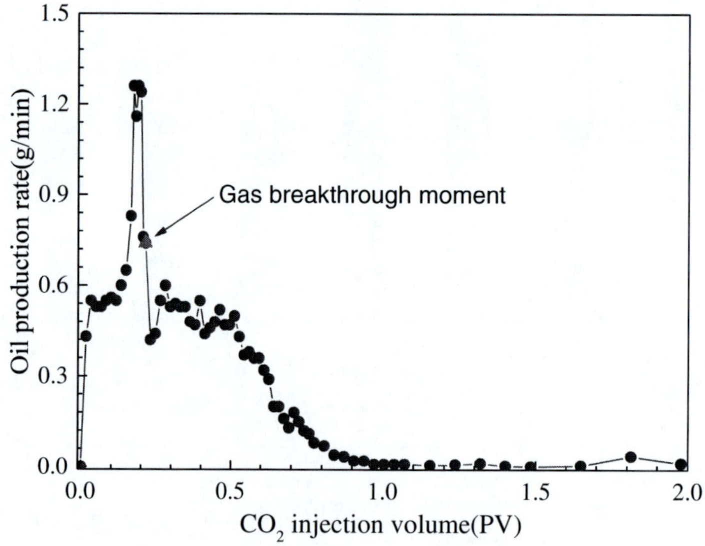

(a) Experiment # 1 (60°C, 2.2 MPa)

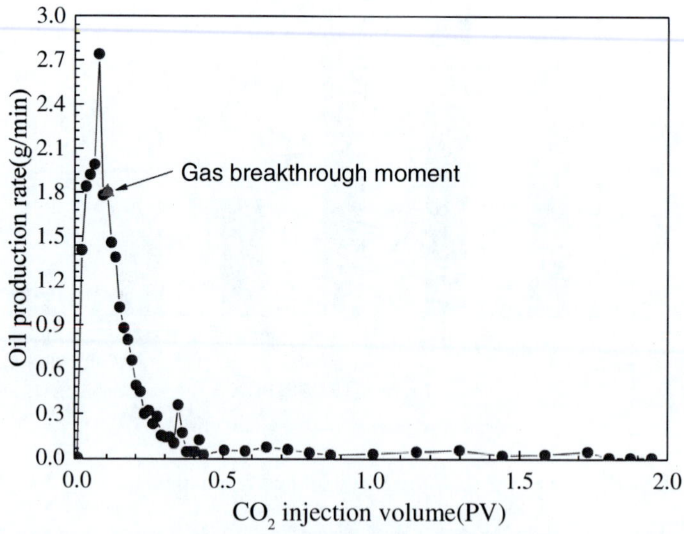

(b) Experiment # 2 (60°C, 5.2 MPa)

Figure 3.23. (Continued)

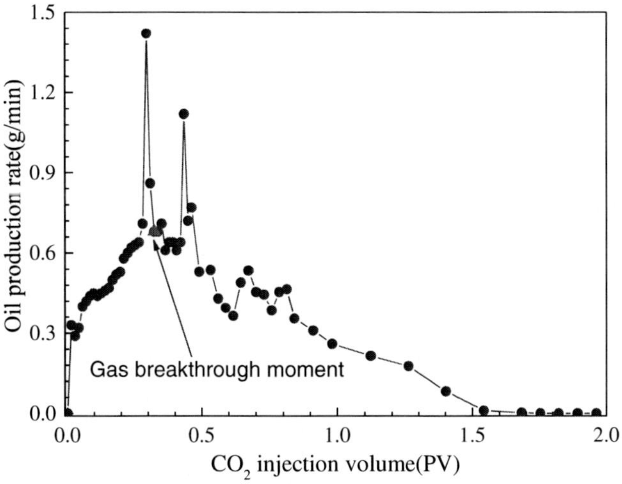

(c) Experiment # 3 (60°C, 9.3 MPa)

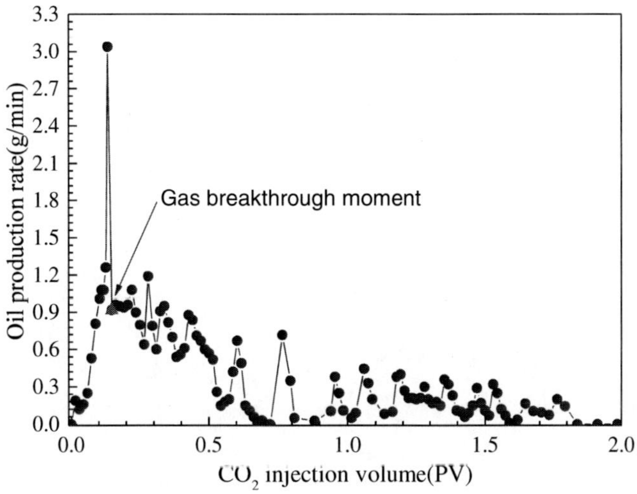

(d) Experiment # 4 (60°C, 14.5 MPa)

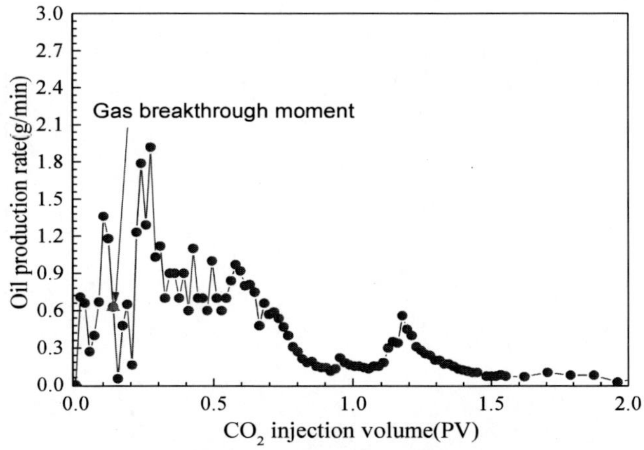

(e) Experiment # 5 (60°C, 20.3 MPa)

Figure 3.23. Production characteristics of CO_2 flooding at 60°C.

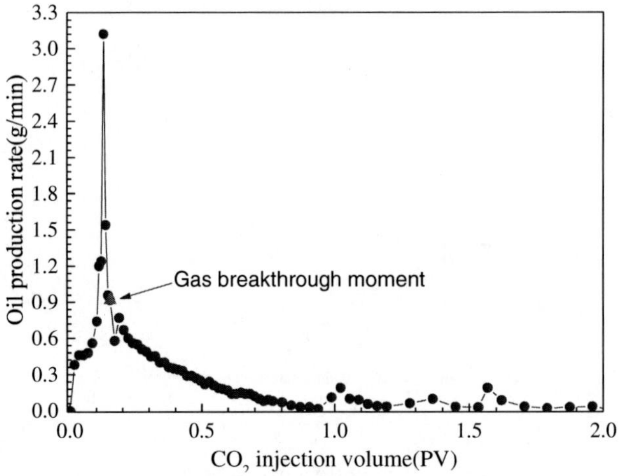

(a) Experiment # 7 (22°C, 2.2 MPa)

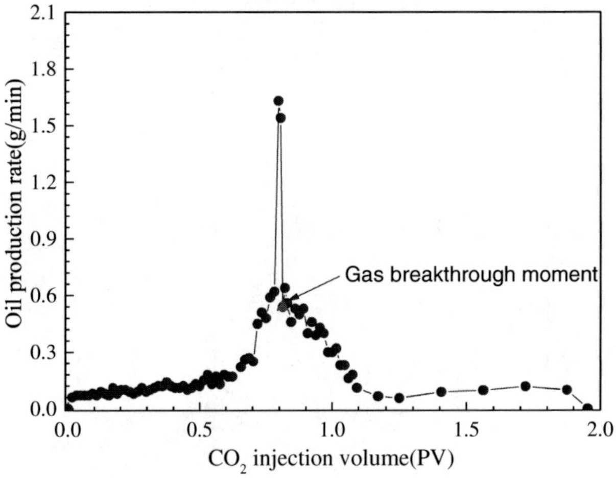

(b) Experiment # 8 (22°C, 5.2 MPa)

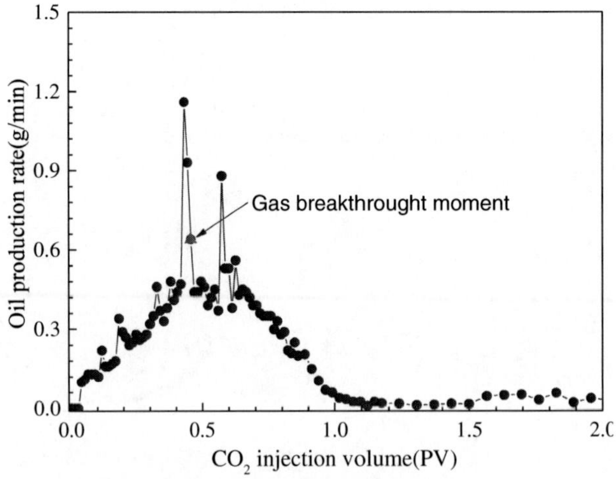

(c) Experiment # 9 (22°C, 9.2 MPa)

Figure 3.24. Production characteristics of CO_2 flooding at 22°C.

Figure 3.24 shows that the oil production rate will suddenly rise before gas breakthrough occurs, which is followed by a phase of rapid decline in the oil production rate. The curve of oil production rate looks similar to an inverted "V".

Figure 3.25 shows that the maximum oil production rate can be 3.09~9.18 times that of the average oil production rate and can be 1.52~3.39 times of the gas breakthrough oil production rate. Therefore, there are two conditions that can be used to predict the gas breakthrough in CO_2 flooding in this paper: (1) the oil production rate of CO_2 flooding suddenly rises and the curve of oil production rate looks similar to an inverted "V"; (2) the maximum oil production rate is over three times bigger than the average oil production rate [49-51].

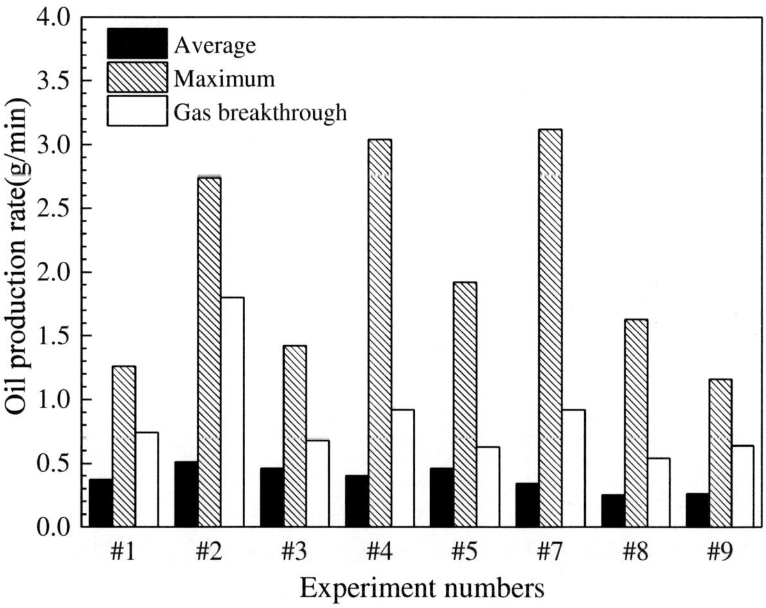

Figure 3.25. Comparison of different production rates of CO_2 flooding.

References

[1] Rao, D. N. (2001). Gas injection EOR-a new meaning in the new millennium. Journal of Canadian Petroleum Technology, 40(2), 11-18.

[2] Pebdani, F. N., & Shu, W. R. (1986). *U.S. Patent No. 4,565,249*. Washington, DC: U.S. Patent and Trademark Office.

[3] Xie, J. Y., Sun, L., Pu, W. F., Jin, F. Y., & He, S. (2015, January). Physical simulation experiment research of air foam flooding in Da Gang Oilfield. In Advanced Materials Research (Vol. 1073, pp. 2335-2344).

[4] Kaszuba, J. P., Janecky, D. R., & Snow, M. G. (2003). Carbon dioxide reaction processes in a model brine aquifer at 200 °C and 200 bars: implications for geologic sequestration of carbon. *Applied Geochemistry*,*18*(7), 1065-1080.

[5] Vazquez M., Beggs H. D. Correlations for fluid physical property prediction[J]. SPE Reservoir Engineering, 1980, 32(5): 968-970.

[6] Chung Frank T. N., Jones Ray A., Nguyen Hai T. Measurements and correlations of the physical properties of CO_2/heavy-crude-oil mixtures[J]. SPE Reservoir Engineering, 1988, 3(3) : 822-828.

[7] Das, S. K., & Butler, R. M. (1998). Mechanism of the vapor extraction process for heavy oil and bitumen. *Journal of Petroleum Science and Engineering*, *21*(1), 43-59.

[8] Mungan, N. (1991). An evaluation of carbon dioxide flooding. *SPE*, *21762*, 20-22.

[9] Li, L., Zhao, N., Wei, W., & Sun, Y. (2013). A review of research progress on CO^2 capture, storage, and utilization in Chinese Academy of Sciences.*Fuel*, *108*, 112-130.

[10] Jarrell, P. M. (2002). *Practical aspects of CO2 flooding*. Richardson, Tex.: Henry L. Doherty Memorial Fund of AIME, Society of Petroleum Engineers.

[11] Beggs H. D., Robinson J. R. Estimating the viscosity of crude oil systems[J]. SPE Reservoir Engineering, 1975, 27(9): 1140-1141.

[12] Thomas, S. (2008). Enhanced oil recovery-an overview. *Oil & Gas Science and Technology-Revue de l'IFP*, *63*(1), 9-19.

[13] Zhaowen Li, Mingzhe Dong. Densities and solubilities for binary systems of dioxide+water and carbon dioxide+brine at 59 °C and pressures to 29 MPa[J]. Chem. Eng. Data, 2004, 49(4): 1026-1031.

[14] Stalkup, F. I. (1978). Carbon dioxide miscible flooding: Past, present, and outlook for the future. *Journal of Petroleum Technology*, *30*(8), 1102-1112.

[15] Holm, L. W., & Josendal, V. A. (1982). Effect of oil composition on miscible-type displacement by carbon dioxide. *SPE Journal*, *22*(1), 87-98.

[16] Orr Jr, F. M., Yu, A. D., & Lien, C. L. (1981). Phase behavior of CO2 and crude oil in low-temperature reservoirs. *SPE J*, *21*(4), 480-492.

[17] Emera, M. K., & Sarma, H. K. (2008). A genetic algorithm-based model to predict CO2-oil physical properties for dead and live oil. *Journal of Canadian Petroleum Technology*, *47*(2), 52-61.

[18] Enick, R. M., & Klara, S. M. (1990). CO2 solubility in water and brine under reservoir conditions. *Chemical Engineering Communications*, *90*(1), 23-33.

[19] Halmann, M. M., & Steinberg, M. (1998). *Greenhouse gas carbon dioxide mitigation: science and technology*. CRC press.

[20] Linn, D. M., & Doran, J. W. (1984). Effect of water-filled pore space on carbon dioxide and nitrous oxide production in tilled and nontilled soils. *Soil Science Society of America Journal*, *48*(6), 1267-1272.

[21] Holm, L. W., & Josendal, V. A. (1974). Mechanisms of oil displacement by carbon dioxide. *Journal of petroleum Technology*, *26*(12), 1427-1438.\

[22] Klins, M. A. (1984). Carbon dioxide flooding: Basic mechanisms and project design.

[23] Pope, G. A. (1980). The application of fractional flow theory to enhanced oil recovery. *Soc. Pet. Eng. J*, *20*(3), 191-205.

[24] Carey, J. W., Wigand, M., Chipera, S. J., WoldeGabriel, G., Pawar, R., Lichtner, P. C., & Guthrie, G. D. (2007). Analysis and performance of oil well cement with 30 years of CO_2 exposure from the SACROC Unit, West Texas, USA. *International Journal of Greenhouse Gas Control*, *1*(1), 75-85.

[25] Holm, L. W., & Josendal, V. A. (1982). Effect of oil composition on miscible-type displacement by carbon dioxide. *SPE Journal*, *22*(1), 87-98.

[26] Baker, L. E., Pierce, A. C., & Luks, K. D. (1982). Gibbs energy analysis of phase equilibria. *Soc. Pet. Eng. J*, *22*(5), 731-742.

[27] Emberley, S., Hutcheon, I., Shevalier, M., Durocher, K., Mayer, B., Gunter, W. D., & Perkins, E. H. (2005). Monitoring of fluid–rock interaction and CO 2 storage through produced fluid sampling at the Weyburn CO 2-injection enhanced oil recovery site, Saskatchewan, Canada. *Applied Geochemistry*,*20*(6), 1131-1157.

[28] Rogers, J. D., & Grigg, R. B. (2001). A literature analysis of the WAG injectivity abnormalities in the CO2 process. *SPE Reservoir Evaluation & Engineering*, *4*(05), 375-386.

[29] Stalkup, F. I. (1978). Carbon dioxide miscible flooding: Past, present, and outlook for the future. *Journal of Petroleum Technology*, *30*(8), 1102-1112.

[30] Rogner, H. H. (1997). An assessment of world hydrocarbon resources.*Annual review of energy and the environment*, *22*(1), 217-262.

[31] Collett, T. S. (2002). Energy resource potential of natural gas hydrates. AAPG bulletin, 86(11), 1971-1992.

[32] Dunn, S. G., Nenniger, E. H., & Rajan, V. S. V. (1989). A study of bitumen recovery by gravity drainage using low temperature soluble gas injection.*The Canadian Journal of Chemical Engineering*, *67*(6), 978-991.

[33] Pourmortazavi, S. M., & Hajimirsadeghi, S. S. (2007). Supercritical fluid extraction in plant essential and volatile oil analysis. Journal of chromatography A, 1163(1), 2-24.

[34] Gittel, A., Sørensen, K. B., Skovhus, T. L., Ingvorsen, K., & Schramm, A. (2009). Prokaryotic community structure and sulfate reducer activity in water from high-temperature oil reservoirs with and without nitrate treatment.*Applied and environmental microbiology*, *75*(22), 7086-7096.

[35] Garcia Alba, L., Torri, C., Samorì, C., van der Spek, J., Fabbri, D., Kersten, S. R., & Brilman, D. W. (2011). Hydrothermal treatment (HTT) of microalgae: evaluation of the process as conversion method in an algae biorefinery concept. Energy & fuels, 26(1), 642-657.

[36] Reverchon, E., & Senatore, F. (1994). Supercritical carbon dioxide extraction of chamomile essential oil and its analysis by gas chromatography-mass spectrometry. *Journal of Agricultural and Food Chemistry*, *42*(1), 154-158.

[37] Yarborough, L. (1977). Multipk Phase Behavior in Porous Media During C02 or Rich-Gas Flooding. *Gas*, *20*, 0-91.

[38] Pope, G. A. (1980). The application of fractional flow theory to enhanced oil recovery. *Soc. Pet. Eng. J*, *20*(3), 191-205.

[39] Kharaka, Y. K., Cole, D. R., Hovorka, S. D., Gunter, W. D., Knauss, K. G., & Freifeld, B. M. (2006). Gas-water-rock interactions in Frio Formation following CO2 injection: Implications for the storage of greenhouse gases in sedimentary basins. *Geology*, *34*(7), 577-580.

[40] Pope, G. A. (1980). The application of fractional flow theory to enhanced oil recovery. *Soc. Pet. Eng. J*, *20*(3), 191-205.

[41] Gozalpour, F., Ren, S. R., & Tohidi, B. (2005). CO2 EOR and storage in oil reservoir. *Oil & gas science and technology*, *60*(3), 537-546.

[42] Wang, X., Zhang, S., & Gu, Y. (2010). Four Important Onset Pressures for Mutual Interactions between Each of Three Crude Oils and CO2†. *Journal of Chemical & Engineering Data*, *55*(10), 4390-4398.

[43] Junior, L. C. R., Ferreira, M. S., & da Silva Ramos, A. C. (2006). Inhibition of asphaltene precipitation in Brazilian crude oils using new oil soluble amphiphiles. *Journal of Petroleum Science and Engineering*, *51*(1), 26-36.

[44] Wang, J. X., & Buckley, J. S. (2001). A two-component solubility model of the onset of asphaltene flocculation in crude oils. *Energy & Fuels*, *15*(5), 1004-1012.

[45] Kawanaka, S., Park, S. J., & Mansoori, G. A. (1991). Organic deposition from reservoir fluids: a thermodynamic predictive technique. *SPE Reservoir Engineering*, *6*(02), 185-192.

[46] Kovscek, A. R., Wong, H., & Radke, C. J. (1993). A pore-level scenario for the development of mixed wettability in oil reservoirs. *AIChE Journal*, *39*(6), 1072-1085.

[47] Morrow, N. R., & Mason, G. (2001). Recovery of oil by spontaneous imbibition. *Current Opinion in Colloid & Interface Science*, *6*(4), 321-337.

[48] Okafor, J. C. (2014). *Characterization of Non-Distillable Crude and Refined Oil Fractions for Asphaltene Precipitation Modeling* (Doctoral dissertation, University of Calgary).

[49] Huu, N. B., Denner, E. B., Ha, D. T., Wanner, G., & Stan-Lotter, H. (1999). Marinobacter aquaeolei sp. nov., a halophilic bacterium isolated from a Vietnamese oil-producing well. *International journal of systematic bacteriology*, *49*(2), 367-375.

[50] Banapurmath, N. R., Tewari, P. G., & Hosmath, R. S. (2008). Performance and emission characteristics of a DI compression ignition engine operated on Honge, Jatropha and sesame oil methyl esters. *Renewable energy*, *33*(9), 1982-1988.

[51] Asps, J. J. (1945). Analysis of decline curves. *Transactions of the American Institute of Mining, Metallurgical and Petroleum Engineers*, *160*, 228-247.

In: Enhanced Oil Recovery
Editor: Alicia Knight
ISBN: 978-1-63463-917-0

Chapter 4

SYNTHESIS AND ASSESSMENT OF A NOVEL AM-CO-AMPS POLYMER FOR ENHANCED OIL RECOVERY (EOR)

Li Qi* and Wang Yabo
CNOOC EnerTech-Drilling and Production Co., Tianjin, China

ABSTRACT

This paper focuses on the polymerization and property assessment of water-soluble AM-co-AMPS polymers for enhanced oil recovery. The AM-co-AMPS polymers are synthesized by acrylamide(AM) and 2-acrylamido-2-methyl-1-propanesulfonic acid(AMPS) in aqueous solution. The synthesis is initiated by (NH4)2S2O8 and NaHSO3 redox system. The effects of initiator concentration, monomers' concentration, AMPS/AM ratio and the polymerization temperature on the synthetic polymer viscosity are studied based on the single factor experiment. The optimal synthesis conditions of AM-co-AMPS polymer are determined according to golden section method. Property assessment is carried out for AM-co-AMPS polymers under different salinity and temperature, and results are compared with Daqing's HPAM. The optimal synthesis conditions are initiator concentration of 3mmol%, monomers' concentration of 20%, AMPS/AM ratio of 20%:80%, and the polymerization temperature of 45.5°C. What's more, The AM-co-AMPS polymer sample employed in the IR characterization was synthesized under the optimal condition discussed above. Then, the property of viscosity of the AM-co-AMPS polymer is carried out in distilled water and a set of formation brines at the temperature of 30°C and 90°C. The results show that under the temperature of 30°C, the viscosity of the target copolymer is 80.5mPa.s and the intrinsic viscosity is 4780mL/g. And the AM-co-AMPS polymer's somehow ability of long term heat-resistance and salt-tolerance is observed: 1. after 32d of aging at 90°C, it maintained a viscosity higher than 20%; 2. compared with the retention values of the solution in salinity of 1%,1.5% and 2%, after 28d of aging at 90°C, this of the solution in salinity of 0.5% show the maximum retention. The AM-co-AMPS polymer has a better thickening

* llsq@sina.com.

properties in the salt water. The salt resistance of AM-co-AMPS polymer has better than DaQing's HPAM at the temperature of 30°C.

Keywords: polymer, AM/AMPS, enhanced oil recovery, redox polymerization

1. INTRODUCTION

In recent years, oil production has been moved toward deeper, and higher salt reservoirs. This trend has produced a challenge for polymer flooding, as in the polymer flooding process synthetic polymers can be chemically and thermally degraded at high temperature and salinity [1]. Polyacrylamides, the most widely used products for mobility control, suffer extensive thermal hydrolysis at high temperatures and, as a result, may precipitate in the presence of divalent cations [2]. Partially hydrolyzed polyacrylamide(HPAM) has acquired some popularity, but is also unstable at high temperature and is probably not suitable for enhanced oil recovery applications in the presence of dissolved oxygen and divalent cations [3]. The use of stabilizing packages might raise this limit somewhat [4, 5]. A.M. Amghi, and P.H. Dee et al[6,7,8] introduced a new stable polymer which in principle contains two or more of acrylamide(AM), N-vinyl-2-pyrrolidone(NVP), and 2-acrylamido-2-methyl-1-propane-sulfonic acid (AMPS). Although the AM-co-NVP polymer showed an exceptional resistance to thermal hydrolysis for the amide groups, its viscosity is not preferred. As reported, its viscosity was as low as 1.8mPa.s at 30°C and 0.8mPa.s at 90°C. Therefore, AM-co-AMPS polymer showed a better performance comprehensively.

The objective of this study is to synthesize polymers that can tolerant the harsh combination of temperature, salinity and hardness for at least 30 days without a substantial loss in viscosity. Polyacrylamides can be modified with various monomers. According to the principle of molecule design of polymer, AMPS is chosen as the second monomer. Sulfo group contained in AMPS can not only increase the stability of the main chain, but also control the degradation of AM. It is found that copolymer of AM and AMPS has a good thermal stability. These systems demonstrate stabilities beyond what would be expected from a simple dilution of AM with AMPS monomer. This paper reports the properties of AM-co-AMPS polymer produced in the laboratory. Details are discussed in later sections.

2. EXPERIMENTAL

2.1. Material

AM monomer reported in the paper was produced by the means of biological catalysis, and was further repurified by ethylalcohol before using. Partially hydrolyzed polyacrylamide (HPAM, Mw=12 million Dalton, degree of Hydrolysis (25%~35%) examined in this paper was provided in powder form by the industrial production workshop of DaQing Refining and Chemical Company. Distilled water was made in our laboratory. The salts used in this paper were reagent grade.

2.2. Polymerization Processes

The full sets of polymers were prepared based on aqueous solution polymerization method. Solution polymerization was carried out in distilled water not containing various amounts of electrolytes as the preferred medium. This process leads to chain polymers containing AM chain link and AMPS chain link. The synthesis condition was investigated in detail in the study, namely mole fraction of initiator, all monomer's mass concentration, monomer ratio of those and the polymerization temperature. An optimal synthesis condition has been presented. Under the optimal synthesis condition, AM-co-AMPS polymer which is heat-resistant and salt-tolerant was developed.

2.3. Polymerization Initiator

Free radical initiation was used to prepare these polymers. Redox system can initiate the polymerization at a low temperature. The range of temperature can be various from 0 to 30°C. In this study, (NH4)2S2O8 and NaHSO3 redox system was screened out as the initiator. The polymerization was initiated in an oxygen-free atmosphere.

3. Sample Preparation and Aging

3.1. All of the Polymers Examined in This Paper Were Prepared As Solutions

All solutions tested were made up with distilled water and a set of formation brines. For each polymer a stock of 10,000 mg/L was prepared, diluted later to 2,000 mg/L. Then, dilution to 200, 400, 600, 800, 1000, 1200, 1500 mg/L were prepared. For each formation brines, a run was made at each of the concentrations (5000, 10000, 30000, 50000 mg/L). Samples (25ml) of 2000 mg/L diluted water solution and a set of brines were placed in 50ml ampule in oven. The oven temperature was set at 90°C. Determination of Polymer Structure Infrared spectroscopy(IR) which conducts as an experimental approach to character molecular composition was employed. The AM-co-AMPS polymer sample employed in the IR characterization was synthesized under the optimal condition discussed above. In this paper, the infrared spectrum was recorded on Paragon1000 infrared spectrometer. The synthesized AM-co-AMPS polymer sample in powder form has been identified by pressing KBr slice of its infrared spectroscopy(IR). Paragon1000 Infrared spectrometer, similar in principle to the UV-Visible spectrometer, permits us to obtain absorption spectra of compounds that are unique reflections of AM-co-AMPS polymer.

3.2. Determination of Apparent Viscosity

Apparent Viscosities measured at particular temperature were obtained using Brookfield DV-III viscometer (manufactured by Brookfield Corporation, USA) fitted with an UL adaptor at room and reservoir temperature using a Thermosel thermostatic device to keep temperature constant during measurement. A shear rate/viscosity curve was derived for each solution using spin speeds of 0.3 to 60 rpm. As a routine the viscosity-shear rate curve was measured twice for each polymer solution with increasing and decreasing values of spin speed [9].

4. Result and Discussion

4.1. Optimization of Polymerization Processes

The effect of mole fraction of initiator, all monomer's mass concentration, monomer ratio of those and the polymerization temperature on the synthetic polymer-viscosity were studied based on the single factor experiment, and the optimal experimental conditions of synthesized AM-co-AMPS was calculated by golden section. Mole fraction of initiator = 0.005 mol%, all monomer's mass concentration = 30%, monomer ratio of those = 20%:80%, and the polymerization temperature =30°C is in common use. A certain condition was studied, it was defined as the variable, the others were as the constant. Table 1 - Table 4 summarize the optimal condition: mole fraction of initiator is 3mmol%, total concentration of monomer is 20%, monomer ratio of AMPS and AM is 20%:80%, the reaction temperature is 45.5°C. Under the temperature of 30°C, the viscosity of the target copolymer is 80.5mPa.s and the intrinsic viscosity is 4780mL/g.

Table 1. Effect of Mole Fraction of Initiator

Item	Mole fraction of initiator, mol%				
	0.1	0.25	0.5	1	2.75
Viscosity, mPa.s	6	24.2	12.6	8	7.1

Table 2. Effect of All Monomer's Mass Concentration

Item	All monomer's mass concentration %			
	10	15	20	25
Viscosity, mPa.s	8.3	20	33.3	22

Table 3. Effect of Monomer Ratio of AM and AMPS

Item	Monomer ratio of AM and AMPS			
	0.27	0.48	0.68	0.89
Viscosity, mPa.s	7.6	8.1	9	12.6

Table 4. Effect of Polymerization Temperature

Item	Polymerization temperature °C				
	30	35	40	45	50
Viscosity, mPa.s	12.6	7.6	17.9	43.5	24.3

4.2. IR Characterization of AM-co-AMPS Polymer

IR spectrum of AM-co-AMPS polymer sample was delineated in Figure 1. The strong absorption peak at 3399.9 cm^{-1} and 3274.3cm^{-1} are assigned to stretching and bending vibrations of V-H. Bands near 1650cm^{-1} yield an intense peak due to stretching >C=O. The bands observed in the regions 650cm^{-1} to 900 cm^{-1} are attributed to wagging vibrations of V-H. The peak at 1186.0 and 1039.4 are assigned to the antisymmetric and symmetric S=O stretching modes. Because of the board peak belonging to wagging vibrations of V-H, the peak of >CH_2 at 725 cm^{-1} is hard to recognize. Meanwhile, the absorptions of stretching >CH_2 occur at 2933.2 cm^{-1} and 2788.6 cm^{-1}.All these absorption peaks are characteristic of a structure type based on AM link chain and AMPS link chain. Therefore, it can be identified as the AM-co-AMPS polymer.

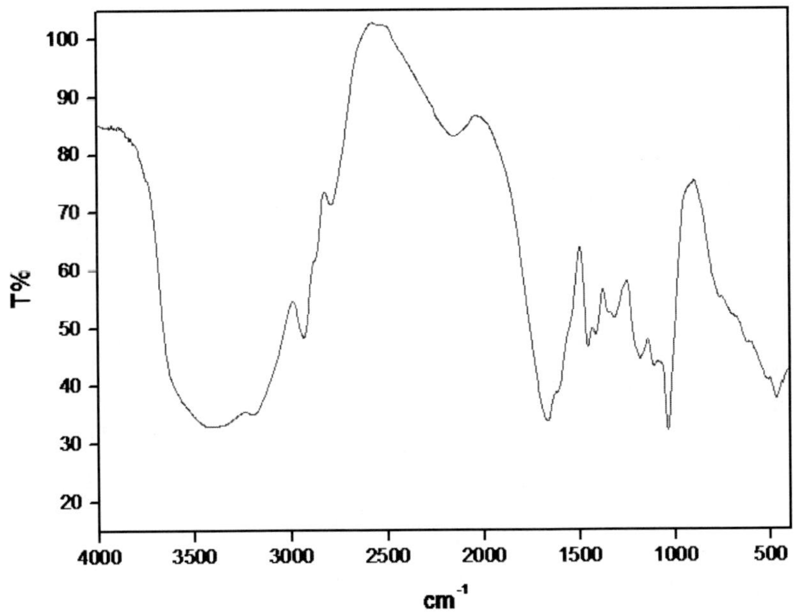

Figure 1. IR Spectrum of AM-co-AMPS Polymer.

4.3. Polymer Rheology

Polymer solution viscosity for a range of polymer concentrations was measured in distilled water and a set of formation brines at a shear of 7.34S^{-1} as shown in Figure 2, Figure

3 and Figure 4. The AM-co-AMPS polymer has a lower molecular weight than the pure hydrolyzed polyacrylamide that we tested. The molecular weight of the HPAM is around 12 million Dalton, while the AM-co-AMPS polymer has a weight around 10 million Dalton. Due to its lower molecular weight, the AM-co-AMPS polymer has a lower viscosity yield in distilled water solution than the HPAM. Otherwise, the viscosity of HPAM is greatly enhanced in the concentration range between 600 and 1000mg/L.

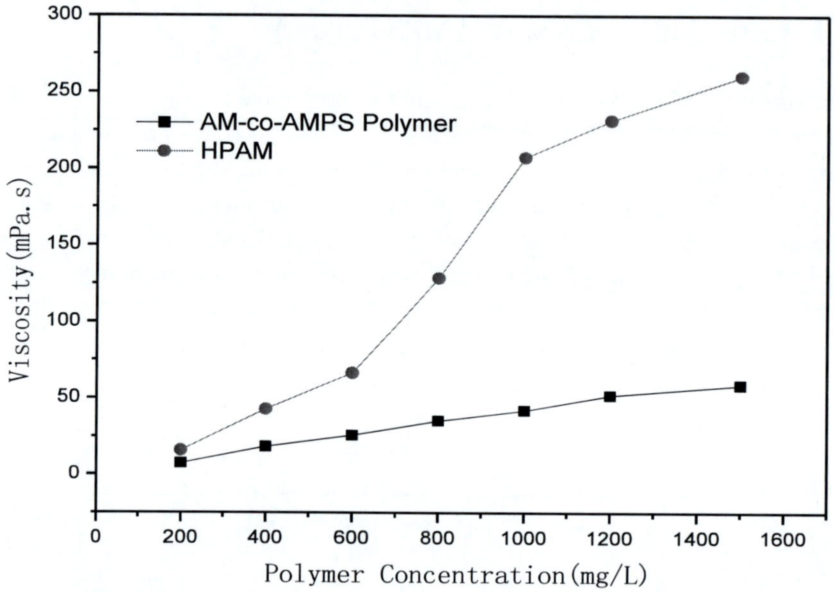

Figure 2. Viscosity vs. polymer concentration in distilled water solution at a temperature of 30 °C and a shear rate of 7.34 S^{-1}.

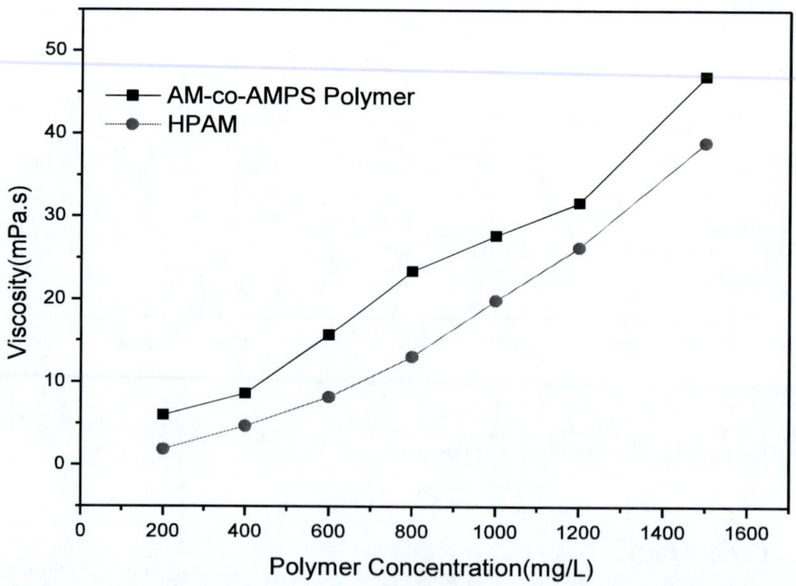

Figure 3. Viscosity vs. polymer concentration in 10000mg/L salt solution at a temperature of 30 °C and a shear rate of 7.34 S^{-1}.

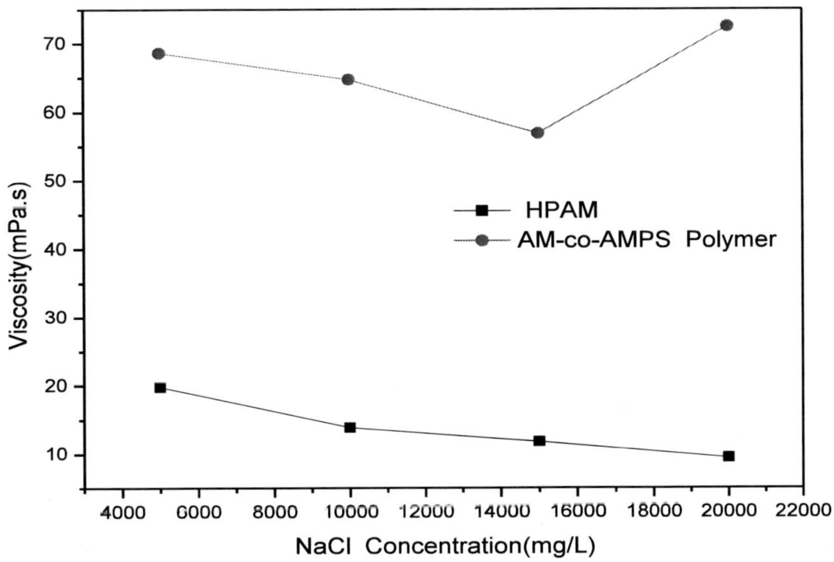

Figure 4. Viscosity vs. NaCl concentration in formation brines at a temperature of 30 °C and a shear rate of 7.34 S^{-1}.

It is interesting to note that the relationship of the viscosity of HPAM and AM-co-AMPS polymer is reversed in formation brines. Analysis combined with Figure 4 suggests that the viscosity of AM-co-AMPS polymer is as much as three times than HPAM's. As expected, the viscosifying power decreases with brine salinity.

4.4. Polymer Viscosity As a Function of Temperature

The viscosity as a function of temperature, the equation solutions were heated in the rheometer while measuring the viscosity at a shear rate of 7.34S^{-1}. The result is depicted in Figure 5. Upon cooling down, the polymer solution viscosity increased again. For HPAM, the viscosity sharply decreased at 50°C, and a cross with AM-co-AMPS polymer around 65°C was plotted. It demonstrates that at a comparatively high temperature, the performance of HPAM is inferior to AM-co-AMPS polymer. It is reasonable to deduce that at high temperature, the use of AM-co-AMPS polymer is recommended. This can be attributed to the steric hindrance given by the AMPS group that protects the main polymer chain.

4.5. Polymer Stability Tests at 90°C

The viscosity retention of 2000mg/L AM-co-AMPS polymer aged in distilled water solution as a function of aged days is shown in Figure 6.

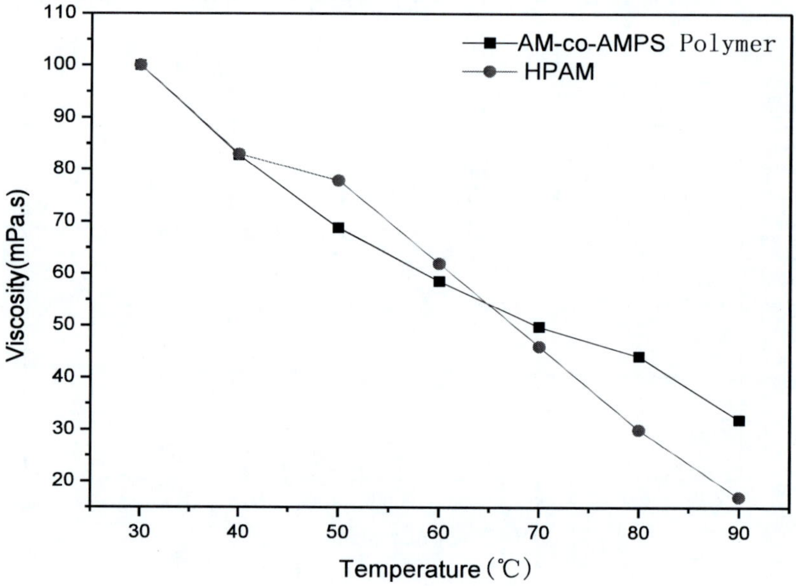

Figure 5. Viscosity vs. temperature in distilled water solution at a shear rate of 7.34 S-1.

After 32d's aging, the retention is about 22%. It demonstrates that molecular weight degradation occurs. If the aging were continued, an increase in retention would be promised to observe. For comparison, our results show a good consistency with the previous work made by R.S. Seright et al. [3], A.M. Araghi et al. [7, 8]. Hydrolysis of polyacryamide introduces negative charges on the backbone of the polymer chain that have a large effect on the rheological properties of the polymer solution. The negative charges on the polymer backbone repel each other and cause the polymer chain to stretch. Each polymer molecule occupies more space in solution, and the apparent viscosity increases accordingly [10].

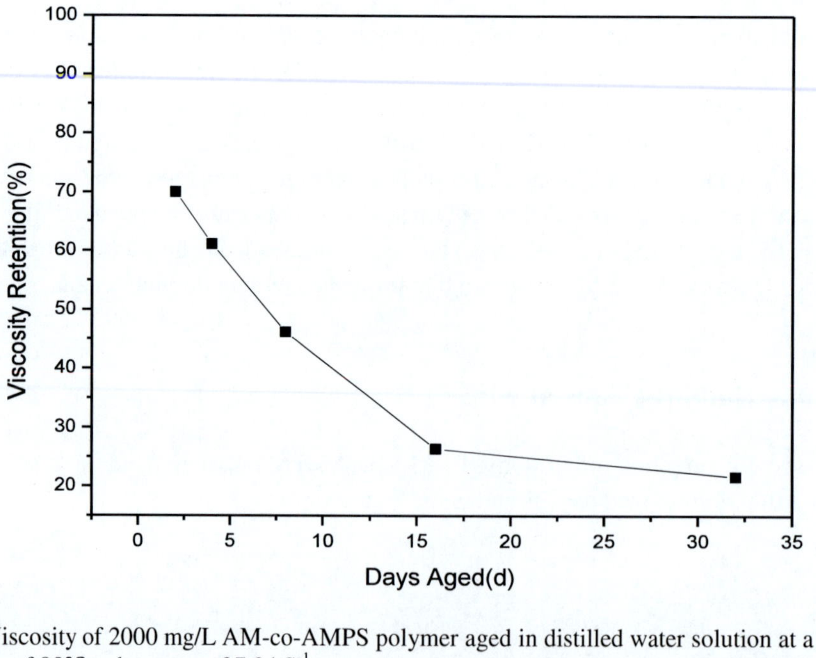

Figure 6. Viscosity of 2000 mg/L AM-co-AMPS polymer aged in distilled water solution at a temperature of 90°C a shear rate of 7.34 S^{-1}.

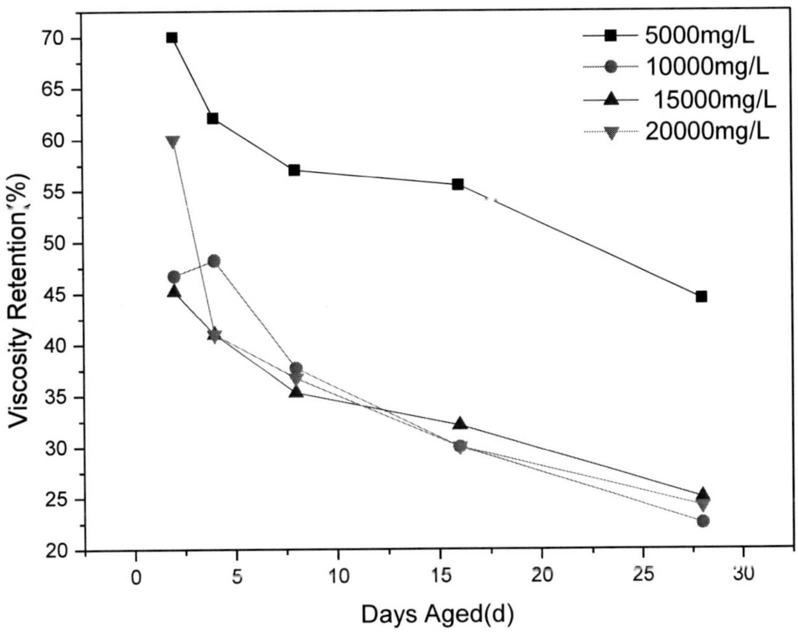

Figure 7. Viscosity of 2000 mg/L AM-co-AMPS polymer aged in formation brines at a temperature of 90°C and a shear rate of 7.34 S-1.

To evaluate the long-term stability in formation brines at 90°C, 2000mg/L solution of AM-co-AMPS polymer were aged for 32d in various brines. Figure 7 shows the plots of viscosity retention vs. ageing days. For all the solutions show a similar decrease trend, while there is little difference between the values in the three high saline brines. Only AM-co-AMPS polymer in the 5000mg/L salt solution reveals a high viscosity retention. At the highest salinity of 2%, however, the contribution of salts to the viscosity becomes very significant.

CONCLUSION

- The optimal experimental condition of the AM-co-AMPS polymer polymerization is determined: mole fraction of initiator is 0.002888mol%, total concentration of monomer is 20%, monomer ratio of AMPS and AM is 20%:80%, the reaction temperature is 45.52°C.
- Under the temperature of 30°C, the viscosity of the target copolymer is 80.5mPa.s and the intrinsic viscosity is 4780mL/g.
- On the same polymer concentration condition, the viscosity of the AM-co-AMPS polymer in salt water solution is higher than the viscosity of the DaQing's HPAM, it shows that the AM-co-AMPS polymer has a better thickening properties in the salt water. The salt resistance of AM-co-AMPS polymer has better than DaQing's HPAM at the temperature of 30°C.
- The AM-co-AMPS polymer's somehow ability of long term heat-resistance and salt-tolerance is observed: 1. after 32d of aging at 90°C, it maintained a viscosity higher

than 20%; 2. compared with the retention values of the solution in salinity of 1%,1.5% and 2%, after 28d of aging at 90°C, this of the solution in salinity of 0.5% show the maximum retention.

REFERENCES

[1] E. C. M. Vermolen, M. J. T van Haasterecht, S. K. Masalmeh, M.. J. Faber, D. M. Boersma, M. Gruenenfelder, "Pushing the Envelope for Polymer Flooding Towards High-Temperature and High-Salinity Reservoirs with Polyacrylamide Based Ter-Polymers," *SPE* 141497.

[2] A. M. Araghi, and P. H. Dee, "Hydrolysis and Precipitation of Polyacrylamides in Hard Brines at Elevated Temperatures," *SPERE*, May 1987, pp.189-98.

[3] R. S. Seright, A. R. Campbell, and P. S. Mozley, "Stability of Partially Hydrolyzed Polyacrylamides at Elevated Temperatures in the Absence of Divalent Cations" Presented at at the 2009 SPE International Symposium on Oilfield Chemistry held in The Woodlands, Texas, USA, 20–22 April 2009; paper 121460.

[4] S. L.Wellington, "Biopolymer Solution Viscosity Stabilization–Polymer Degradation and Antioxidant Use," *SPEJ*, Dec 1983, pp.901-12.

[5] M. K. Abdo, "*Waterflood Oil Recovery Process Employing Stabilized Biopolymers*," U.S. Patent No.4,141,842, Feb. 1979.

[6] H. L Hsieh, A. M. Araghi, G. A. Stahl et al., "Water-Soluble Polymers for Hostile Environment Enhanced oil Recovery Application" Makromol. *Chem., Macromol. Symp*, 1992, pp 121-135.

[7] P. H. Dee, A. M. Araghi, E. S. James et al., "Development and Evaluation of EOR polymers Suitable for Hostile Environments-Part 1: Copolymers of Vinylpyrrolidone and Acrylamide" *SPE Reservoir Engineering*, Nov 1987, pp 461-467.

[8] A.M. Araghi, D.H. Cleveland, W. W. James et al., "Development and Evaluation of EOR polymers Suitable for Hostile Environments: II- Copolymers of Acrylamide and Sodium AMPS" Presented at SPE International Symposium on Oilfield Chemistry held in San Antonio, Texas, USA, 4-6. February1995, paper 16273.

[9] M. A. Leveratto, J. Lauri, C. Sanz, J. Sigal and S.M. Farouq Ali, "Eor Polymer Screening for an Oil Field with High Salinity Brines" *SPE* 26989.

[10] Don W Green. Enhanced oil recovery (SPE textbook series) [M]. Henry L. Doherty *Memorial Fund of AIME*, Society of Petroleum Engineers, 1985.103.

In: Enhanced Oil Recovery
Editor: Alicia Knight
ISBN: 978-1-63463-917-0

Chapter 5

IMPROVED OIL RECOVERY VS. ENHANCED OIL RECOVERY

Nikolay P. Zapivalov*
National Research Tomsk Polytechnic University,
Institute of Petroleum Geology and Geophysics,
SB RAS, Novosibirsk, Russia

ABSTRACT

In this chapter, the author presents a principally new scientific-technological paradigm of exploitation, preservation, and rehabilitation of petroleum resources.

An oil deposit in the Earth crust is considered as a self-organizing, living fluid-saturated system capable of restoring its balance. Active works during development of an oilfield produce a strong perturbation in the near-equilibrium system and substantially deform its natural parameters. In case of a moderate perturbation, the self-organizing system restores its balance; a prolonged or intensive perturbation considerably exceeding the critical threshold will destroy the system.

Nowadays, the problem of enhanced oil recovery (EOR) and improved oil recovery (IOR) is a key problem in petroleum theory and practice. In world oil-and-gas practice, two different terms are in use: EOR (enhanced oil recovery) meaning intensive, forcible methods; and IOR (improved oil recovery) – advanced and moderate methods.

INTRODUCTION

The enhanced oil recovery methods do not provide a scale effect. There are about 1500 active projects in the world using various EOR technologies, which only results in additional production of about 100-120 MT/Y. It is approximately 2% of the total of produced oil in the world – an equivalent of transporting and other losses.

* Corresponding author: Tel.: 7 (903) 935-87-25, e-mail: ZapivalovNP@ipgg.sbras.ru.

In the USA, the "additional" oil recovery remains at the level of 30-35 MT/Y and has not exceeded this threshold value since 1986. In 1986, there were 512 active projects and in 2008 there were 184, so there is an obvious tendency of a decrease. The situation in other regions of the world is much the same. In terms of the final oil recovery, at many oilfields the efficiency of EOR is very low or zero.

However at present, development of oil-and-gas resources is usually aimed at immediate commercial return; therefore, all possible methods are used for accelerated intensification of oil and gas extraction, which in the longer term leads to the destruction of the oil deposit as a living system in the thermodynamic rock-fluid interconnection.

The only way out is to rehabilitate and revitalize such deposits as a whole or as separate productive zones. This will permit to restore the natural energy parameters of fluid saturated systems and also to provide a balanced proportion of hydrocarbons in fractures and porous matrix. Such individual fluid-dynamic cycles, based on self-organization of natural systems, can essentially increase the oil recovery.

Also, a review is presented of promising modern trends in technologies for sparing and moderate development of oilfields (Improved Oil Recovery instead of Enhanced Oil Recovery) aimed at longevity and high oil recovery.

Among the present-day innovative technologies for exploration and development of oilfields, such trends are worth noting as geofluid-dynamical, seismic-geophysical, geomechanical, and technological; besides, there are some ideas and projects not implemented so far.

Geofluid-Dynamical Aspects. Critical Threshold of Perturbation

It is necessary to distinguish two states of an oil deposit in the earth crust: natural state formed before human intervention, and naturally-technogenic state during the process of active exploration and development. In any state, an oil deposit can be considered as a living fluid-saturated system

The natural state of any accumulation of hydrocarbons is unsteady in the geological time-scale. Depending on various fluctuations and bifurcations, it can be equilibrium and non-equilibrium. The natural accumulations of hydrocarbons can enlarge or grow smaller, or even fully collapse (disappear) in relatively short geological (human) time periods. In fact, it is a self-organizing system.

The active works during the development of an oilfield produce a strong perturbation in the near-equilibrium system and substantially deform its natural parameters. In case it is a moderate perturbation, the self-organizing system restores its balance. It is especially important at the late stages of the oilfield life.

A prolonged or intensive perturbation considerably exceeding the threshold will destroy the system. As a result, the formation pressure drops, production rate falls off, the layer is watered out, and even its mineral composition changes.

The perturbation threshold can be estimated through the draw-down pressure. The author has established that the optimal draw-down pressure (formation pressure FP – bottom-hole pressure BHP) must not exceed 5÷8 MPa (FP – BHP $\leq$ 5÷8 MPa). This value is almost universal and applicable for all types of reservoirs and many oilfields. It is apparently seen in Figure 1.

In the figure 1c, S=f(ΔP) crosses the abscise axis at ΔP = 5 MPa. At lower pressures, S is negative, which indicates the fractured type of reservoir. Increasing pressure gives positive values of S, showing that the matrix works through porous reservoir.

The main features of any living system are its energy potential and working capacity. Any oil reservoir is an open fluid-dynamic system with variable exergy and versatile gradients of mass-and-energy transfer. Their threshold values are determined by the boundary parameters of the system at any given moment of time [1].

Taking into account more than century of world experience in the active developing of oilfields, the stages of their development are determined as shown in Figure 2.

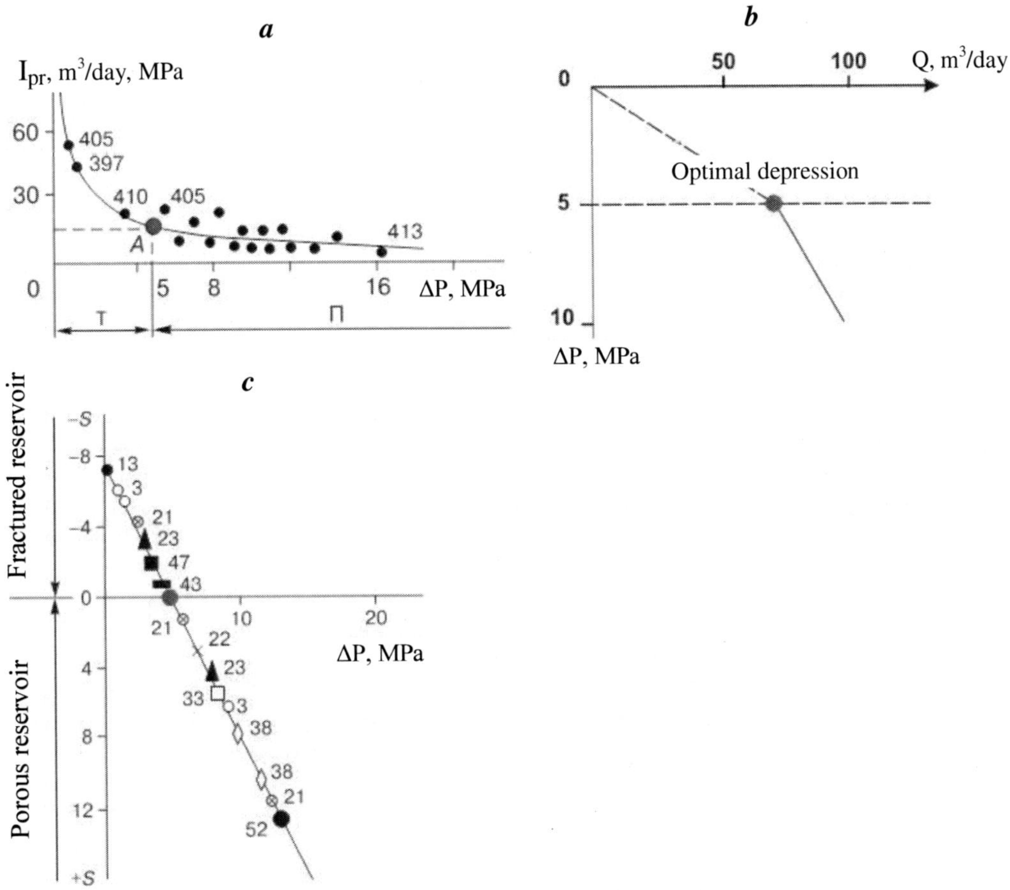

Figure 1. a) Indicator diagram and dependence of productivity index (I_{pr}) from draw-down pressure in wells. Nizhnevartovsky oilfields, West Siberia. b) Indicator diagram for well 43 in Barsukovsky oilfield, Belarus. c) Description of fluid-dynamic parameters of carbonate reservoirs of porous-fractured type. Correlation between skin-factor and depression (drawdown) at the wells of Beshtentyak oilfield, Kirgizstan.

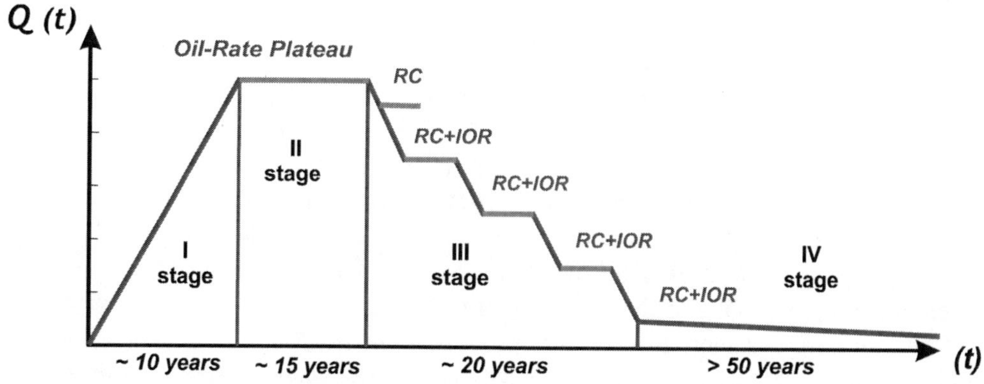

Figure 2. Generalized chart of oilfield "life". Along the *y*-axis, the dynamics of oil production is shown as ***Q***(t). RC is rehabilitation cycles, IOR is the methods of improved oil recovery. Oil-rate plateau is the stable state of the system (optimal level of oil production). The duration of the oil-rate plateau is determined by a scientifically-reasonable monitoring project of developing and by professional skills of the industrial specialists.

Rehabilitation cycles in the process of developing an oilfield and gentle (moderate) methods of oil recovery are the basis of oilfield life extension [2]. They allow longtime development of an oilfield and getting maximum oil recovery.

It is established that the actual reserves of oil and gas can be replenished in the process of oilfield development. It is possible in two cases.

1) An active present-day process of hydrocarbon generation is proceeding in the layer. This case was proved in the Mexican bay, Eugene Island oilfield, and other regions. An additional feeding (inflow) with newly formed portions of hydrocarbons is possible from both inside and outside the system.
2) An individually adjusted moderate development is performed, combined with regular periodic rehabilitation works. In this case, a balanced exchange of fluids proceeds between the matrix (block) and filtration channels of the layer. Also, equilibrium is observed between the rock pressure and formation pressure [1, 3].

There are many examples throughout the world demonstrating a renewal of activity in the well after some period of rehabilitation (rest) of the whole oilfield or its separate blocks [4].

We can see many long-livers among oilfields:

- Production of oil at the Digboy oilfield in India (Assam region) began in 1890 and proceeds until now.
- There are more than 100 deposits in the USA that have been developed for over 50 years. For example, the giant East-Texas sandstone oilfield (Upper Cretaceous) was estimated at 1 milliard tons of oil. Its development began in 1930 and still proceeds.
- A model example is Azerbaijanian oilfields.
- In the Grozny oil district (Russia) some oilfields are under development for more than 60 years: October oilfield, Tashkalinsky, and Oysungursky oilfields.
- In 2007, R. Muslimov introduced a new vision of the prospects of development of the unique Romashkinsky oilfield in Tatarstan (Russia), which is under development

for 65 years (see Figure 3). Now the oil production has the level of 12-15 million ton per year, and oil recovery exceeds the projected value. R. Muslimov considers that this oilfield will be under development until the end of the 21st century; thus, its lifetime will exceed 150 years. He even predicts the lifetime of this oilfield as long as until 2285, due to an inflow of new portions of hydrocarbons [5-6].

On the contrary, Verch-Tarsky oilfield in Novosibirsk region (West Siberia) is an example of cruel exploitation of a living oil geosystem. The actual chart of "life" of Verch-Tarsky oilfield (Figure 4) can be compared with Figures 2 and 3. Apparently, an oil-rate plateau is absent here. Also, an alarming point is that the content of water in the product of the Verch-Tarsky oil reservoir is now more than 70%. It is a result of intensive hydraulic fracturing and water injections. Rather than exploring or at least developing the oilfield, the oilmen just extract the actual reserves in a supercritical mode. All new wells undergo hydraulic fracturing from the very beginning of their exploitation.

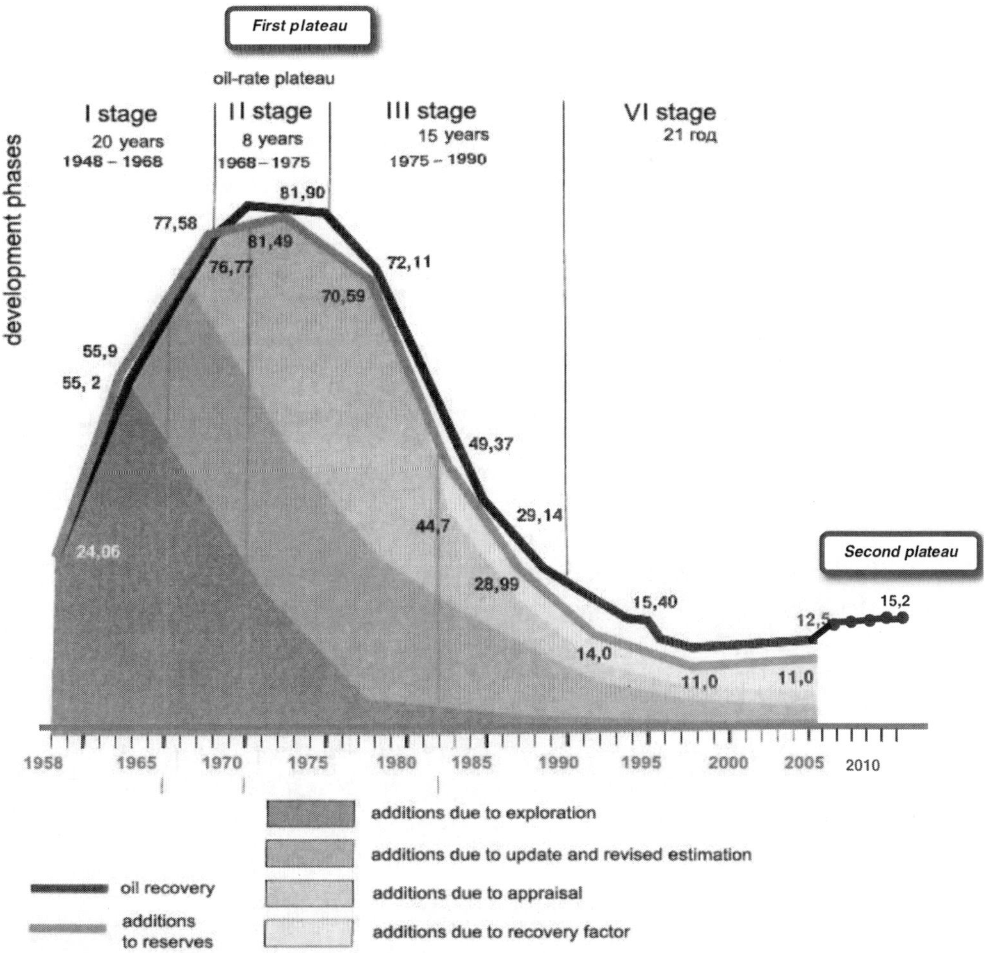

Figure 3. Chart of "life" of Romashkinsky oilfield, Tatarstan (after R. Muslimov). Oil recovery and accretion to reserves, million tons.

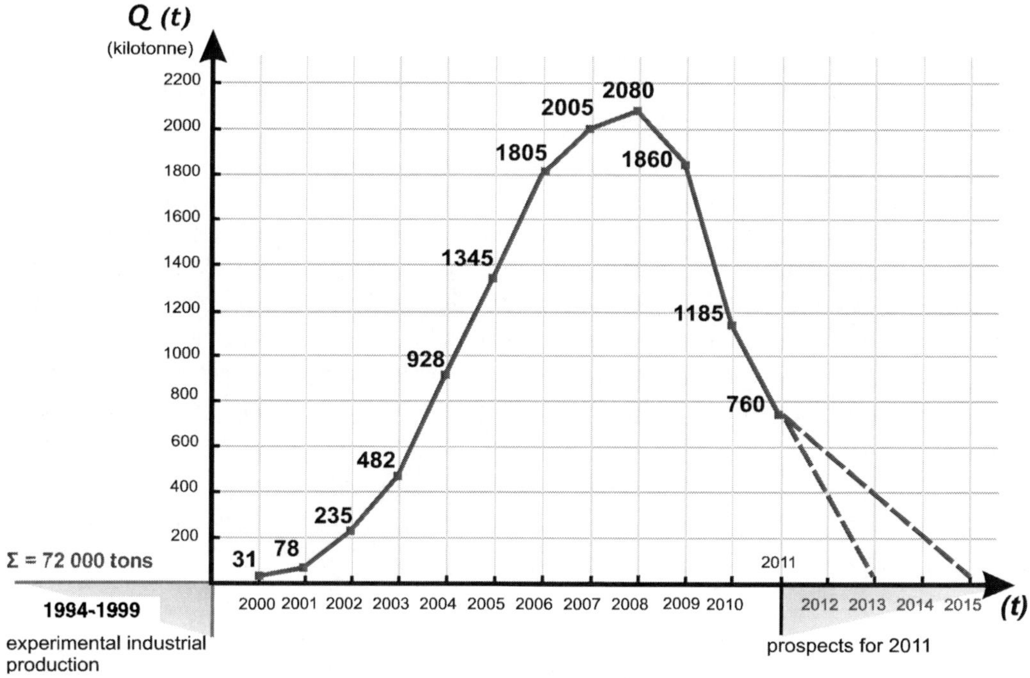

Figure 4. Chart of "life" of Verch-Tarsky oilfield. There are only two stages, rise and falling.

To keep the formation pressure, the layer is intensively "flushed" with water in volumes many times exceeding those of the oil product. In April, 2010 there were 92 operating wells and 74 water-injected wells at the oilfield. The average daily flow rate was 41 tpd, and water content in the product was 80 tpd, i.e. twice as large. Pressurizing in such a way is also destructive for an oil-saturated reservoir [7].

Unfortunately, many oilfields in West Siberia are being developed in a similar way.

"Piston-type" oil displacement in a layer will not provide an effective percolation. It should be taken into account that a large-scale injection of foreign water into the layer will change its mineral composition and consequently all its natural properties in an abrupt and irreversible way; thus, active technogenic metasomatosis will occur, with a negative effect.

It should be noted that in 2010, there were about 200 oil recovery projects in the USA and no one hydraulic fracturing project. They use CO_2 injections widely. By the way, near Verch-Tarsky field there is West-Mezhovsky (Veselovsky) gasfield, where in one of the wells a blowout of CO_2 with the flow rate of 200 000 m^3/d was obtained.

It is necessary to spare the natural properties of petroleum reservoirs, especially their mineral compositions. Oilfields should be developed in an optimal mode, so as not to exceed the critical threshold of perturbation indicated by the value of depression. For effective long-term development of oil and gas fields, it is necessary to observe the critical threshold (5-8 MPa), use rehabilitation cycles, and apply moderate individual methods to increase both current and final oil recovery.

Environmental issues are of a special importance here. Forcible, boosted methods of oilfield exploitation strongly perturb the geological medium and create multiple bifurcation points both in the depth and on the surface. It is extremely essential, because our planet is thus being turned into a large oil production site. In a short period of time, even the seascape has been changed greatly, in some places really beyond recognition.

A gentle, sparing long-term exploitation of oilfields will provide gradual mutual adjustment of the natural-technogenic objects (field facilities) and the environment. The economic effect of long-term oilfield exploitation will be definitely positive in a short time.

(N.P. Zapivalov, Novosibirsk, Russia)

SEISMIC-GEOPHYSICAL METHOD. DYNAMIC-FLUID MODELS (DFM)

Parameters of a fluid-saturated medium of a discrete structure are, by their physical character, functions of elastic modules and of the current stressed state. Therefore, the most appropriate method for estimating the gradient pressure may be based on a complex analysis of the seismic parameters and other geologic and geophysical data [8].

The results of applying DFM-technologies at various world basins showed the possibility to define (prognosticate) the location of the maximum fluid-saturation zones (foci) with a sufficient reliability. The application of DFM-technologies is illustrated in Figure 5.

DFM-technology can be effectively applied at all stages of oilfield exploration and development including the process of monitoring; it has already proved successful in various regions. The technology is aimed at mapping the productive zones (foci) with active fluid cross-flows.

(V.B. Pisetsky, Yekaterinburg, Russia)

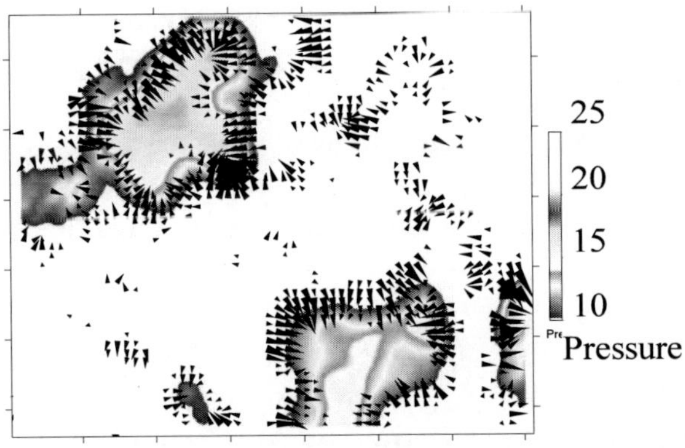

Figure 5. Map of differences of fluid-dynamic parameters [9]. The raster color shows the changes in fluid pressure gradients; the arrows show differentiated fluid flow vectors.

GEO-MECHANICAL TECHNOLOGY. METHOD OF DIRECTIONAL UNLOADING OF THE RESERVOIR (ROCK LOOSENING)

The idea of the rock-loosening method [10-11] is to create stresses in the well vicinity so as to form multiple new micro- and macro-fractures, by way of non-uniform directional unloading of the reservoir. The unloading is performed through a pressure relief in the well and choosing a special bottom-hole design. This newly generated system of fractures functions as a new network of filtration channels, with a permeability considerably exceeding the initial one.

It should be noted that the rock-loosening method impacts the bottom-hole area in the range up to ten well radii. This accounts for the high effectiveness of the method in development of the producing wells and repair works at the injection wells.

In 1954-1955, S.A. Khristianovich developed a theory and methodology of hydraulic fracturing of oil reservoirs, which up to now remains one of the most effective methods of oil production intensification. In 1990, Khristianovich suggested a totally new approach to the problem of rock deformation, with a gradual decrease of the formation pressure.

Rock sample tests were carried out at the Institute of Problems in Mechanics RAS, at an ITCTS experimental station (Independent Triaxial Compression Testing System). The results demonstrated a remarkable phenomenon concerning the influence of the bottom-hole depression on the rock permeability of the well vicinity. Depression of 6 to 9 MPa proved to decrease the rock permeability. It is especially evidently shown in sandstones with high clay content. It presumably results from plastic deformation of clays due to the tangential stresses appearing in the reservoir: they tend to «close» the filtration channels. This corroborates the thesis of geofluid-dynamic critical perturbation threshold 5÷8 MPa suggested above [12].

This deterioration in the rock permeability of the well vicinity lowers the well productivity. Actually, the effect of a sharp decrease in the well flow-rate due to higher depressions has been frequently observed in practice; there even appeared a special term for it: *collapsing of a reservoir.*

The mechanism of the DUR method (directional unloading of the reservoir) is that of the hydraulic fracturing method, but inverted vice versa.

With the rock-loosening method, development of a well can be conducted simultaneously with rehabilitation of the reservoir permeability in the bottom-hole area. No additional equipment or additional up-and-down operations are required. Thus, both time and expenses needed for the well development essentially decrease while the quality of the works increases.

The rock-loosening method is applicable at any depth of formation. It can produce a considerable economic effect at oilfields where the costs of well drilling and development are high, say at the sea shelf. It can be supposed to be a most sparing and effective method.

(S.A. Khristianovich, Yu.F. Kovalenko, V.I. Karev, Russia)

GASGUN®: TECHNOLOGY OF STIMULATING WELLS WITH SOLID PROPELLANTS

The necessity to optimise the impact on the bottom-hole area resulted in creating an effective technology with the use of solid propellants – Gasgun® technology. The technology is developed by the research team of *The GasGun Inc.* Co. headed by Dr. R.A. Schmidt [13].

Three various ways of stimulating a reservoir are compared at Figure 6a: explosion, hydraulic fracturing, and the Gasgun® method. The experimental and field tests have shown the Gasgun® method to be the most effective one. Figure 3b shows schematically a typical fracture pattern produced by the Gasgun® technology in the bottom-hole area of a reservoir. A principal feature of the Gasgun® technology is the use of solid propellants creating oscillating gaseous jets in the perforated well or even in the open hole.

With the use of solid propellants, large quantities of high-pressure gas can be produced. The burning characteristics of solid propellants can vary in a wide range; through appropriate adjusting of the burning intervals, multiple radiating fractures can be produced in the restricted treatment zone. Their structure will have a number of obvious advantages as compared with the results of the conventional hydraulic fracturing method.

For the last ten years, the Gasgun® method has been already used over 4,000 times in the USA, Canada, Europe, Africa, and Middle East, with the well depth 70 to 3,000 m. Promising results are shown for sandstone, limestone, dolomite, shale, coal, chalk, marlstone, and diatomite. The developers claim that Gasgun® stimulation has the following advantages as compared with hydraulic fracturing:

- Minimal vertical fracture growth out of the productive zone;
- Multiple circumferential and radiating fractures;
- Selected zones stimulated without the need to set packers off;
- Minimal formation damage from incompatible fluids;
- Homogeneous permeability for injection wells;
- Minimal onsite equipment needed;
- Much lower costs.

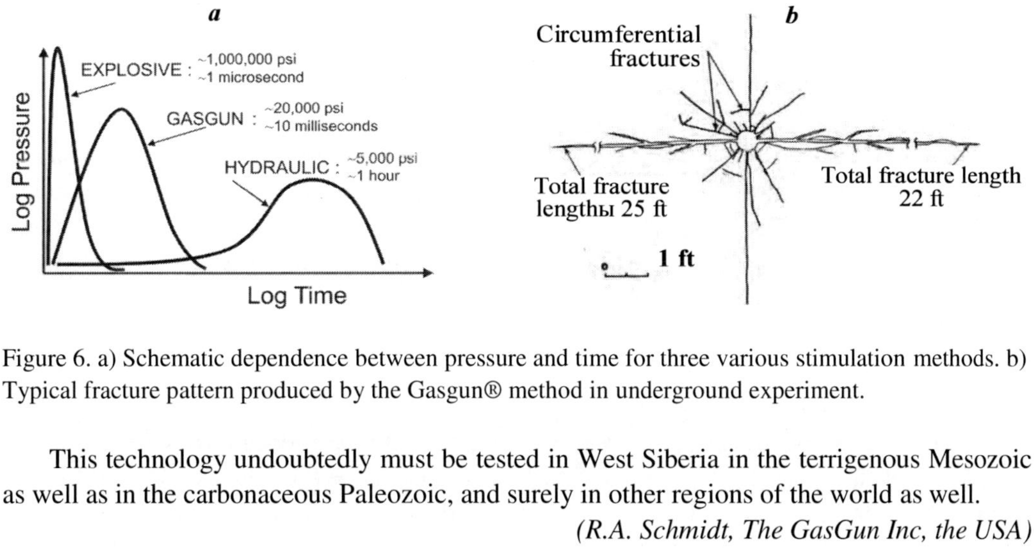

Figure 6. a) Schematic dependence between pressure and time for three various stimulation methods. b) Typical fracture pattern produced by the Gasgun® method in underground experiment.

This technology undoubtedly must be tested in West Siberia in the terrigenous Mesozoic as well as in the carbonaceous Paleozoic, and surely in other regions of the world as well.

(R.A. Schmidt, The GasGun Inc, the USA)

Novel Ideas and Projects Not Implemented So Far

1. Metasomatic Dolomitisation: A Possibility of Applying Nano-Technologies to Form Highly-Productive Reservoirs (Artificial Metasomatosis)

Oil and gas occur in various natural reservoirs including dolomites, which contain 40% of all the world oil reserves. By changing the architectonic pattern of voids and cavities, dolomitisation can increase the pore volume in compact limestones. This increases not only porosity but also permeability. It is known that the radius of a calcium ion (Ca^{2+}) is 0.99 Å, or

99 nm, while the size of a magnesium ion (Mg^{2+}) is 0.66 Å, or 66 nm. So, when Calcium is replaced with Magnesium, an additional empty space appears (fractures, caverns etc.). Hence, natural nano-dimensional metasomatic processes facilitate the formation of high-rate collectors, especially in the Phanerozoic carbonate rocks. The Maloichskoe oilfield (West Siberia) discovered in 1974 is one of the best-studied oilfields. The main productive horizon is at the depth 2794-2850 m and consists of carbonaceous rocks, namely limestones and dolomites. This oilfield clearly shows the focal character of dolomitisation, which is eventually the key factor in forming the well productivity. A specific feature of such foci (western and south-western parts of the oilfield, wells 9, 6, 117, 2) is active secondary dolomitisation across the Middle Devonian reef. The areas with high-rate wells apparently prove to coincide with the western facial-tectonic zone (Figure 7). It should be noted that metasomatic foci have no definite stratigraphic referencing, and their morphology normally cannot be viewed in terms of folded forms analysis and the superposition law. It is possible to initiate an accelerated technogenic process of metasomatic dolomitisation so as to create (renew) highly-productive foci at an oilfield. In fact, it will allow managing the process of an oilfield development and increase the oil recovery factor.

For these purposes, we should first determine the composition of the carbonate materials and formation water. The technology of injection of magnesium-containing liquid, or nano-particles of granular magnesium, is unlikely to prove too difficult. As a result, the specific surface of pores and cavities will increase; the fluid cross-flow from the block matrix to the fractures will intensify, and even active formation of new hydrocarbon masses will start. Also, this will stimulate the percolation processes; the well productivity and the actual oil recovery factor will increase. In certain cases, the process of induced and accelerated dolomitisation (metasomatosis) may be supported with certain wave or thermal stimulation.

In case it is successfully applied, this technology can considerably improve the duration of oilfield development and the total oil recovery factor.

(N.P. Zapivalov, Novosibirsk, Russia)

2. Laser-Based Technologies

These days, the attention is focused on novel breakthrough technologies for studying subsurface resources, geological prospecting, and oil-and-gas production. In a longer prospect, laser-based technologies are supposed to allow almost 100% oil extraction at any oilfield, without environmental pollution [14].

A. Lenetsky, Director of the Research and Production Company «Bereg», claims that the method suggested by the Company is a conceptually new one and has no world analogues. They suggest using laser instead of conventional rock drilling tool. A laser tool does not destroy the rock but rather fuses it; besides, it can pass at an angle, and therefore penetrate into hard-to-reach places. The technology will make it possible to restore old oilfields from which oil cannot be extracted any more through conventional methods.

In 2012 an American start-up company, *ForoEnergy*, also announced that they have developed a technology of laser application in oil production. Their method involves rapidly cracking the surface of hard rocks with a high-energy laser. *ForoEnergy* representatives assert that during the test run they managed to send a laser beam of 20 kW through a fibre-optical

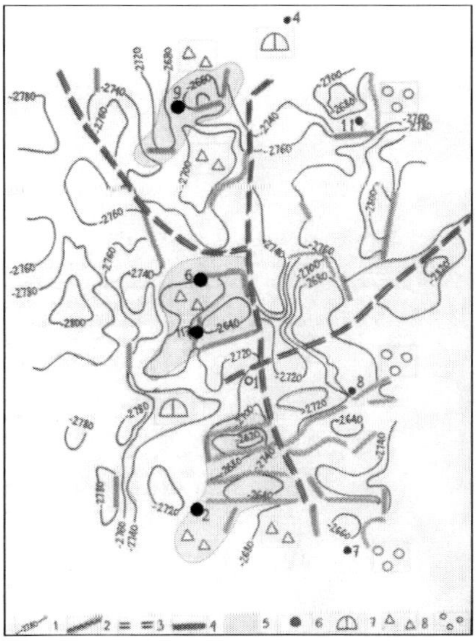

1 – Hypsographic curves of the surface of carbonaceous Paleozoic rocks, m;
2 – Subvertical zones of erosional-tectonic protrusions;
3 – The presumable abyssal fractures;
4 – Tectonic deformations;
5 – Foci of secondary dolomitisation;
6 – The wells which have given oil inflow;
Lithofacies:
7 – of organogenic reeves;
8 – of foreslopes;
9 – of back-reef lagoons.

Figure 7. 3D-seismic survey map of the Maloichskoe oilfield (West Siberia).

channel for a distance of 1.5 km. With the *Foro* technology, the intensive laser strike destructs the hard rock in such a way that no further difficulties arise afterwards when a mechanical drilling tool is used. This method can 10 times improve drilling effectiveness. Industrial tests of the technology will take place as soon as in 2014.

Saudi Aramco Company reports [15] that their Advanced Research Center has developed a new perforation method using laser technology. This method is supposed to make Saudi Aramco the first to introduce in-situ laser perforation to the petroleum industry.

A direct application of laser perforation will facilitate hydraulic fracturing in open-hole horizontal wells (oriented fracturing), which can greatly enhance the wells' productivity. This technology may show significant advantages as compared with conventional perforation technologies, because it does not involve compaction in the rock.

Specialists of *Saudi Aramco* are sure that the in-situ tests of laser perforation in the well will make a ground for further research in petroleum engineering, including laser drilling.

(ForoEnergy, USA; Saudi Aramco, Saudi Arabia)

CONCLUSION

1. A principally new scientific-technological paradigm is suggested for developing, saving, and replenishing oil-and-gas resources, with the account of the critical stability threshold of the system in order to provide for the fluid-dynamic balance. This approach will make it possible to save for long the active oil resources; they can be replenished at the expense of newly formed hydrocarbon masses and soft *sparing* methods of improving oil recovery without destroying the composition, structure, and properties of the fluid-dynamic productive strata.
2. For successful prognosis, effective prospecting and long-term development of oilfields, it is necessary to take account of the latest geo-dynamics of the Earth crust including its behavior in the gradient terms. For these purposes, satellite survey and mapping should be used in various modifications as well as other monitoring technologies.
3. Optimisation of the sparing methods and techniques should be adjusted depending on the individual features of the object.

REFERENCES

[1] Zapivalov, NP; Popov, IP. Fluid-dynamic models of oil-and-gas deposits. – Novosibirsk: Geo, 2003. – 198 p. (*in Russian*)

[2] Zapivalov, NP. Rehabilitation cycles are the basis of compensation of actual reserves at commercial oilfields. – Proceedings of International Conference "Degassing of the Earth", May 20-24, 2002, Moscow, 330-332 (*in Russian*).

[3] Zapivalov, NP; Lobov, VI. Geofluid-dynamic methods in managing the stressed-deformed state of oil-saturated reservoirs and well productivity. – Geodynamics and Stress State of the Earth's Bowels: *Proceedings of International Conference*, Institute of Mining Engineering, Novosibirsk, Oct. 6-9, 2003. 447-454. (*in Russian*)

[4] Reznikov, AN. Geosynergetics of oil and gas. – Rostov-on-Don: Publishing house «ЦВВР» («Centres of valeology of Russian higher educational establishments»), 2008. (*in Russian*)

[5] Muslimov, RKh. A new look at the prospects of development of the supergiant Romashkinsky oilfield. – *Geology of oil and gas Journal*, 2007. 3-12 (*in Russian*).

[6] Muslimov, RKh. Features of exploration and development of oilfields in conditions of market economy. – Manual for students. Kazan, Academy of Sciences of Tatarstan Republic, 2009. P. 690 (*in Russian*).

[7] Zapivalov, NP. Novosibirsk oil-2010 as a mirror of Russian petroleum industry. – EKO: All-Russian Economic Journal, Novosibirsk, 2010, №9. – 31-49 (in Russian).

[8] Pisetski VB. Prognosis of fluid-dynamic parameters of a basin based on seismic data. – Yekaterinburg: Ural State Mining University, 2011 (*in Russian*).

[9] Zapivalov, NP; Pisetski, VB. New geo-fluid dynamics method for mapping active fluids in oil-and-gas saturated strata // DEW: Drilling and Exploration World. – India. – August 2012. – Vol. *21*, No 10. – P. 55-60.

[10] Khristianovich, SA; Kovalenko, YuF; Kulinich, YuV; Karev, VI. Improvement of productivity of oil wells using the geo-loosening method // Oil&GasEURASIA. – 2000. – № 2. – P. 90-94.

[11] Karev, VI; Kovalenko, YF. Stress condition control as a method for a perfect well drilling//Oil&GasEURASIA. – November 2012. – №11. –16-19. (*in Russian*)

[12] Zapivalov, NP. Dynamics of oilfield life // Bulletin of Tomsk Polytechnic University. – Tomsk: 2012. – V. 321. – No. 1. –206–211. (*in Russian*)

[13] Schmidt, RA. Fracturing with solid propellants offers advantages over traditional stimulation // DEW: Drilling and Exploration World. – India. – October 2009. –47-51.

[14] «Kapital» of December 11, 2013. http://kapital.kz/economic/24434/v-rossii-nauchilis-dobyvat-neft-s-pomocshyu-lazera.html (*in Russian*)

[15] http://theenergyinfo.com/ServiceDetails.aspx?ID=60(InformationAgency *theenergyinfo .com*)

In: Enhanced Oil Recovery
Editor: Alicia Knight
ISBN: 978-1-63463-917-0

Chapter 6

OPTIMIZATION OF POLYMER FLOODING IN RESERVOIRS UNDER BOTTOM-WATER CONDITION TO MAXIMIZE ECONOMIC PROFIT

Byungin Choi and Kun Sang Lee*
Department of Natural Resources and Environmental Engineering,
Hanyang University, Seoul, South Korea

ABSTRACT

Bottom-water condition has been considered as a challenging issue for implementing Enhanced Oil Recovery (EOR) techniques. In the application of polymer flood which is known as one of effective EOR methods, the existence of strong bottom aquifer leads to severe polymer loss to the aquifer, water fingering, and water production. Because these factors affect economics of the polymer flooding operation, injection design of the polymer flooding should be optimized to maximize profit. This study aims for the evaluation of polymer flooding under bottom water condition considering economic feasibility. Quantitative comparisons considering oil recovery, water production, and economics of polymer injection were made to obtain maximum net present value. The results indicated that oil recovery was increased 5% and water breakthrough was delayed about 40 days to maximize the net present value by 40%. After the optimization of injection designs, impacts of geological uncertainties on net present value were analyzed. According to sensitivity analysis, vertical permeability had most negative impact on the net present value. This study showed that injection of polymers under bottom water condition could be optimized to be an effective EOR technique for maximizing profit.

Keywords: Polymer flooding, bottom-aquifer, blocking process, sweep efficiency, water coning

* Department of Natural Resources and Environmental Engineering, Hanyang University, Seoul 133-791, S. Korea. (Phone) 82-2-2220-4428. (Fax) 82-2-2220-4427. (E-mail) kunslee@hanyang.ac.kr.

INTRODUCTION

The operation of oil reservoirs under bottom-water conditions has been recognized as a problem of considerable interest. Due to rapidly decreasing oil recovery and extremely large water influx in the early production time, such reservoirs could be forced to suspend or abandon the operation despite the significant amount of remaining oil. A number of projects have been focused on the problems and remediation of the reservoir condition [1-3]. The presence of bottom-water aquifer makes oil production more complicated. Generally, the aquifer is located under the relatively shallow reservoir column bearing the oil. It is because the specific gravity of the water is greater than that of oil [4]. Maintaining the economic oil production rate by controlling the excessive producing water under such condition was a major concern for the operator.

Given the poor efficiency of primary recovery procedure, the reservoirs were considered as candidates for Enhanced Oil Recovery (EOR) techniques as a means by which to improve the ultimate recovery. Among many EOR options, the chemical EOR techniques, especially polymer and gel treatment, were implemented to control the mobility ratio and water production. Yeung and Farouq [5] used a term of "single" and "simultaneous" fluids injection for elaborating the difference. The "single-fluid" is desired for the blocking mode to take place in the bottom-water zone and the displacing mode to take place in the oil zone. The "simultaneous-fluid" injection consists of injecting displacing and blocking fluids into the corresponding layers simultaneously. They claimed that better control of the displacement process is obtained by using simultaneous injection process.

The polymer flooding is recognized as the "simultaneous-fluid" injection process. It is expected to improve the oil recovery by maximizing the cross flow of oil under such reservoir condition. Results from the experiments conducted by Barnes [6] and Hodaie and Bagcl [1] indicated that the larger the quantity of viscous slug injected, the greater the cross flow of oil ahead of the displacing front and the lower WOR during that period of a displacement. They reported that the breakthrough of oil was delayed compared with that of reservoir without water zone.

The recovery efficiency is also affected by the well configuration. Popa and Clipea [7] reported that vertical wells as injectors and horizontal wells as producers are more preferable than use of horizontal wells as injectors and vertical wells as producers. Algharaib and Ertekin [8] investigated the effects of various parameters on the performance of horizontal well. They suggested that use of vertical wells as injectors and horizontal wells as producers for reservoirs with bottom water conditions is effective. In addition, Shirif [9] stated that a certain well combination (horizontal production and vertical injection) gave better oil recovery due to the increase in the swept area. Many experimental and simulation studies have investigated the recovery efficiency of polymer floods with regard to injection design parameters [10, 11]. The flow behavior of polymer solution is strongly affected by the excessive water from reservoir and aquifer. Because the recovery efficiency does not guarantee the economic profit under such condition due to the significant operating cost related with excessive production water. Not only recovery efficiency of polymer flooding, but water breakthrough from bottom aquifer are necessarily analyzed at a same time. However, most of previous researches focused only on the flow of oil, whereby blocking efficiency for water breakthrough was regarded as relatively less important aspect. Thus, based on both important parameters in this work, profit and cost are comprehensively calculated in terms of NPV.

This work tried to find the optimum design considering the location of injection wells and concentration of polymer solution simultaneously. The results from the various simulations were compared in terms of objective function of Net Present Value (NPV). By using the index of NPV, the oil recovery, water production, and operating cost were quantitatively considered during the evaluations process. Moreover, to investigate the effects of key uncertainties from geological conditions on the NPV, a sensitivity analysis (SA) was conducted.

BACKGROUND THEORY

Polymer flooding is a kind of water

flooding which uses mobility control agents to improve the sweep efficiency. In this process, the key parameter is viscosity of polymer solution. The viscosity of polymer solution is determined by the Flory-Huggins Equation [12], which is followed:

$$\mu_p^o = \left[\mu_w 1 + (A_{p1}C_p + A_{p2}C_p^2 + A_{p3}C_p^3)C_{sep}^{S_p}\right] \tag{1}$$

where μ_w is the water viscosity (cp), μ_p^o the polymer solution viscosity at zero shear rate (cp), C_p the polymer concentration in water (gmol/m^3), A_{p1}, A_{p2}, A_{p3} and S_p the fitting constants, and C_{sep} the effective salinity for polymer (meq/ml).

The NPV is composed of unit price and/or cost of produced oil, produced water, injected water, and consumed polymer. NPV has been used to evaluate the profits or economics of the projects. The comparison in terms of NPV was more quantitative and accurate than other objective parameters. A number of studies investigated the performance of chemical flooding regarding to the NPV [13-16].

In this work, not only the operating conditions, but the geological properties which have far more uncertainty due to the limitation in the quality and quantity of the data were investigated. In practice, given the subjective nature of probability assessment, there are uncertainties associated with the actual input probabilities. It is, therefore, important to determine the sensitivity of the simulation results to changes in the estimates of the input parameters [17]. The calculated result at a given reservoir conditions was always uncertain value which provokes risk on the estimation of that projects. Considering these key uncertainties, the effects of the undetected geological event in the reservoir on the NPV of polymer flooding was analyzed using sensitivity analysis. For sensitivity analysis task, the proxy models are then used to determine the main (linear) effects, interaction effects, and quadratic (nonlinear) effects.

MODELING

The productivity of polymer floods is affected by operating conditions such as injected polymer concentration, injector location, and well configuration as well as reservoir properties such as porosity, permeability, and saturation. For the well configuration, Ertekin [8] pointed out the use of horizontal wells as producers and vertical wells as injectors showed the greatest performance under bottom-water condition. In this work, the well patterns for every case were simulated reflecting the result of well application from Ertekin's research.

The oil reservoir used in this simulation work had dimensions of 1,000ft×10ft×300ft above infinite aquifer which was 500 times larger than the size of oil-bearing reservoir. The simulation domain was on the x-z cross-sectional area. Vertically, the oil reservoir is comprised of 15 layers; and each layer is 100×1 grid blocks. Each grid block has dimensions of 10, 10, and 15 ft for x, y and z directions, respectively. Figure 1 shows the reservoir model above the infinite bottom-aquifer. A vertical injector is located at grid (1, 1, 1-3) in the base case. A producer is horizontally completed at top of the reservoir (91-100, 1, 1). Water saturated zone was contacted with the bottom of reservoir. This aquifer zone consists of five layers; upper four layers has same dimensions with the reservoir grid blocks, while the last bottom layer was made 500 times larger than the size of reservoir grid blocks for representing the infinite aquifer zone.

The viscosity of polymer solution depending on the polymer concentration was shown in the Figure 2. According to the Figure 2 the maximum soluble concentration was set to be 2,000 ppm. At the concentration, the viscosity was 60 cp without salinity effect. To eliminate the salinity degradation, the make-up brine was assumed as pure water having little salinity ion.

Increasing the viscosity of the displacing fluid has two disadvantages: It reduces the maximum rate of injection into the reservoir, and it costs more than water. Polymer concentration is the one of the key factors that affects the displacing efficiency of polymer flood. To examine the effect of operating conditions such as location of well location and concentration of polymer solution on the productivity of polymer floods applied on the reservoir in communication with bottom-water aquifer, the extensive numerical simulations were carried out in various concentrations and well locations for 2 years. The polymer flooding was applied in the duration of the entire operating period, during which about 1 pore volume of polymer solution was injected.

For the reservoir conditions such as porosity, permeability and unwanted geological events, the effects on the recovery efficiency was quantitatively investigated using sensitivity analysis. Effects of polymer adsorption capacity were also examined. Input data including reservoir conditions were listed in the Table 1. Additionally, the results of each case were analyzed with respect to the NPV as shown in Table 2.

RESULTS

This study aims to verify the efficiency of polymer flooding under bottom-water aquifer condition. By comparing with other injection designs, the optimum design was deduced with regard to the oil recovery, water cut, and NPV at two year after polymer injection.

Polymer Flooding vs. Waterflooding

To compare the efficiencies between the waterflooding and polymer flooding, the cumulative oil recovery of each case was calculated as shown in Figure 3. In the case of waterflooding, the cumulative oil recovery is about 40,000 bbl during the water injection. As the polymer flood was applied in the same reservoir, the oil recovery factor increases from 45% up to53%. Because of water breakthrough in the relatively early time, the swept efficiency of waterflooding is significantly lowered up to 24% while polymer flooding still shows stable sweep efficiency of 79% as shown in the Figure 4. Especially, the unswept oil remains high in the bottom zone contacted with water saturated area. Polymer injection modifies the poor performance of waterflooding. It results from the better sweep efficiency by the mobility control. So far, these cases were calculated for the base case where the injection well located at top of the layer. The oil recovery can be improved by applying the optimum polymer injection design. In the following section, a series of cases associated with polymer concentrations and injector locations were compared to find optimum design of polymer flooding under bottom-water aquifer.

Sweep Efficiency

During oil recovery, the injected fluids were disturbed by the water influx from the bottom-water zone. As shown in Figures 5 and 6, the oil displacement exhibited the close relationship with the sweep efficiency. To analyze the effects of various operating conditions on the sweep efficiency, the shape and location of front between the injected solutions and oil were examined. Also, the area under 40% of residual oil saturation was calculated to compare sweep efficiency between polymer flooding and waterflooding. For the waterflooding, the displaced oil flows fast into the producer located in the top layer. The oil remained in bottom zone is little pushed from the injection water. The shape of interface implied that only the oil in the upper layers moved into the producer resulting in poor sweep efficiency. In the case of waterflooding, the front of injection fluid was shown as ambiguous, while the front of polymer flooding was more distinct between the displacing and displaced fluids. It results from the unfavorable mobility ratio of waterflooding. After applying the waterflooding, 60 % of the OOIP remained in the swept zone. The oil saturation after the waterflooding is about 0.2 higher than that after polymer flooding. According to the results shown in Figure 5, the upper interface move faster toward the production well compared to the interface in bottom zone in the case of injector located in the top of reservoir. The remaining oil was mainly distributed near the bottom of the reservoir, which leads to the poor vertical sweep efficiency. As the injection well located in the deeper zone, the poor vertical sweep efficiency was improved. For the cases of injectors at layers10-12 and layers13-15, the front flowed evenly along the vertical cross-section. As the location of injection well was deeper, the oil zone moves with maintaining the piston-like shape in the reservoir. Especially, the mobilization of oil remaining in the bottom of reservoir exhibited a noticeable difference depending on the injector location.

Table 1. Reservoir and fluid properties

	parameters		values
reservoir	size (ft^3)		1,000 × 10 ×300
	porosity, ϕ		0.25
	permeability, k (md)		800
	depth (ft)		1,500
	pressure (psi)		3,000
	compressibility, c_f (psi^{-1})		0
	initial saturation	water, S_{wi}	0.3
		oil, S_{oi}	0.7
fluid	viscosity (cp)	water	0.6
		2,500 ppm polymeric solution	60
		oil	12

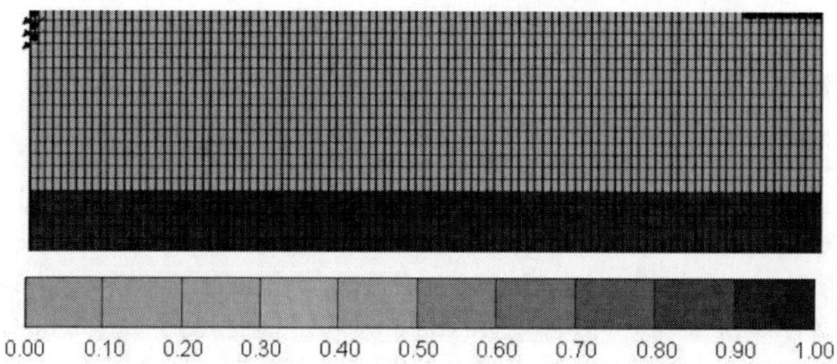

Figure 1. Initial water saturation condition of the reservoir model with bottom-water aquifer.

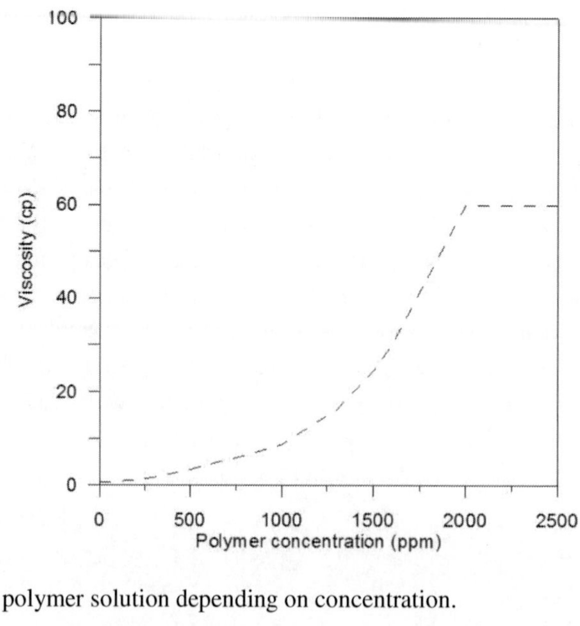

Figure 2. Viscosity of polymer solution depending on concentration.

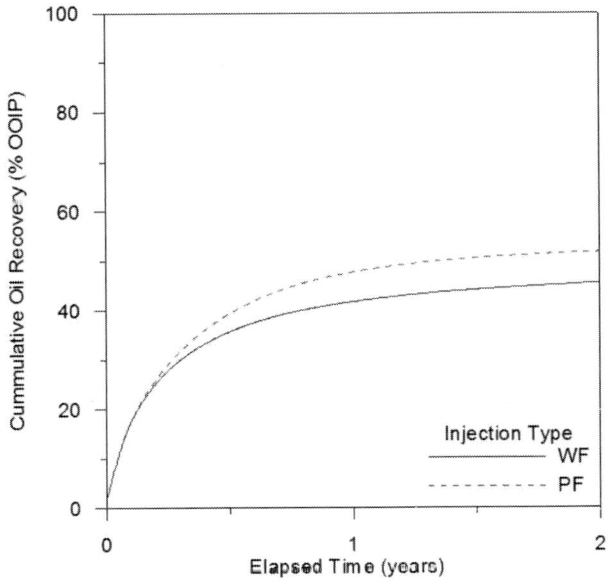

Figure 3. Cumulative oil recovery from waterflooding and polymer flooding.

Table 2. Terms of objective function

parameters	oil production	water production	water injection	polymer consumption
unit cost	+$ 50/bbl	-$ 4/bbl	-$ 2/bbl	-$ 1.5/lb

Table 3. Cumulative oil recovery (% OOIP)

injection design		injection well location (perforation location)				
		1-3	4-6	7-9	10-12	13-15
polymer conc. (ppm)	500	47.18	47.63	48.00	48.40	48.75
	750	47.87	48.50	49.07	49.62	49.96
	1,000	48.38	49.16	49.92	50.59	50.87
	1,250	48.78	49.71	50.61	51.41	51.51
	1,500	46.54	50.01	51.04	51.92	51.93
	2,000	47.29	50.71	51.83	52.75	52.52

Table 4. Elapsed time for reaching water cut of 90% (days)

injection Design		injection well location (perforation location)				
		1-3	4-6	7-9	10-12	13-15
polymer conc. (ppm)	500	157	151	142	132	124
	750	164	159	150	137	126
	1,000	168	164	156	141	127
	1,250	172	167	160	144	129
	1,500	169	168	162	146	129
	2,000	173	171	166	150	132

Table 5. The NPV at 2 years for waterflooding

injection design	NPV (USD)
waterflooding	670,756

Productivity of Polymer Flooding

The oil recovery factor was compared to find the optimum design among a series of combinations. The flow of oil and water influx was significantly affected by mobility of displacing fluid and injected vertical location. According to the table 4 and 5, high concentrated polymer flooding with 2,000 ppm of polymer concentration shows that oil recovery increases about 5% and water breakthrough is delayed about 40 days depending on the injection location. As a result, it is the most efficient case that injector is in the middle of the oil column. As regards moderate mobility ratio cases such as 1,000 ppm and 1,250 ppm, the maximum oil recovery is almost same.

Table 4 shows the results of elapsed time for reaching water-cut of 90% from each polymer flooding case. Under bottom-water aquifer condition, the water-cut is as important as the oil recovery because of its impacts on operating cost. The water-cut of waterflooding is higher than that of polymer flooding. For the waterflooding, the oil breakthrough appeared relatively early in the operating process, while the amount of oil flow is relatively smaller than that of polymer flooding. It is because of poor sweep efficiency and water coning. After the bottom-water aquifer affected, the excessive water inflows into the producer. It causes the economic operation to be stopped earlier than the operators expected. The water-cut of 0.9 in the case of waterflooding occurred at 123 days after the water injection. For the 500 ppm polymer flooding, the time to reach 0.9 is delayed about 20 days. Furthermore, the time was delayed up to 100 days in the case of 2,000 ppm polymer injection. It affects operating cost significantly and is closely related to the economic profit of each case.

Optimum of Polymer Concentration and Injection Well Location

With respect to the NPV, the optimum injection design conditions were examined. The comparison of results was made in terms of NPV at two year after polymer flooding. The maximum value of each case during the total polymer injection was also investigated. The NPV at the 2 years after the polymer injection for all the 30 cases is shown in the Figure 7. With respect to the polymer concentration, the NPV for the over 1,250 ppm is greater than 1 million USD. Results imply that the NPV increases with the polymer concentration due to the improved sweep efficiency. For the case of injecting low concentrated polymer ranging from 500 ppm to 1,000 ppm, representing less favorable mobility ratio, the optimum location of injection well was shown to be the lowest three layers (layers 13-15) with contact to the bottom-water aquifer. As the injected polymer concentration increases, the optimum location for injecting polymer solution moves upward. The maximum NPV was shown at 10-12 layers for 1,500 ppm and 7-9 layers for 2,000 ppm. One reason is that excessive water from the aquifer affects poor efficiency in polymer flooding. Also water from the aquifer hinders polymer solution from reaching producer.

Maximum NPV

Maximum NPV shows that result for each case is greater from 0.1 to 0.3 million USD than the previous cases. The maximum NPV at certain elapsed time for cases are shown Figure 9. This figure indicates that the maximum values do not exhibit large difference depending on the polymer concentration. From 500 ppm to 2,000 ppm, the difference in maximum values is about USD 20,000. The difference is at least USD 100,000 depending on the injection locations. It also indicates that profit from optimum design approaches about USD 180,000 which was 15 % improvements. The optimum location for maximum NPV moved upward in comparison with the previous NPV cases. These differences result from the excessive water influx from the aquifer. The excessive water influx disturbs the flow of polymer solution. It causes not only the poor efficiency of polymer flooding, but also delayed flow of polymer solution. Under unfavorable mobility ratio, the time to reach maximum NPV was delayed with the increasing depth of the injector location. For the cases of 500 to 1,250 ppm, the delayed time caused by the deeper location was about 50 days. There is change of time that reaches the maximum NPV. As can be seen, the time was delayed with increasing viscosity under unfavorable mobility ratio. However, if the polymer concentration is over 1,500 ppm, time of the maximum NPV was related with the results of the optimal location. It did not show any relationship with concentrations or locations. It means that the profits from the polymer flooding in aquifer reservoir condition are determined by the combination of many associated parameters. These differences were good clue for operators or decision makers to use optimization process. It is more useful when many parameters were considered simultaneously, especially, the many parameters were associated with cost.

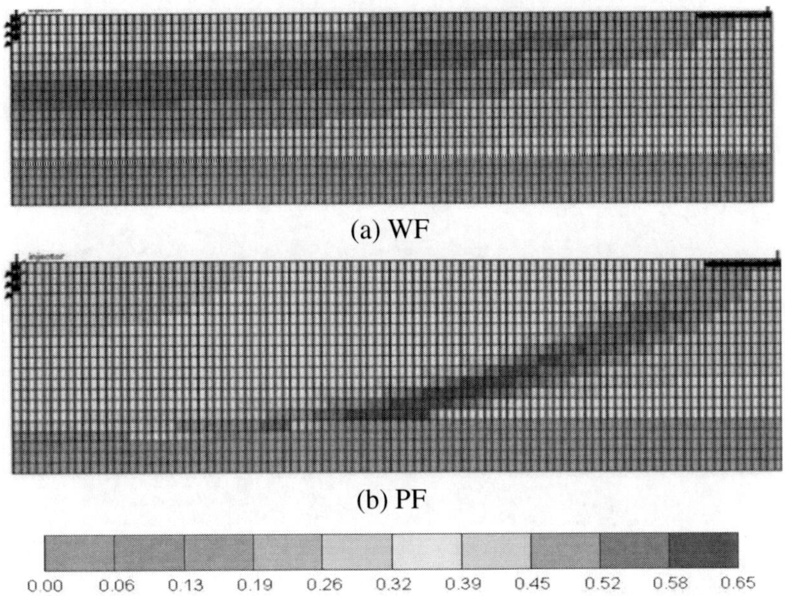

Figure 4. Oil saturation at one year after the application.

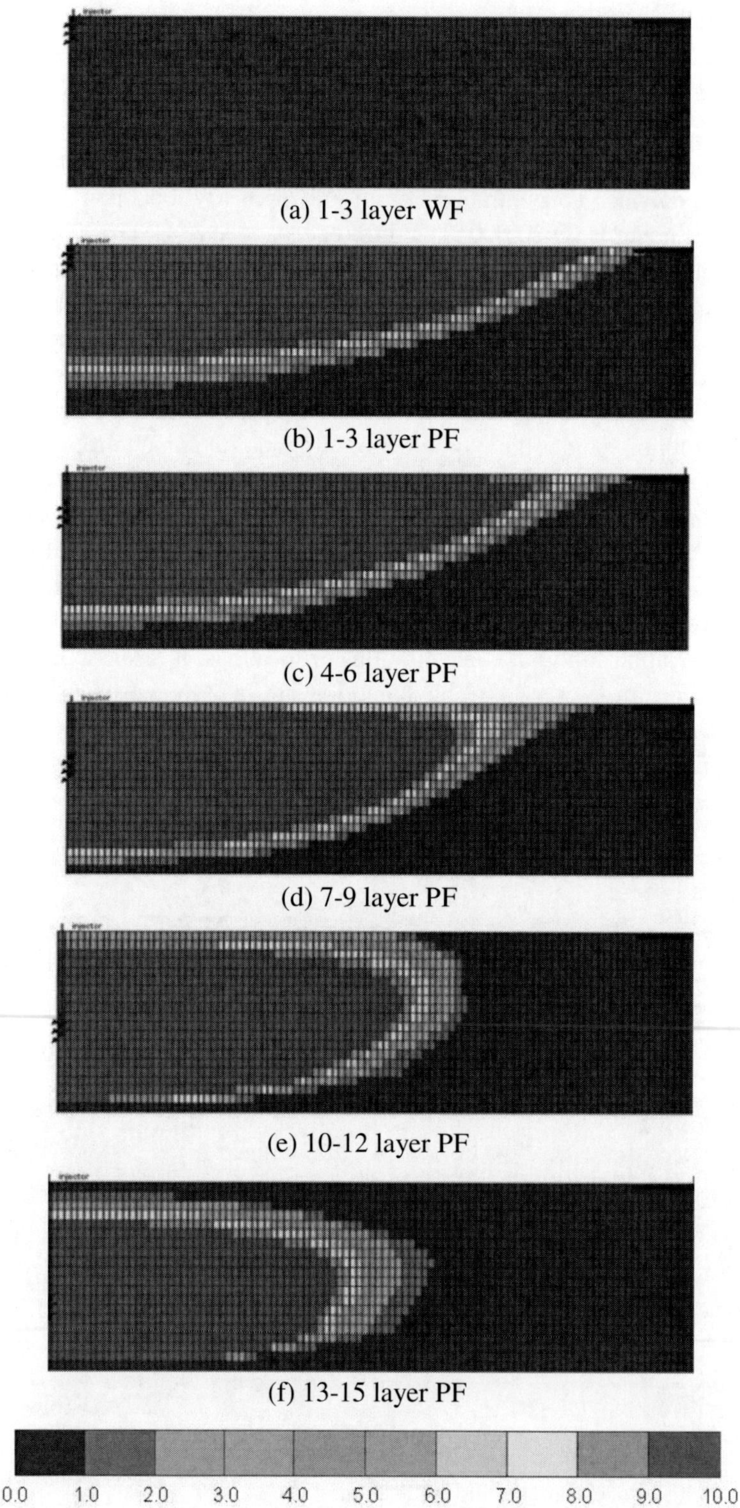

Figure 5. Viscosity of injected aqueous phase depending on the injection well location at one year after 1,250 ppm polymer injection.

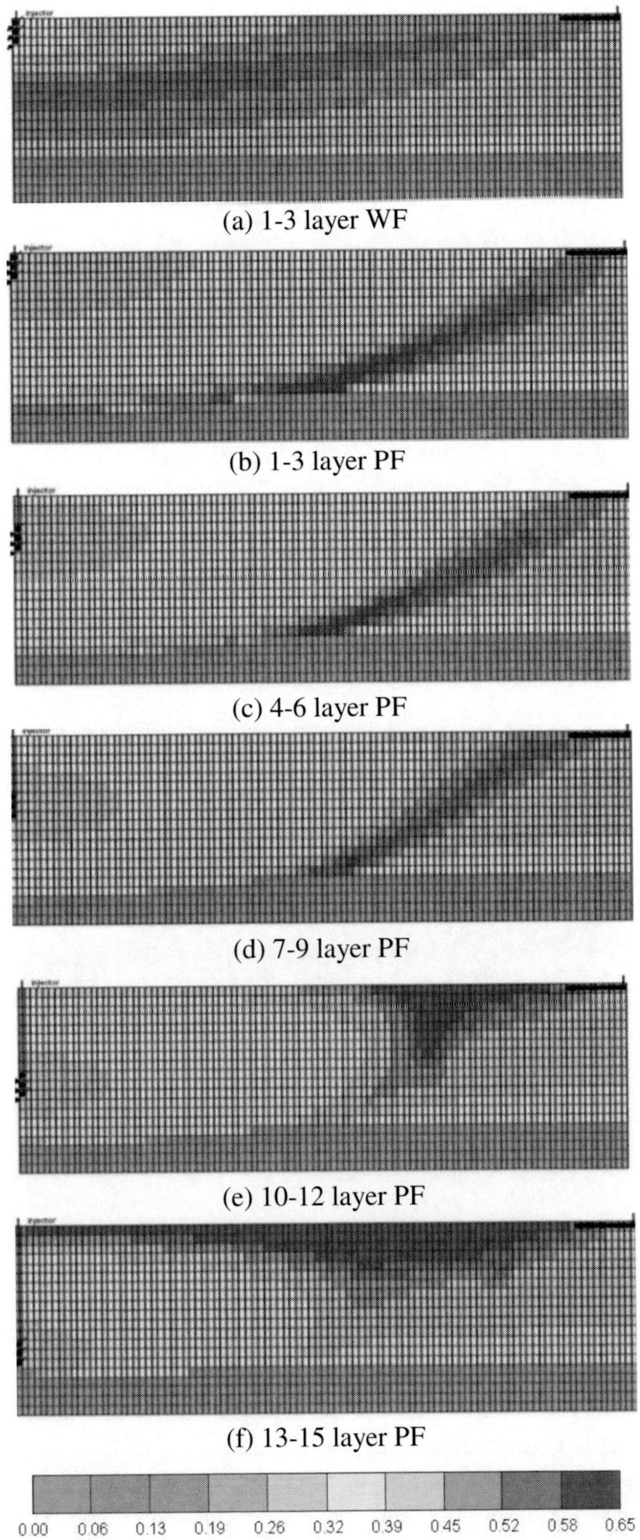

Figure 6. Oil saturation depending on the injection well location at one year after waterflooding and 1,250 ppm polymer injection.

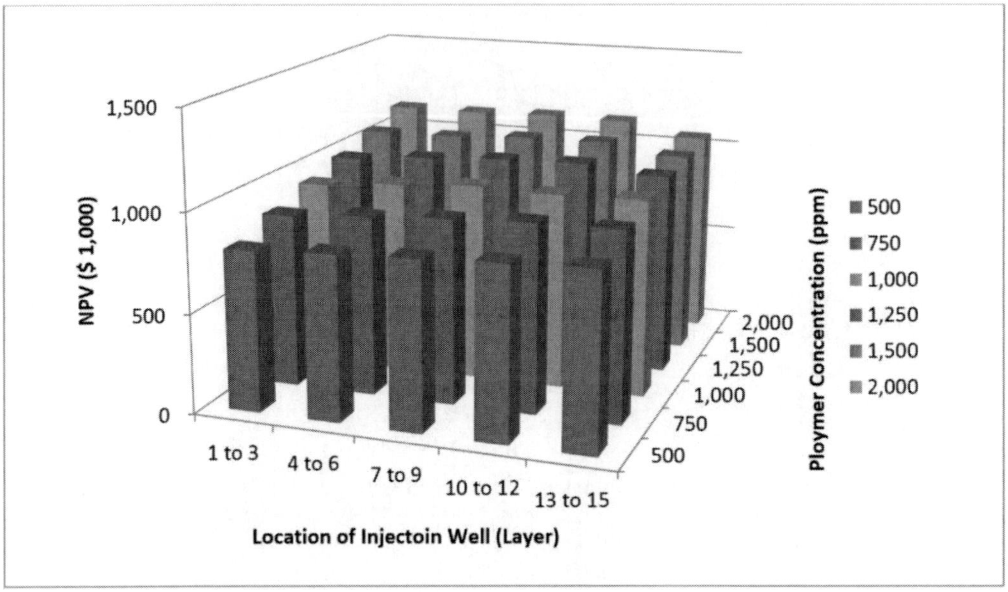

Figure 7. Net Present Value (NPV) at 2 years after polymer injection with operating design variables.

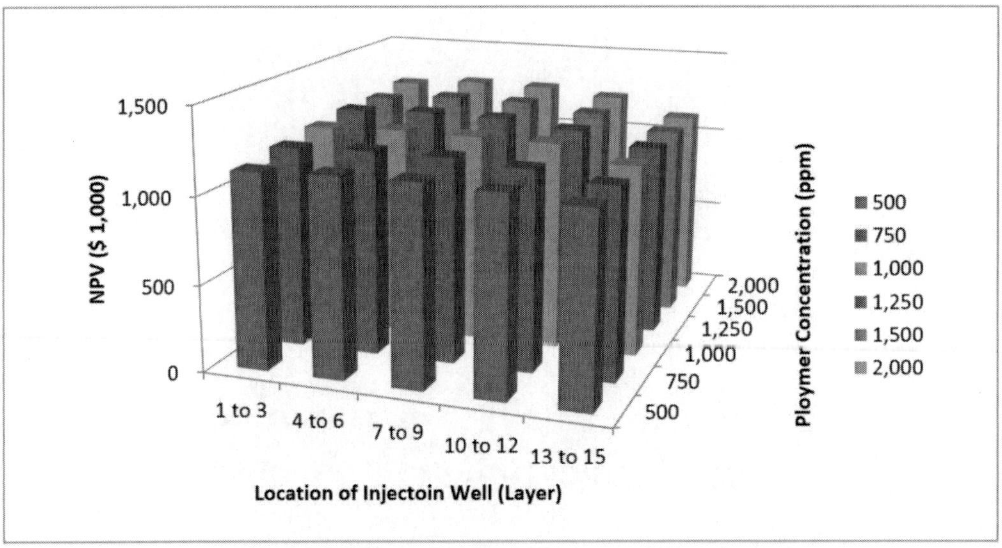

Figure 8. Maximum Net Present Value (NPV) with operating design variables.

Table 6. Maximum NPV at elapsed time for waterflooding

injection design	NPV (USD)
waterflooding	1,041,435 (155 days)

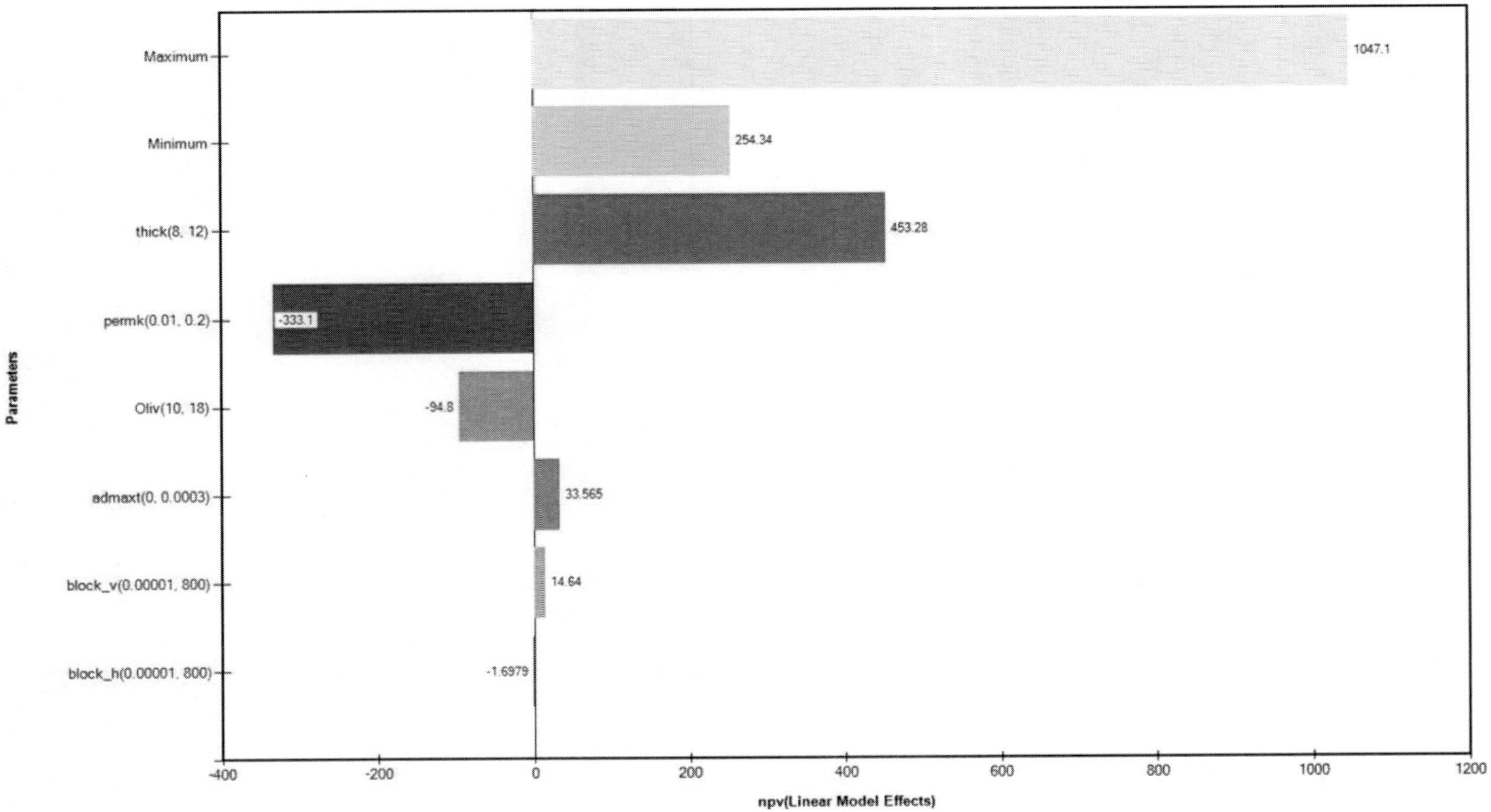

Figure 9. Tornado plots of sensitive analysis for key uncertainties associated with geological events.

Sensitivity Analysis of Geological Properties

Even though the optimum design was applied in the real oil field, the results from the project were always different from the prediction. It results from the key uncertainties which induced a considerable risk. The important geological uncertainties, which might be inherent and undetected in bottom-water aquifer reservoir condition, need to be summarized. The key parameters and their possible ranges were taken into accounts for in the process of sensitivity analysis. The sensitivity analysis was conducted with regard to the NPV. The effects of the uncertainties on the NPV are summarized in the tornado diagram in Figure 10. The results show that the polymer flooding process is negatively sensitive to ratio between vertical permeability and horizontal permeability ($\frac{k_v}{k_h}$). The higher the ratio, the more polymer lost to the aquifer and more excessive water production. The polymer flooding is also negatively affected by higher viscosity of oil. If the oil viscosity increases up to 18 cp and the mobility ratio becomes more unfavorable, about USD 300,000 is lost in terms of the NPV by reduced the sweep efficiency of the polymer flooding. Whether shale barrier influences positively or negatively on the NPV is determined by its orientation. For the vertical impermeable barriers even though the magnitude of changed NPV was relatively small, it has an adverse impact on the NPV. If the orientation of baffle was changed to the horizontal, it has a positive impact on the NPV because it acted like as a blocking curtain delaying the water breakthrough from the bottom aquifer. As a result, the NPV increases up to USD 15,000 due to the presence of horizontal baffle. Polymer adsorption causes the NPV increasing because of increasing efficiency of blocking material. The dominant parameters among many geological events were proved to be the thickness of oil column and the ratio of the vertical to horizontal permeability. The thickness of oil zone is always key parameter thorough entire operation, but the ratio becomes an important issue in reservoir with aquifer. This is related with the previously mentioned problem in the bottom-water conditions. The excessive water production is regarded as more important parameter than the oil recovery, which significantly affects the NPV of whole operating process. Besides, the operating time of polymer floods was determined to maximize the NPV. Therefore, to increase the NPV from the polymer floods, operators or decision-makers should mainly focus on the reducing the risk associated with the water production. Additionally, the efforts to identify the risks associated with the polymer flooding development have to be taken into account.

CONCLUSION

This study presented a systematic and efficient approach to design and optimize a polymer flooding taking into account technical and economic uncertainties. From a number of simulation performed in order to estimate the productivity of a polymer flooding applied in reservoir with bottom-water aquifer, the important conclusions were derived in this work.

1) By implementation of polymer flooding, the oil recovery increases and water production decreases. At the 2 years after polymer injection, the NPV of polymer flooding was indicated to increase up to 80 % compared to that of waterflooding.

This emphasizes that the implementation of polymer flooding under such conditions is efficient treatment.

2) With optimal combination of design variables in this work, the incremental of maximum NPV is on a par with 15% of total NPV of the optimal one compared with the poorest case.
3) For the unknown geological conditions we studied, the most dominant parameter is proven as the ratio of the vertical and horizontal permeability excepting the thickness of oil column. The results of sensitivity analysis indicate that the controlling excessive water production is major concern to determine if the project is economic or not.

REFERENCES

[1] Hodaie, H., Bagcl, A. S. *Proc. SPE Middle East Technical Conference & Exhibition,* Manama, Bahrain, 3-6 April, 1994.

[2] Elkaddifi, K., Shirif, E., Ayub, M., Henni, A. *J. CPT.* Bottom-water Reservoirs, Simulation Approach 2008, 47(2), 38-43.

[3] Al Azri, N., Jaml, E., Murshidi, A., Al Mahrouqi, A., Al Busaidi, I., Kazzaz, W., Ramalingam, S., de Kruijf, A., Al Kharusi, B., Brooks, D. *Proc. SPE EOR Conference at Oil & Gas West Asia,* Muscat, Oman, 11-13 April, 2010.

[4] Dusseault, M. B. *The Future of Oil: Canadian Viscous Oil Resources and New Production Technologies,* Univ. of Waterloo, Waterloo, Ontario, Canada, 2006.

[5] Yeung, K., Farouq Ali, S. M. *J. CPT.* Waterflood Reservoirs with a Water Leg: Dynamic Blocking Process 1995, 34(7), 50-57.

[6] Barnes, A. L. *J. Pet Tech.* The Use of a Viscous Slug to Improve Waterflood Efficiency in a Reservoir Partially Invaded by Bottom Water 1962, 14(10), 1147-1153.

[7] Popa, C. G., Clipea, M. *Proc. SPE International Conference on Horizontal Well Technology,* Calgary, Alberta, Canada, 1-4 November, 1998.

[8] Algharaib, M., Ertekin, T. *Proc. SPE Mid-Continent Operations Symposium,* Oklahoma city, OK, USA, 28-31, 1999.

[9] Shiriff, E., *Proc. SPE Asia Pacific Oil and Gas Conference and Exhibition,* Brisbane, Australia, 2000.

[10] Lake, L. W. *Fundamentals of Enhanced Oil Recovery,* Prentice Hall Inc., Upper Saddle River, NJ, USA, 1986.

[11] Sorbie, K. S., *Polymer-Improved Oil Recovery,* CRC Press Inc., Boca Raton, FL, USA, 2000.

[12] Flory, P. J., Fox, T. G. *Principles of Polymer Chemistry,* Cornel Univ. Press, Ithaca, New York, USA, 1953.

[13] Zhang, J., Delshad, M., Sepehrnoori, K., Pope, G. A. *Proc. SPE Latin American and Caribbean Petroleum Engineering Conference,* Rio de Janeiro, Brazil, 20-23 June, 2005.

[14] AlSofi, A. M., Blunt, J. M. Proc. *SPE Enhanced Oil Recovery Conference,* Kuala Lumpur, Malaysia, 19-21 July, 2011.

[15] Van Doren, J., Douma, S., Wassing, B., Kraaijevanger, H., de Zwart, B. R. *Proc. SPE Enhanced Oil Recovery Conference,* Kuala Lumpur, Malaysia, 19-21 Junly, 2011.

[16] Mollaei, A., Lake, L. W., Delshad, M. *Proc. SPE Reservoir Simulation Symposium,* The Woodlands, TX, USA, 21-23 February, 2011.

[17] Bratvold, R. B., Begg, S. H. *Making Good Decisions,* SPE, Richardson, TX, 2010.

In: Enhanced Oil Recovery
Editor: Alicia Knight

ISBN: 978-1-63463-917-0

Chapter 7

OPTIMIZED DESIGN OF WATER-ALTERNATING-GAS EOR PROJECTS IN HETEROGENEOUS RESERVOIRS

Moon Sik Jeong and Kun Sang Lee *

Department of Natural Resources and Environmental Engineering,
Hanyang University, Seoul, South Korea

ABSTRACT

To overcome reduced oil recovery resulting from poor mobility ratio and sweep efficiency by the low viscosity of CO_2, water-alternating-gas (WAG) process has been commonly used as an advantageous alternative. This study examines how to optimize WAG processes in heterogeneous reservoirs by rigorous simulation of WAG injection. Compositional simulations of areally heterogeneous reservoirs were used to quantify the effects of process parameters such as WAG ratio and CO_2 slug size, well parameters including well length and type, and reservoir heterogeneity characteristics such as Dykstra-Parsons' coefficient and correlation lengths on the oil recovery and economic value. Optimization was carried out with net present value (NPV) as an objective function. The main conclusions of this study show that more WAG cycles are required for maximum NPV at a given operation period. Higher water fraction in WAG injection is needed to improve sweep efficiency in more heterogeneous reservoirs. The longer horizontal wells maximize NPV as their orientation and permeability correlation direction is parallel. The proposed methodology has significant improvement on the oil recovery up to 43% and hence yields higher NPV up to 47%. The importance of using the reservoir simulation as a tool for optimum design before implementing WAG injection in practice is evidenced. It also demonstrated that the approach is viable to give the optimistic scenario of oil recovery prediction under WAG injection into heterogeneous reservoirs.

Keywords: optimization, WAG cycle, well pattern, well length, NPV

---

* (Phone) 82-2-2220-2240. (Fax) 82-2-2220-4427. (E-mail) kunslee@hanyang.ac.kr.

INTRODUCTION

Injecting fluids in oil reservoir is a common manner to increase oil production. The injected fluids are usually water and gas, which have potential to displace and sweep the oil remained in the reservoir to production wells. CO_2 gas injection is an attractive method to enhance oil recovery in oil industry and to reduce greenhouse gas at the point of environmental issue [1].

Gas injection technique has higher microscopic displacement efficiency than water injection since the interfacial tension between oil and gas has lower than water and oil. When the interfacial tension is close to zero, oil and gas are in miscible state. If the miscibility occurs, the oil can be completely recovered in the contact area. Although miscibility is not accomplished, mass transfer between the oil and gas components still occur. This mass transfer causes lower interfacial tensions compared to water. In this immiscible condition, gas injection is still effective method to recover the remained oil (independently of its saturation). Application of gas injection is effective in various production stages, independently of the mass transfer mechanism. Uncertainties of reservoir influence to gas injection more than water injection because more parameters exist for production and they can affect the response of reservoir. The mass transfer which occurs during miscible process must be created appropriately by EOS model. In heterogeneous reservoirs, the gas injection has poor sweep efficiency due to the unfavorable mobility ratio [2].

Mobility control in gas injection is of significantly important factor for successful oil recovery process. In reservoir condition, the viscosity of hydrocarbon gas or CO_2 usually has about 1/10 lower than that of oil [3]. Caudle and Dyes [4] suggested the water-alternating-gas (WAG) process to improve macro sweep efficiency of the gas injection process. The purpose of WAG process is to improve gas sweep efficiency through water injection. The injected water stabilizes the gas front and controls the displacement efficiency. According to Christensen, Stenby, and Skauge [5], this process integrates advantageous aspects of gas and water flooding to oil recovery. Gas flooding has better microscopic displacement efficiency than water whereas water injection has better macroscopic sweep efficiency. WAG process also involves other important aspects such as swelling, composition change, viscosity reduction caused by dissolved gas in oil and decrease of the residual oil saturation caused by the three phases flow and effects of relative permeability hysteresis [6]. The WAG method also includes economic profits because of reduction in the injected gas volume into the reservoir [7].

The WAG process has been successfully applied in many EOR projects. However, it is a difficult procedure to design its numerical simulation [1]. In the WAG simulation, the technique of alternately shutting-in and opening the injection wells is necessary in order to change the injected fluid. Furthermore, effects of parameters such as WAG ratio, various well patterns, well length and reservoir heterogeneities should be considered. For optimization of the simulation processes with different scenarios over wide range of reservoir and design parameters with reduce computational time, DECE (Designed Exploration and Controlled Evolution) method was utilized. It is possible to maximize computational efficiency by providing the evaluation of parameters for maximum NPV with a limited number of simulation runs.

The main objectives of this study are to compare the oil recoveries in immiscible WAG process for different well patterns, to optimize the processes and to maximize net present value (NPV) under different reservoir heterogeneity index. Four patterns of wells and various number of WAG cycles are implemented in different heterogeneous reservoirs. Moreover, the length of horizontal well is changed for optimal recovery design. Comparison of NPVs shows the optimum design reflecting the best WAG properties and well pattern.

METHODOLOGY

Optimization Method

In this study, DECE (Designed Exploration and Controlled Evolution) method that is CMG's proprietary optimization method was applied. To be brief, it performs Designed Exploration and Controlled Evolution processes, iteratively and sequentially. Designed Exploration is a method for structure and organization. It based on statistical principles which identify the impacts of various parameters. The designed exploration stage selects a reasonable number of runs efficiently. In this stage, with experimental design and Tabu search techniques, parameter values are selected and datasets of simulation are created. In the Controlled Evolution process performs statistical analysis on the simulation results deducted from designed exploration stage. On the basis of DE and CE process, all of the parameter values are scrutinized. Then, the algorithm judges if there is better opportunity to improve the quality of solution from the rejected values. The system remembers the rejected values and will not use them in the next stage. In order to avoid the local minima trapping, rejected values are checked again by DECE algorithm. The algorithm confirms rejected values are still reasonable. If it determines they are not reasonable, the values are recalled and reused in optimization process.

Objective Function

In CO_2-EOR optimization, the net present value (NPV) is commonly used as the objective function. The NPV is calculated with

$$\mathrm{NPV}=\sum_{t=1}^{T}\frac{C}{(1+r)^t} \tag{1}$$

where t is the time step, C is the cash inflow for the time step of t, r is the annual or periodic discount rate, and T is the cumulative investment or production period.

The annual discount rate is converted to daily discount rate using

$$\mathrm{Daily\,DiscountRate}=e^{\frac{\ln(1+\mathrm{Annual\ Discount\ Rate})}{365}}-1 \tag{2}$$

The cash inflow is given by

$$C = (\text{Oil Rate} \cdot \$/\text{bbl}) - (CO_2 \text{ Mass Rate} \cdot \$/\text{lb}) - (\text{InjectedWaterRate} \cdot \$/\text{bbl}) - (\text{ProducedWaterRate} \cdot \$/\text{bbl}) \quad (3)$$

Eq. (3) can be easily altered to reflect temporal variation of fluid prices and costs. The purpose of this study is to maximize the NPV of the WAG process.

SIMULATION

Model Description

Compositional simulator is used in this study. The reservoir size is described as $200 \times 200 \times 15$ ft^3 ($61 \times 61 \times 4.5$ m^3) and the grid number is composed by $20 \times 20 \times 1$ (Fig. 1). Average permeability is 50 md (4.9×10^{-14} m^2) and vertical permeability is 5 md (4.9×10^{-15} m^2). The porosity is 25%. These values were referred to the Midale unit in Weyburn field. The Midale unit is consisted of dolomitic unit (Marly) and limestone unit (Vuggy). The ranges of permeability and porosity are distributed from 0.5 (4.9×10^{-16} m^2) to 500 md (4.9×10^{-13} m^2) and from 15 to 37% at Marly unit [8]. The initial oil and water saturations are found 64% and 36%, respectively. Weyburn field fluid data were used for oil components and the specific compositions are exhibited in the Table 1. The initial pressure and temperature are 3,000 psi (20,684 kPa) and 145°F (63°C), and the bottom hole pressure of producer is constant by 1,400 psi (9,653 kPa) and injection rates of CO_2 and water are constant. Production was simulated for 5 years.

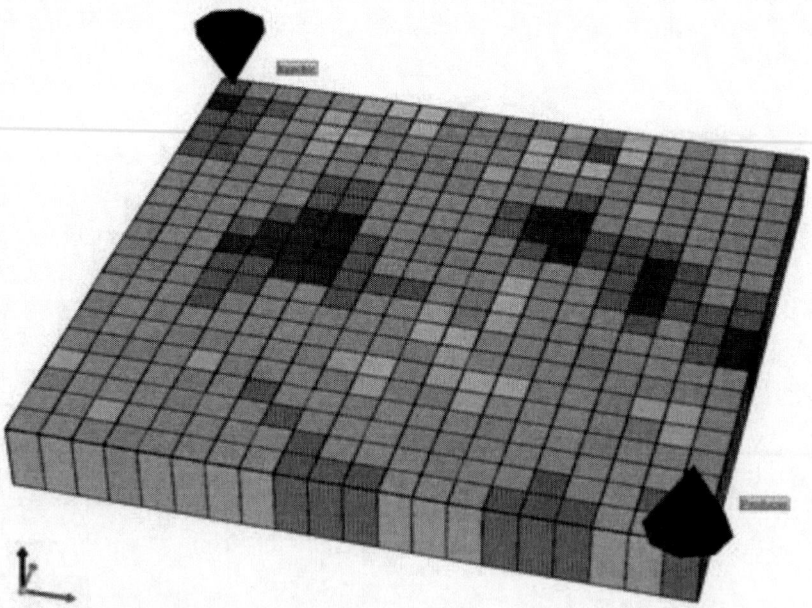

Figure 1. Heterogeneous reservoir in $V_{DP} = 0.3$.

Table 1. Oil components and mole fractions

Component	Mole fraction
CO_2	0.0074
N_2 to CH_4	0.0956
H_2S to C_2H	0.0434
C_3H_8	0.0785
IC_4 to NC_4	0.0655
IC_5 to NC_5	0.0459
C_6 to C_9	0.2156
C_{10} to C_{17}	0.2202
C_{18} to C_{27}	0.1027
C_{28} to C_{30}	0.1252
Total	1

The four cases of heterogeneities were used in this study. The level of heterogeneity is represented by Dykstra-Parsons coefficient (V_{DP}) as 0.0, 0.3, 0.5 and 0.7. This coefficient is an effective way to represent the grade of reservoir heterogeneity and it is also named the reservoir heterogeneity index (HI) and the range is $0 < V_{DP}$ (HI) < 1 [9]. The distribution of permeability is used in calculation for it. The equation is:

$$V_{DP} = \frac{k_{50} - k_{84.1}}{k_{50}} \tag{4}$$

where k_{50} is permeability value at 50% probability and $k_{84.1}$ is permeability value at 84.1% from the cumulative sample.

1. $V_{DP} = 0$, ideal homogeneous reservoir
2. $0 < V_{DP} < 0.25$, slightly heterogeneous
3. $0.25 < V_{DP} < 0.50$, heterogeneous reservoir
4. $0.50 < V_{DP} < 0.75$, very heterogeneous reservoir
5. $0.75 < V_{DP} < 1$, extremely heterogeneous reservoir

In this study, k_{50} and $k_{84.1}$ were 50.8 and 35.2 md (5.0×10^{-14} and 3.5×10^{-14} m^2) in cases of V_{DP} 0.3. It means that the variation of permeability is not too large. The k_{50} and $k_{84.1}$ were 51.3 and 24.9 md (5.1×10^{-14} and 2.5×10^{-14} m^2) respectively in V_{DP} 0.5 reservoirs, and in V_{DP} 0.7 cases, they were 52.3 and 15.5 md (5.2×10^{-14} and 1.5×10^{-14} m^2). In these very heterogeneous reservoirs, the not only differences between k_{50} and $k_{84.1}$ but also those of minimum and maximum permeability are larger than the other cases. For example, minimum and maximum permeability of V_{DP} 0.7 reservoirs is 3.5 and 376.8 md (3.5×10^{-15} and 3.7×10^{-13} m^2).

The parameters for maximum NPV calculation are CO_2 injection period, water injection period and the length of horizontal wells. The periods of CO_2 and water injection are increased from 60 to 720 days. Four patterns of wells were used in this study (vertical injector – vertical producer, vertical injector – horizontal producer, horizontal injector – vertical producer, horizontal injector – horizontal producer). The horizontal wells have no branch wells, and their orientations are parallel to *x*- or *y*-axis. The lengths of them are changed from 20 to 200 ft (6.1 to 61 m) long and the interval is 20 ft (6.1 m).

For the NPV estimation, conservative annual discount rate of 10% was used in this study. Cash inflow for the time step is calculated with the oil price, the cost of injected CO_2 and the cost of injected and disposal cost of produced water. The price of oil is $70 per barrel ($440/m^3) for the entire production period. The cost of water disposal is considered to be $1.5 per barrel ($9.4/m^3) while the water injection cost is $0.25 per barrel ($1.57/m^3). The CO_2 purchase cost is assumed to be $60 per metric ton of CO_2 [10]. All the parameters are expressed in Table 2.

Table 2. Values for NPV calculation

Parameters for NPV calculation	CO_2 injection period	60~720 days (12 steps)
	water injection period	60~720 days (12 steps)
	injector length	20~200 ft (6.1~61 m) (10 steps)
	producer length	20~200 ft (6.1~61 m) (10 steps)
Costs for NPV calculation	annual discount rate	0.1/year
	oil price	$70/bbl ($440/m^3)
	injected water cost	$0.25/bbl ($1.57/m^3)
	disposal water cost	$1.5/bbl ($9.4/m^3)
	injected CO_2 cost	$60/ton

Random Field Generation

The Dykstra-Parsons coefficient is a measurement of sample variability or dispersion in dimensionless. Even if the coefficient has constant value, distributions of permeability can be different. In this study, with the same Dykstra-Parsons coefficient, various correlation lengths

(λ_x, λ_y) were applied to make different permeability distributions (Fig. 2, Table 3). If the correlation length of x-direction is larger than that of y-direction, the permeability has stronger correlation in x-direction than y-direction.

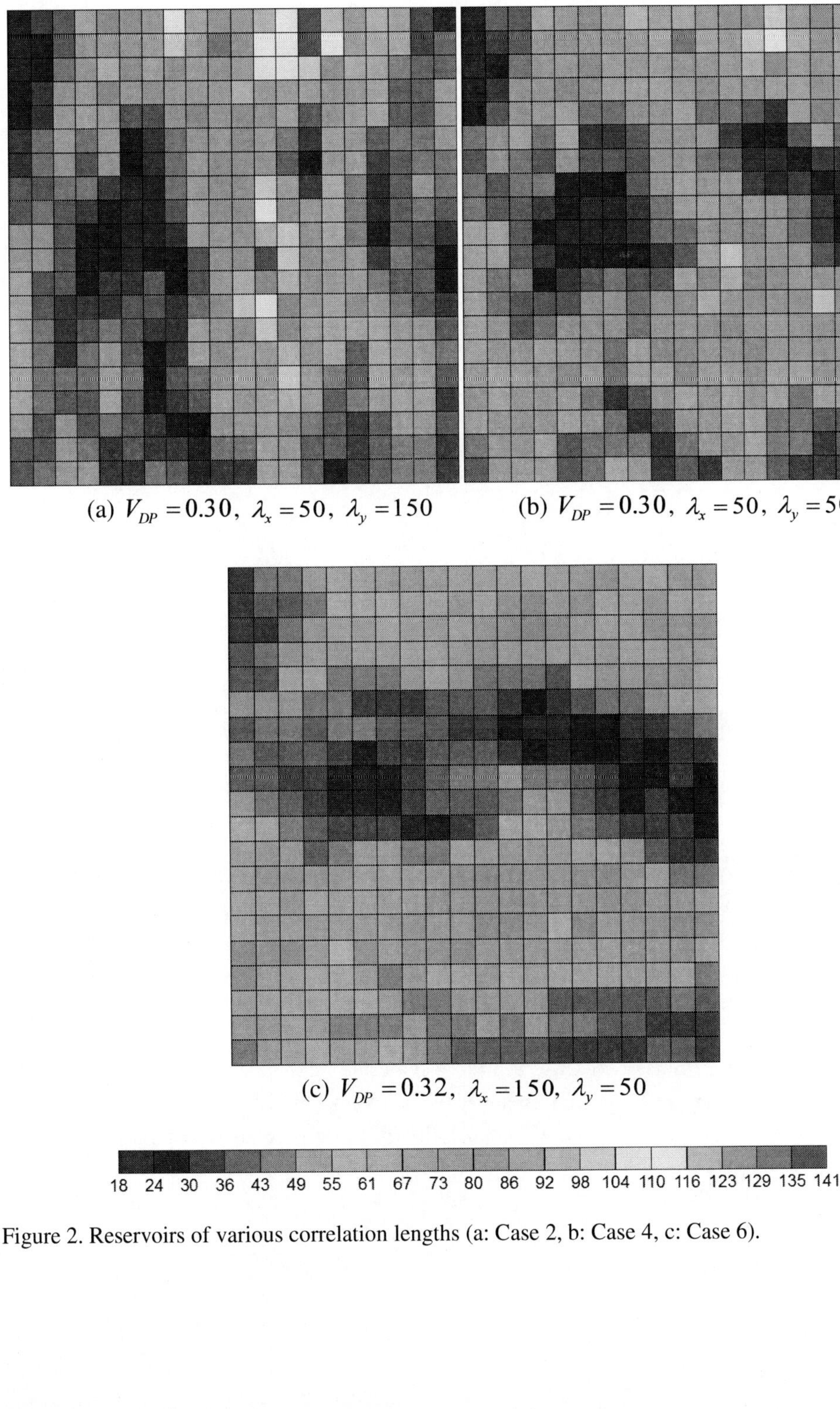

(a) $V_{DP} = 0.30$, $\lambda_x = 50$, $\lambda_y = 150$

(b) $V_{DP} = 0.30$, $\lambda_x = 50$, $\lambda_y = 50$

(c) $V_{DP} = 0.32$, $\lambda_x = 150$, $\lambda_y = 50$

Figure 2. Reservoirs of various correlation lengths (a: Case 2, b: Case 4, c: Case 6).

Table 3. Cases of random field generations

case No.	HI (V_{DP})	λ_x , λ_y	remarks
1	0.00	-	homogeneous reservoir
2	0.30	50, 150	heterogeneous reservoir
3	0.31	50, 100	
4	0.30	50, 50	
5	0.31	100, 50	
6	0.32	150, 50	
7	0.52	50, 150	very heterogeneous reservoir
8	0.50	50, 100	
9	0.51	50, 50	
10	0.53	100, 50	
11	0.51	150, 50	
12	0.70	50, 150	
13	0.68	50, 100	
14	0.72	50, 50	
15	0.72	100, 50	
16	0.70	150, 50	

RESULTS

NPV

Tables 4 and 5 show every optimization results. Table 4 indicates the optimal CO_2 slug sizes and water injection amount in each WAG cycle for maximum NPV. The optimal well patterns, orientations of horizontal well and well lengths for each heterogeneous reservoir are shown in Table 5. Four patterns of wells, which are combined with horizontal and vertical injection and production wells, are used in this study. The optimum oil production strategies for the maximum NPVs exist in Table 4. In Fig. 3, black dot represents base case and red diamonds denote optimum cases. This figure indicates that the optimum cases which are estimated by optimization algorithm have increased NPV comparing to the base case. As is well known, NPV decreases with increasing of reservoir heterogeneity. For Cases 4, 9 and 14, these reservoirs have the same correlation length but different heterogeneity index. Case 4 has the maximum NPV, and in order of Case 9 and 14 have the next maximum NPV. The NPV has the tendency of increasing with increment of oil recovery (Fig. 4). This trend proves that the oil recovery significantly affects NPV because the oil price is relatively higher than other costs. In this circumstance, highly heterogeneous reservoirs generate more severe CO_2 channeling and earlier breakthrough (Figs. 5~6), and these cause reduction of the oil recovery.

Table 4. Results of WAG ratio

case no.	injected water (HCPV)	injected CO_2 (HCPV)	recovery factor (%)	NPV ($\$10^5$)
1	0.07	0.13	65.2	5.6685
2	0.07	0.13	62.5	5.4224
3	0.07	0.13	61.8	5.3463
4	0.07	0.13	62.9	5.4187
5	0.07	0.13	64.6	5.5970
6	0.07	0.13	65.0	5.6472
7	0.07	0.13	60.0	5.2045
8	0.07	0.13	59.6	5.1622
9	0.07	0.13	60.8	5.1835
10	0.07	0.13	63.1	5.4346
11	0.07	0.07	64.1	5.5437
12	0.07	0.07	56.5	4.7905
13	0.07	0.07	56.8	4.8312
14	0.07	0.07	57.3	4.8567
15	0.07	0.07	62.8	5.3795
16	0.07	0.07	63.0	5.4163

Table 5. Results of well patterns

case No.	well pattern	orientation	injector length (ft)	producer length (ft)
1	V-V	-	-	-
2	H-H	Y	200 (61 m)	120 (37 m)
3	H-H	Y	200 (61 m)	120 (37 m)
4	V-V	-	-	-
5	V-V	-	-	-
6	V-V	-	-	-
7	H-H	Y	200 (61 m)	120 (37 m)
8	H-H	Y	100 (30 m)	120 (37 m)
9	H-V	Y	20 (6.1 m)	-
10	H-V	Y	20 (6.1 m)	-
11	V-V	-	-	-
12	H-V	Y	200 (61 m)	-
13	H-V	Y	100 (30 m)	-
14	H-V	Y	40 (12.2 m)	-
15	H-V	Y	20 (6.1 m)	-
16	V-V	-	-	-

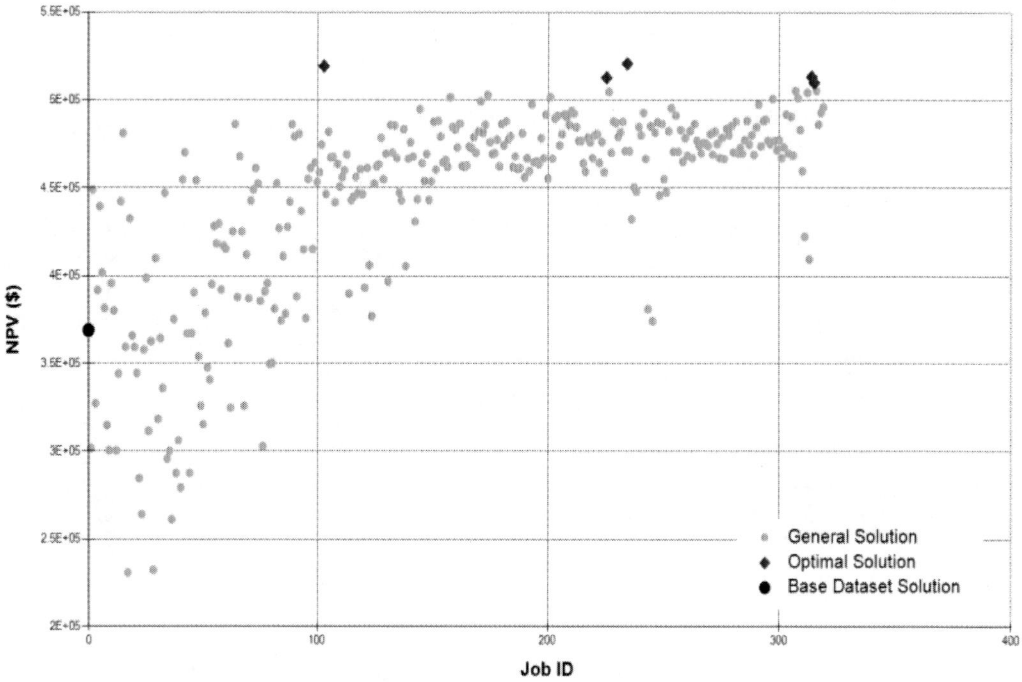

Figure 3. NPVs obtained by optimization algorithm.

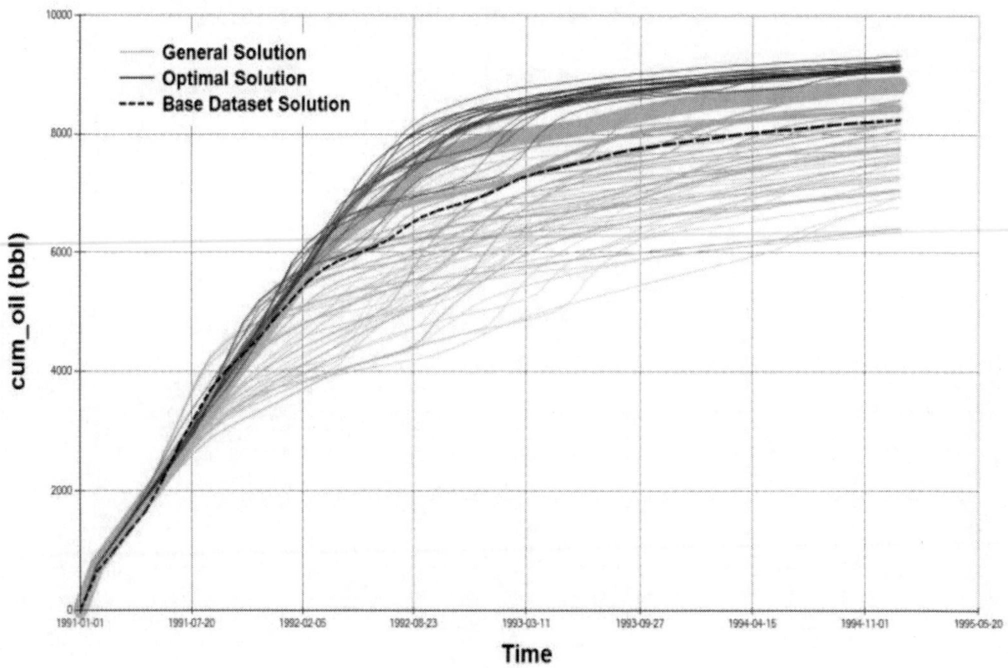

Figure 4. Cumulative oil recoveries of general, optimal and base solutions.

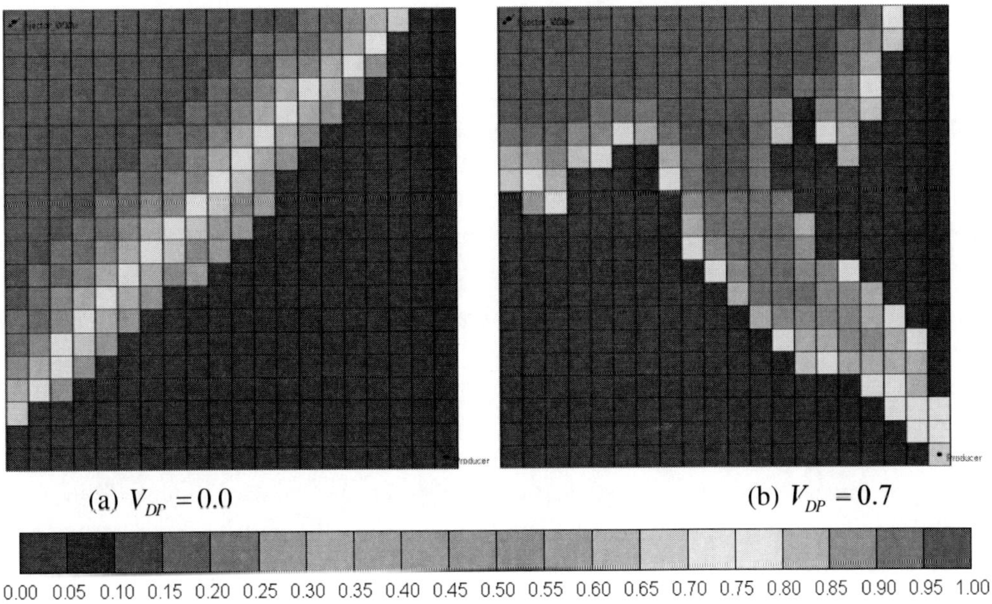

Figure 5. CO_2 mole fraction after 0.4 HCPV injected.

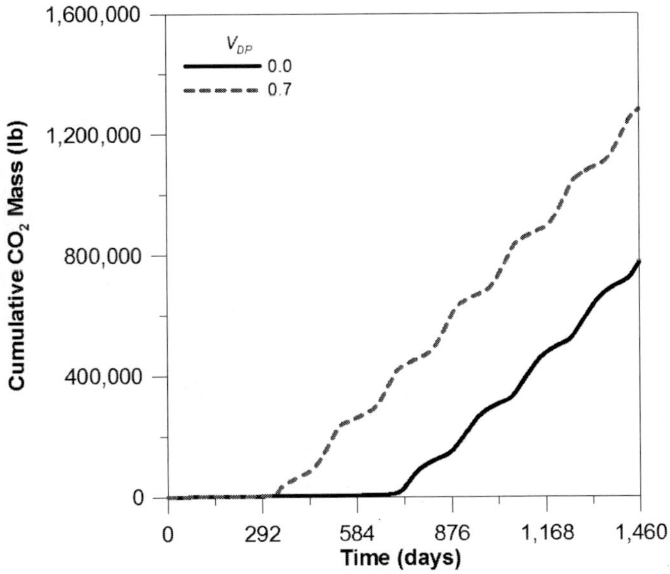

Figure 6. Cumulative CO_2 mass recoveries of homogeneous and heterogeneous reservoirs.

WAG Ratio and Well Patterns

In this study, in order to optimize the design of WAG process at these random fields, various WAG ratios and well patterns are considered as design parameters. In homogeneous reservoir (Case 1), the optimal WAG ratio is 1:2 and vertical injector – vertical producer (V-

V) well pattern has the maximum NPV of 5.7×10^5. Horizontal well is generally known as an effective technique to produce oil with a large contact area. However, in constant injection rate, the horizontal well could induce low oil recovery. The reason is that injected CO_2 mole fraction in unit pore volume is lower at near injector than vertical well's case (Fig. 7). If the injected CO_2 mole fraction in unit volume is low, displacement efficiency becomes poor. Fig. 8 is the oil saturation near well block. This graph indicates that the smaller CO_2 mole fraction in unit volume can cause the larger volume of remained oil.

Cases 2 and 3 are heterogeneous reservoirs ($V_{DP}=0.3$) which permeability correlation length of y-direction is relatively higher than x-direction's. In these cases, optimal WAG ratio is 1:2 and well pattern is horizontal injector – horizontal producer wells parallel to y-direction. The optimal injector length is 200 ft (61 m) and producer length is 120 ft (37 m). It indicates the importance of the optimal well length design for the best performance in heterogeneous reservoir. Heterogeneous reservoir has both high permeability and low permeability zones. After the injected fluids arrive at the producer through the high permeability zone, larger amounts of following fluids flow through the high permeability zone more easily. In contrast, the low permeability area is still left as unswept zone. This phenomenon can be improved by the well length optimization (Fig. 9). As a result, cumulative oil recovery and NPV increased by 5.5% and 6.2% compared with base case.

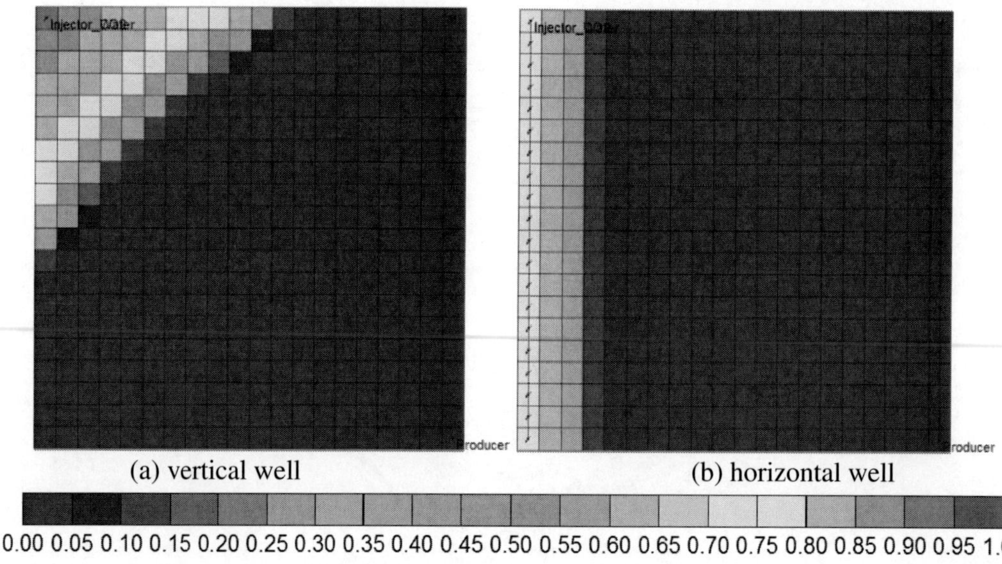

Figure 7. CO_2 mole fraction after 0.1 HCPV injected.

The change of correlation length ratios (λ_y/λ_x) can affect optimum well patterns. Cases 7 through 11 describe it well. These cases have different ratio of correlation length but same heterogeneity index. When the ratios are large (Cases 7 and 8), combination of the horizontal injector and producer which orientation is parallel to y-direction has the best performance among various well patterns. The optimal WAG ratio is 1:2 for all cases except Case 11. As the ratio of correlation length decreases, the optimum length of horizontal injector shortens and vertical producer has better performance. The Case 11, which correlation ratio is the lowest, obtains the maximum NPV when WAG ratio is 1:1 and well pattern is vertical injector – vertical producer. NPV increased by 30% in the optimum case compared to base case. In the most heterogeneous reservoirs (Case 12 through 16), when WAG ratio is 1:1 and optimum well pattern is horizontal injector – vertical producer, the maximum NPV is expected except Case 16.

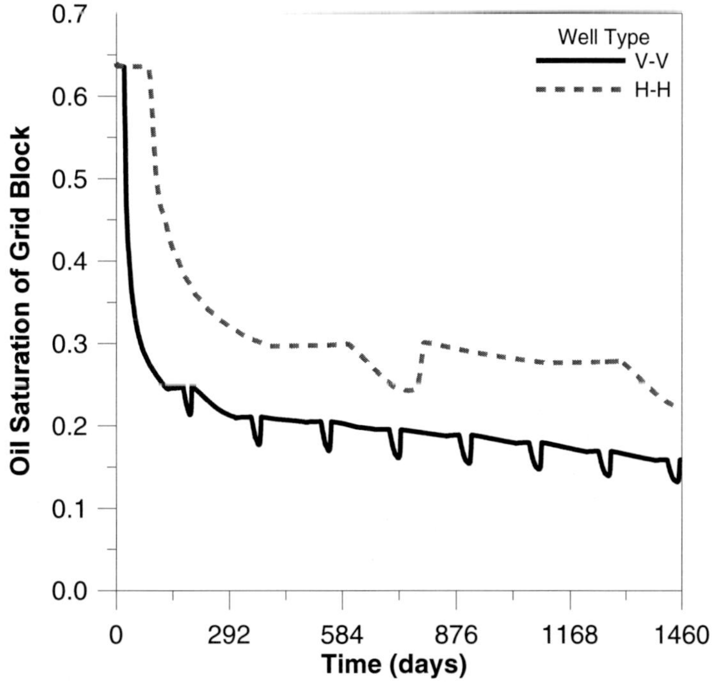

Figure 8. Oil saturation at near well bore block.

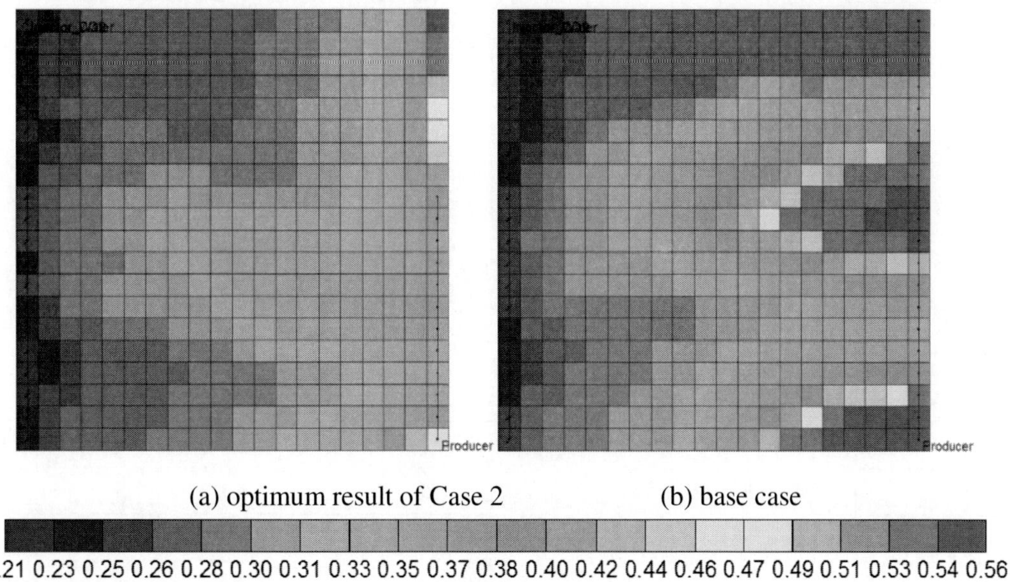

Figure 9. Oil saturation of Case 2 after three years.

All of the optimal cases have the maximum oil recovery to make the maximum NPV. WAG ratio and well pattern are optimized to improve sweep efficiency and displacement efficiency. The sweep efficiency is the ratio of contact area between the injected fluid and reservoir oil and total area. The CO_2 mole fraction at the contact area affects the displacement efficiency. The displacement efficiency and sweep efficiency correlate with each other. Therefore, through the appropriate WAG ratio and well pattern, maximum oil recovery can be obtained.

Computational Efficiency

The computational efficiency which is the number of calculations for optimization can be improved by the DECE optimization algorithm. This algorithm uses the selected values by optimization method. This procedure can significantly reduce the simulation running time. If all parameters used in optimization are considered, total number of calculation is about 34,000 including useless jobs. However, when optimization algorithm is applied, it only calculates the objective function with 4~5% of all the jobs. The utilization of the calculations is in the Table 6.

Table 6. Calculation efficiency of optimization process

case no.	HI	total no.	task no.	utilization
1	$V_{DP} = 0.0$	17,424	680	3.90%
2	$V_{DP} = 0.3$	34,560	1,771	5.12%
3			1,457	5.08%
4			1,936	5.60%
5			1,623	4.50%
6			1,668	4.83%
7	$V_{DP} = 0.5$		1,679	4.86%
8			1,730	5.01%
9			1,706	4.94%
10			1,651	4.78%
11			1,675	4.85%
12	$V_{DP} = 0.7$		1,674	4.84%
13			1,730	5.01%
14			1,881	5.44%
15			1,641	4.75%
16			1,613	4.67%

CONCLUSION

In this study, various heterogeneous reservoirs which have different correlation lengths are used for WAG process design optimization. The optimum WAG ratio is different with reservoir heterogeneities. If the heterogeneity indexes are 0.0 to 0.5, optimal WAG ratio is 1:2. However, the optimal WAG ratio decreases to 1:1 as heterogeneity of reservoirs increase to 0.7. The most heterogeneous reservoirs can have a severe channeling effect. Therefore, injected CO_2 mole fraction should be reduced for better sweep efficiency. The optimized well patterns, orientations and lengths are affected by permeability distributions. Although the reservoirs have same heterogeneity index, these three parameters could be changed due to correlation length to increase sweep and displacement efficiency. These optimal strategies for oil production contribute to optimize the NPVs. The NPVs calculated by optimization algorithm increase by 5 to 30%. Especially, in very heterogeneous reservoirs, the increases in NPVs are more significant. Furthermore, the DECE optimization algorithm can help to improve computational efficiency about 94~95%. The optimum WAG design which is obtained by optimization algorithm can achieve the best economic profit in heterogeneous reservoirs.

REFERENCES

[1] Ligero, E. L.; Mello, S. F.; Munoz Mazo, E. O.; Schiozer, D. J. *Proc. SPETT 2012 Energy Conference and Exhibition* Trinidad, Spain, 11-13 June, 2012.

[2] Egermann, P.; Robin, M.; Lombard, J-M.; Modavl, A.; Kalam, M. Z. *SPE Reservoir Eval. & Eng.* 2006, 9(6), 621-629.

[3] Rao, D. N.; Ayirala, S. C.; Kulkarni, M. M.; Sharma, A. P. *Proc. SPE/DOE Symp. on Improved Oil Recovery* Tulsa, OK, USA, 17-21 April, 2004.

[4] Caudle, B. H.; Dyes, A. B. *J. of Society of Petroleum Engineers* 1958, 213, 281-284.

[5] Christensen, J. R.; Stenby, E. H.; Skauge, A. *SPE Reservoir Eval. & Eng.* 2001, 4(2), 97-106.

[6] Namani, M.; Kleppe, J. *Proc. SPE Enhanced Oil Recovery Conference* Kuala Lumpur, Malaysia, 19-21 July, 2011.

[7] Pariani, G. J.; McColloch, K. A.; Warden, S. L.; Edens, D. R. *J. of Pet. Tech.* 1992, 44(9), 984-988.

[8] Bogatkov, D. *Integrated Modeling of Fracture Network System of the Midale Field*, MS Thesis, University of Alberta, Canada, 2008, 23.

[9] Tiab, D.; Donaldson, E. C. *Petrophysics Ed 3*, Elsevier, Waltham, MA, USA, 2012, 176-178.

[10] Jahangiri, H. R.; Zhang, D. *Int. J. of Greenhouse Gas Control* 2012, 8, 22-33.

In: Enhanced Oil Recovery
Editor: Alicia Knight

ISBN: 978-1-63463-917-0

Chapter 8

PREDICTION OF BREAKTHROUGH TIME IN THE VAPEX PROCESS WITH THE NUMERICAL MODELING

***Morteza Sabeti*[1], *Mohammad R. Ehsani*[1], *Mohammad Nikookar*[2,*], *Leili Sahranavard*[2] *and Amir H. Mohammadi*[3,4]**

[1]Department of Chemical Engineering of Isfahan University of Technology, Isfahan, Iran

[2]IOR Research Institute, National Iranian Oil Company (NIOC), Tehran, Iran

[3]Institut de Recherche en Génie Chimique et Pétrolier (IRGCP), Paris Cedex, France

[4]Thermodynamics Research Unit, School of Engineering, University of KwaZulu-Natal, Howard College Campus, Durban, South Africa

ABSTRACT

In recent years, there has been an increasing interest in modeling miscible flow in petroleum reservoirs. Besides, there are several compositional simulators in order to estimate fluid variables at each time step, however this paper attempts to acquire a simple numerical method based on the compositional simulation. Therefore a two-dimensional, three-phase numerical model reservoirs with formulation details, numerical solution method, and computational results was described for simulation of VAPEX process in heavy oil. The model included the effects of three-phase relative permeability and pressure-dependent fluid properties and also interphase mass transfer of solvent in heavy oil was considered. The three-phase mass balance equations were solved sequentially using finite difference method. Some VAPEX laboratory data were used for studying the accuracy of this model. Comprehensive and comparative studies together with extensive sensitivity analysis from 0.0178 to 0.00178 cc/s as solvent injection rates were conducted to specify the dependency of the oil recovery, breakthrough time and SOR to injection rate. The simulation results indicated that high injection rates decrease the breakthrough time and lead to more SOR, which is unacceptable from the economical point of view.

[*] Corresponding Authors: M. Nikookar: Email: mohamad_ni@yahoo.com.

Therefore, an interesting plane was suggested to control the solvent waste with a high oil recovery.

Keywords: VAPEX, enhanced oil recovery, mathematical model, heavy oil, compositional model

INTRODUCTION

With the advancement of technology and horizontal drilling, heavy oil reservoir extraction methods were entered into a new phase. The vapor extraction (VAPEX) process was reported (Butler and Mokrys, 1991; 1993) as a new and convenient synthetic procedure for the recovery of heavy oil and bitumen in 1991. Because of many advantages of VAPEX method (James, Rezaei, and Chatzis, 2008), this promising process has received so much interest in recent years. In the VAPEX process, two horizontal wells are drilled inside the reservoir for solvent injection and oil production. The solvent is injected through an upper horizontal well and dissolves and diffuses in heavy oil to reduce its viscosity and makes it mobile, then can be easily extracted from the production bore. The feature of the VAPEX process helps to create several analytical, semi-analytical methods (Sabeti, Nikookar, et al., 2014) to estimate oil production but these methods is not always accurate.

Butler and coworker (1989) were the first researchers to work on hydrodynamic of VAPEX on small scale experiment tests and developed a 1D analytical model which was able to estimate oil production rate. They supposed three simple essential assumptions: (a) one dimensional diffusion perpendicular to solvent-oil interface was considered as the only mass-transfer mechanism (b) Darcy's law was considered to calculate the drainage rate of tilted oil (c) dissolution of solvent was a constant value whereas it could change the rate of drainage. Based on the Butler model some variations of other theoretical models proposed (Das and Butler, 1998; Derakhshanfar and Gu, 2012; Sabeti, Nikookar, et al., 2014; Lin, Zeng, and Gu, 2014; Shi and Leung, 2014) to demonstrate the real drainage mechanism of VAPEX. Among the aforementioned works, Drakhshanfar and Gu (2012) developed a new model for determination of the solvent convective dispersion and then oil drainage in a typical VAPEX process. They supposed a finite control volume with unit thickness in the solvent-oil interface, then wrote the material balance equation for that element. In their study, the molecular diffusion mechanism dominates mass transfer in the normal transition zone direction, while convective dispersion overcomes along the transition zone. They applied several assumptions to the mentioned material balance equation and a linear or quadratic velocity profile for the diluted heavy oil gravity drainage to obtain convective dispersion coefficient as an unknown adjustable parameter. Once the optimal coefficient is determined from the laboratory experiments, their model can be easily used to better predict the field scale VAPEX process. Azad and Chalaturnyk (2012) showed that the rising and growing of the steam chamber during the SAGD process is similar to circular geometry. Later Lin et al. (2014) take account this geometrical idea and apply it in VAPEX (solvent analog of SAGD) process. Instead of diffusion and dispersion coefficients they defined adjustable transition-zone thickness in their model to indirectly include mechanical dispersion mechanism in their calculations and in contrast to constant oil production rate, accurately estimate the real production rate. Sabeti et al. (2014) also introduced a semi-analytical model by considering moving boundary layer (solvent-oil interface) on the base of Butler's theory. They discredited solvent-oil boundary in several segments and specified diffusion thickness for each segment in order to calculate special drainage rates of each part. Even, in their approach, variation of thickness of diffusion as well as angle of drainage along the boundary was influenced on trend of oil production rate. In a novel study on the base of Butler's model, Rasti et al. (2011) derived a dimensionless form of oil production rate correlation with four adjustable constants in order to involve the most effective physical parameters in the oil production rate correlation. The authors demonstrated that dimensionless numbers of *Ns*, *Pe* and *Sc* and also the adjustable constants are variables which determine the value of the correlation in the typical VAPEX process. So, they combined their new model with CMG results to obtain adjustable constants and promote Butler's theory. Although all models introduced here are fast and successful to present realistic trend of drainage vs. time, but the semi-analytical methods (Butler like theories) almost are conditional and can't specify variation of the production respect to injection rate. In most of studies relating to the VAPEX, solvent breakthrough has been prevented from occurring so that only tilted oil was able to flow downwards along the transition zone under gravitational driving force (with the intention of developing analytical formulae) (Lin, Zeng, and Gu, 2014). In this way, you can see the experimental results of Kok et al. (2009) which studied the effects of solvent rates on extraction of three different crude oils in VAPEX. Generally, they demonstrated the injection rate of solvents (especially propane in their works) had a strong impact on the slope of the cumulative produced oil curve and also the analytical rate-calculations were unable to match against their experimental results, mainly because of gas push of VAPEX solvent. On the other hand, in strictly speaking, recent experimental works have indicated that the interplay between drainage height, capillary forces, permeability and the solvent on the oil drainage are complex and unclear to apply in the field scale process with such analytical models (Ahmadloo, Asghari, and Henni, 2014; 2013). For overcoming these shortcomings, sets of reservoir characteristics such as relative permeability, mechanical dispersion and diffusion, capillary pressure and geometrical variables must be considered (Ghesmat, 2014). Indeed, to study the performance of VAPEX process and to apply on the reservoirs, a compositional modeling is needed (Sabeti, Ehsani, et al., 2014).

There are several references available for compositional modeling of VAPEX performances for petroleum engineering by different specifications on fluid and rock properties or initial conditions (Nghiem, Kohse, and Sammon, 2001; Azin et al., 2005; Nourozieh et al., 2009; Haghighat, Maini, and Abedi, 2013) in which the most of these simulations have been done by CMG tools (Pourabdollah, Mokhtari, and Kharrat, 2012). The CMG software can simulate VAPEX process by generalized equation-of-state at compositional reservoir with regard to transport phenomena in the fluid mixtures. In order to simulate VAPEX process, it is necessary to apply a realistic correlation for transport of components in fluid phases respect to characteristic of studied reservoirs in the mass conservation equations of the compositional modeling. Although the CMG software is friendly user and readily available, but, this simulator (or other ones) has (have) a limited choice for diffusion or dispersion and viscosity correlations as well as limited access to the algorithm and governing equations used in its numerical simulation (Tam, 2007). So it is needed to develop a compositional code, including all governing equations and required correlations for better understanding of influence of initial conditions on the performance of the process.

This study provides a simple procedure for solving compositional model in respect to the mass transfer of components with workable developed diffusion and dispersion coefficients for VAPEX. It was tried to illustrate the proven formulations in detail in order to become a foundation for other developments. Although equations of state have advantages for governing of fluid properties, these equilibrium equations are computationally more expensive and less accurate against the experimental data (Young and Stephenson, 1983). For this reason, such equations were not used in the present paper. The VAPEX process can be simulated with less complexity and more accuracy by using lumped oil-solvent system, K-values and reliable correlations for the oil viscosity and the solvent diffusion. Another subject of this study is research on the best solvent injection rate for the optimum SOR and the oil extraction. For the mentioned subject, one Iranian heavy oil reservoir was chosen and several cases were applied at the injection rate to study the effects of solvent injection rate on the production rate.

COMPOSITIONAL FORMULATIONS

The simulator developed here is used to model three-phase flow with the transport of multicomponents in hydrocarbon phases in a porous medium. Whereas, there are several compositional approaches under different conditions for numerical reservoir simulation (Kazemi, Vestal, and Shank, 1978; Young and Stephenson, 1983; Branco and Rodriguez, 1996; Mei et al., 1998), this work provides simple formulation for multicomponent mathematical modeling based on Kazemi's (1978) method. This method and related equations are explained in detail at the Appendix.

In this study a reservoir is considered as rectangular geometry which, according to finite difference form the reservoir is divided into two dimensional grid blocks and subsequently the material balance is taken account for each grid. The presented model consists of conservation of mass for oil and gas, and water. The fluid velocity in mass conservation equations is estimated by the Darcy's law equation. The Darcy equation expresses the fluid velocity within the porous media by pressure difference and gravity force in the defined directions.

Some auxiliary equations are used to reduce the unknown variables, such as K-value for determination of the mole fraction of oil and gas components, total fluid saturations of each grid block and correlation between capillary pressure and fluid saturations. Therefor, all formulations quantitatively create $2N_c$+9 dependent variables: x_m, y_m, $\boldsymbol{u_\alpha}$, P_α and S_α with $2N_c$+9 independent formulas at each reservoir grid block at any given time. Some parameters such as porosity, fluid densities and fluid viscosities can be changed during the process under variation of pressure variable. The molar density and porosity changes are illustrated by the following equations:

$$\xi = \xi_0 \exp(C_F (P - P_0)) \approx \xi_0 (1 + C_F (P - P_0)),$$
$$C_F = \frac{1}{\xi}\frac{\partial \xi}{\partial P} \tag{1}$$

$$\phi = \phi_0 \exp(C_R (P - P_0)) \approx \phi_0 (1 + C_R (P - P_0)),$$
$$C_R = \frac{1}{\phi}\frac{\partial \phi}{\partial P} \qquad (2)$$

According to Das and Butler studies (1998) diffusion of the solvent through the thick oil and reduction of oil viscosity are defined as the major mechanism of the VAPEX process. So, it is necessary to apply a suitable mechanical dispersion and viscosity correlation on the simulator for modeling of the VAPEX process. Here, the oil phase viscosity is calculated by the Suh (1984) and Lederer (1933) correlation as follows:

$$\mu_{mix} = \mu_b^{f_b} \mu_s^{f_s}$$
$$f_b = \frac{\gamma C_b}{\gamma C_b + C_s}$$
$$f_s = 1 - f_b \qquad (3)$$
$$\gamma = \frac{17.04(\rho_b - \rho_s)^{0.537} \rho_b^{3.2745} \rho_s^{1.6316}}{Ln(\frac{\mu_b}{\mu_s})}$$

In these correlations, C_s and C_b are volume fraction of solvent and heavy oil, specific gravity, shown by ρ and γ, is an empirical constant between zero and unity.

The correlations for the mass transport coefficients used in the oil phase are defined by Hayduk (1982) and Boustani's (2001) work.

$$K_{m\alpha,X} = \frac{D_{m\alpha}}{F\phi} + 0.5\frac{u\, dp.\delta}{\phi} + 0.0157\frac{v\, dp.\delta}{\phi},$$
$$K_{m\alpha,Z} = \frac{D_{m\alpha}}{F\phi} + 0.5\frac{v\, dp.\delta}{\phi} + 0.0157\frac{u\, dp.\delta}{\phi} \qquad (4)$$

As it can be seen, the right sides of Eqs. (7) are combined of three terms to represent molecular diffusion, longitudinal and transverse dispersion due to the velocity, where particle diameter, inhomogeneity and formation resistivity factor are introduced by *dP*, δ and *F*, respectively. Of course, it is possible a constant number of about *1m* be used instead of the dispersion coefficient (Nghiem, Kohse, and Sammon, 2001). The molecular diffusion coefficient correlation is:

$$D(cm^2/s) = 13.3 \times 10^{-8} T^{1.47} V_a^{0.71} \mu^{(\frac{10.2}{V_a} - 0.791)}, \qquad (5)$$

which, V_a*(cm2/mol)* is the molar volume at the normal boiling point of solvent, and μ*(mPa.s)* is the oil viscosity at *T(K)*.

Also the measurements on the three-phase relative permeability have been done with *Naar* and *Wygal's* model as follows:

$$k_{rw} = \left(\frac{S_w - S_{wc}}{1 - S_{wc}} \right)^4,$$
$$k_{ro} = \frac{S_o^{\ 3}(1 - S_g + 2S_w - 3S_{wc})}{(1 - S_{wc})^4}, \qquad (6)$$
$$k_{rg} = \frac{S_g^{\ 3}(2 - S_g - 2S_{wc})}{(1 - S_{wc})^4}.$$

When water starts to flow S_{wc} refers to critical saturation of water phase and during drainage it mentions as residual saturation. It should be mentioned in a case of validation, the correlation between relative permeability of water and oil phase has been obtained by a curve fitting from the experimental data.

For numerical solution, the partial differential equations governing the fluid flow of system are solved using implicit pressure- explicit saturation and composition method. In the presented IMPESC method firstly, pressure, saturation and composition of each grid block at $n+1$ are guessed in order that the fluid and rock properties can be calculated and as a results the matrix coefficients of the pressure finite difference equation are computed. Next the pressure of each grid block can be solved by *the Gaussian elimination method*. The following step, the saturation and composition of each of grid blocks at $n+1$ are obtained by the explicit equations.

This procedure is repeated until convergence is obtained. The IMPESC method is efficient and is computationally less expensive than a fully implicit one, however it should be noted the time step in IMPESC must be small enough to obtain a converged answer for practical problems during mathematical calculations. A small step time require a large computation time which makes bored with sensitivity analyses.

VALIDATION OF THE MODEL

Two numerical examples based on the presented compositional simulator are solved to validate the performance of the simulator. The first example is a 1D numerical model with 20 grid blocks which is considered to validate against Buckely- Leverett theory. Properties of a core and its fluid (taken from CMG-STARS template data) are presented in Table 1 and Table 2 (Here the capillary pressure is ignored according to Buckely- Leverett equations). The water is flooded to one side of the core at a constant rate of 0.005cm3/s and the production on the other side of the core is obtained at a constant pressure.

Figure 1 compares the core's water saturation results of the model with the modeling of Buckley- Leverett theory at 7hr after the beginning of the water-flooding. At normalized position 0.75 (according to the figure), there is a difference between the model and the theory results. Buckley-Leverett is an analytical relation for frontal- advanced theory which takes into accounts different simplification assumptions. But this theory gives real results with accurate frontal saturation of water in a horizontal core sample. Always numerical modeling of water-flooding causes deviation from Buckley-Leverett theory results (such as CMG simulator). However, the average of water saturation is the same in the model and the theory.

Table 1. Rock and fluid properties of the studied core sample which is flooded by water

Property	Permeability (mDarcy)	Porosity(%)	Reservoir pressure (kPa)	Oil density (g/cm3)	Initial oil viscosity(cp)	Residual oil saturation	Initial water saturation
Value	226	12.6	4000	0.725	2.24	0.35	0.22

Table 2. Relative permeability of water and oil phase in the studied core sample

S_w	K_{rw}	K_{row}
0	0	0.893
0.22	0	0.893
0.3	0.09	0.721
0.4	0.202	0.439
0.5	0.33	0.234
0.6	0.556	0.05
0.65	0.65	0
1	0.65	0

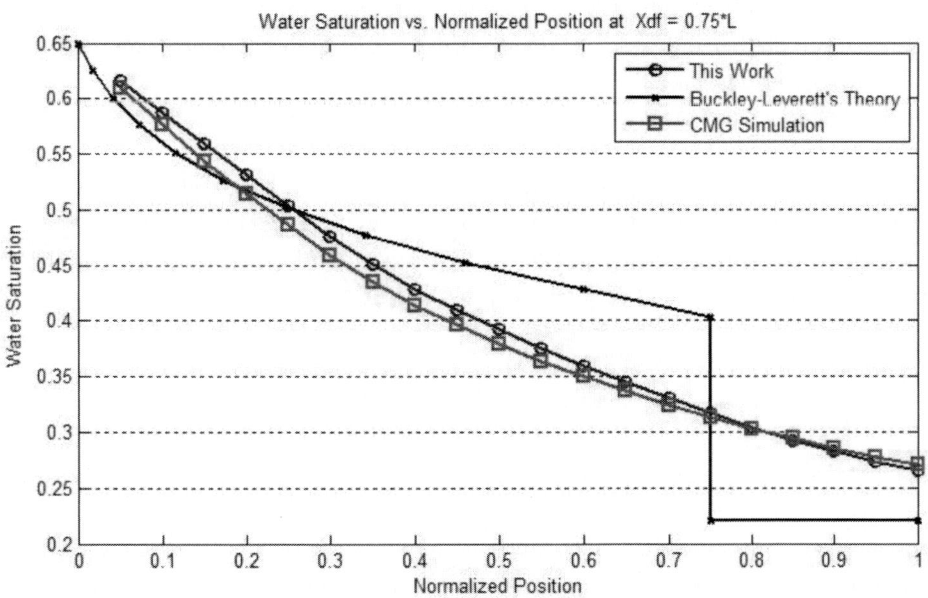

Figure 1. Validation of the model against the theory when the frontal saturation of water is located at $0.75 \times L$ (or at 7hr).

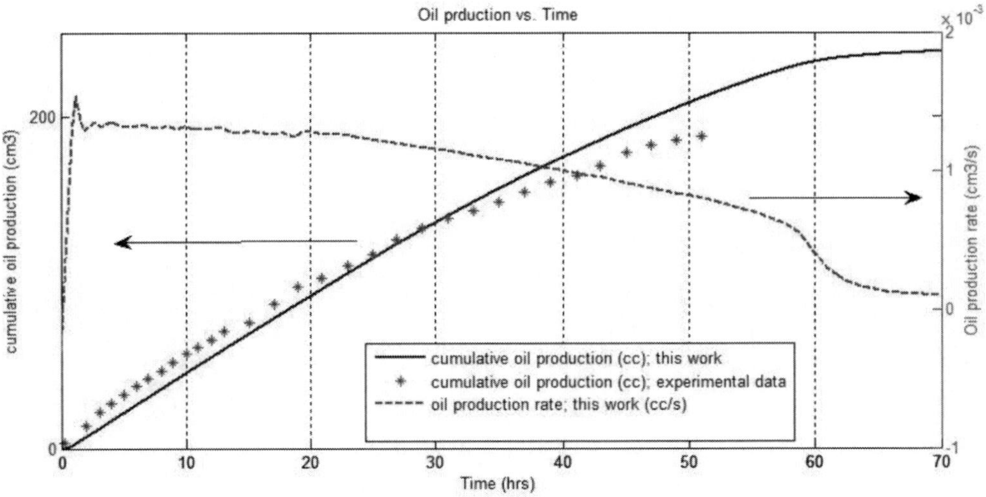

Figure 2. Comparison of accumulative oil production by this work and experimental results.

In the second example, the numerical model is validated to find the performance of mathematical VAPEX simulator. The VAPEX experiments by Moghadam et al. (2009) in a packed cell is chosen to test the simulator model. The packed bed size in their test was 40×10×2 cm^3. Propane at T=20.8 $^{\circ}C$ and P=800 kPa was used as an extracting solvent. The sand packed with 103 Darcy permeability and 32.9% porosity was utilized in order to fill the bed by the Lloydminster heavy oil with *11900 cp* viscosity.

Figure 2 shows the accumulative production rate for the described simulator along with the experimental results for 50 hours. It is clear that there is a good agreement between the experimental data and the modeling results. However, there is a discrepancy between the experimental and predicted results of the process. Among the possible reasons for the differences, the uncertainties in the diffusion and dispersion coefficients, viscosity correlation and neglecting the asphaltene precipitation in the mathematical model can be pointed. It is worthwhile to mention that the asphaltene content of the original heavy oil was about 17% which the most amount of it can be precipitated into the pore layer of experimental system when enough solvent is dissolved. As results it is not unexpected that a tiny lower production with better quality and in-situ deasphalted heavy oil was extracted during the experimental measurement by Moghadam et al.(Luo et al., 2007; 2008).

OPTIMUM SOLVENT INJECTION RATE

Now, by considering the accuracy of the compositional model, VAPEX process in one of the Iranian heavy oil reservoirs (KEM) (Moosavi et al., 2005) will be studied. The dimensions and characteristics of KEM reservoir are presented in Table 3. In order to study the effects of solvent injection rate on the VAPEX performance, several solvent rates are subjected on the system. Solvent selection is one of the important steps for handling of economical aspect of such process.

The criteria for better selection depend on reservoir conditions such as dew point pressure, density difference, solubility, diffusivity, reservoir temperature and pressure

(Ahmadloo, Asghari, and Henni, 2014; Hosseini Pour, Azin, and Yunan, 2011). It should be mentioned the VAPEX process would be more economic if the space of the solvent chamber fill by gas phase of solvent rather than liquid one. Because loads of solvent is needed to occupy the evacuated space by liquid rather than a gas solvent. In addition, the possibility of vapor usage leads to increasing the difference in phase density and high gravity drainage for tilted oil.

Therefore using vapor solvent makes the process more economical. Beside the desired solvent (propane or butane) almost methane and CO_2 have been used as non-condensable carrier gas to adjust the dew point pressure of the solvent mixture to the experimental conditions (Ahmadloo, Asghari, and Henni, 2014). The pure hydrocarbon solvents (propane or butane) are in a liquid state under the KEM conditions. A mixture of light hydrocarbons is the suggested opinion to prevent solvent condensation under the defined conditions. So, the solvent mixture including C_1(40%)-C_3(60%) will be injected into the reservoir to prevent solvent liquefaction under reservoir conditions. In this mixture, methane has been applied as a carrier gas since the injected solvent approaches to its dew-point at T=60 $^{\circ}C$ and P=6391kPa. The lumped oil solvent system is used for reduction of the time and computing errors. So, a rectangular porous medium geometry totally filled with pure heavy oil having a high viscosity is considered and a horizontal production well is located at the bottom center of the system and one injection well above the production in a distance of the formation height. As discussed in previous paragraphs, the solvent is injected through an upper horizontal well and dissolves and diffuses in the heavy crude oil to reduce its viscosity and makes it mobile.

Therefore, mechanical dispersion of solvent-oil is an important term in viscosity reduction and heavy oil production in the process and the choices of suitable correlations of diffusion and dispersion coefficients as well as heavy oil viscosity respect to variation of solvent mole fraction in oil phase is essential during mathematical computations. Unfortunately software packages such as CMG tools don't have a flexible correlation for VAPEX processes. In this stage the molecular diffusion-dispersin correlated by Hayduk and Minhas (1982) and the concentration-dependent viscosity of Shu (1984) are used in studying the KEM oil recovery with the VAPEX process.

The operating conditions explained in the brief words are introduced as input values for the compositional simulator. After a primary period of the injection, the injected solvent dissolves into the heavy oil, then diffuses and disperses in the virgin regions of the heavy oil to reduce the viscosity of the defined heavy oil by several orders of magnitude, then similar to a counter-current, tilted oil drains downward and the vapor solvent starts accumulate above the side of the reservoir. Eventually, over time, the tilted oil is collected nearby the well production by gravitational forces and vapor chamber in up expands laterally in sideways. Figure 3 (a) and (b) show the heavy oil mole fraction and the oil saturation at 120 min after beginning of the process. Definitely after a period of time the regions of the heavy oil and solvent chamber can be detected by the contour plot tools. As can be observed on these figures, the sideways expansion of the solvent chamber and the depth of diffusion are similar to laboratory pictures illustrated by previous authors (Das and Butler, 1998; Etminan, 2007; Moghadam, Nobakht, and Gu, 2009). It should be mentioned the curvature and S-shape of solvent chamber can be changed by a different variety of mechanical dispersions and reservoir characteristics such as relative permeability (Ghesmat, 2014), and this is another subject for future studies.

Table 3. Rock and fluid data for KEM reservoir

Porosity (%)	33
Vertical Permeability (μm^2)	0.85
Horizontal Permeability (μm^2)	1
S_w (%)	0.5
Operational pressure (kPa)	6391.44
Formation width (cm)	10
Model Height (cm)	5
Model Thickness (cm)	2
Initial oil density (gr / cm^3)	0.985
Initial oil viscosity ($mPa.s$)	1758

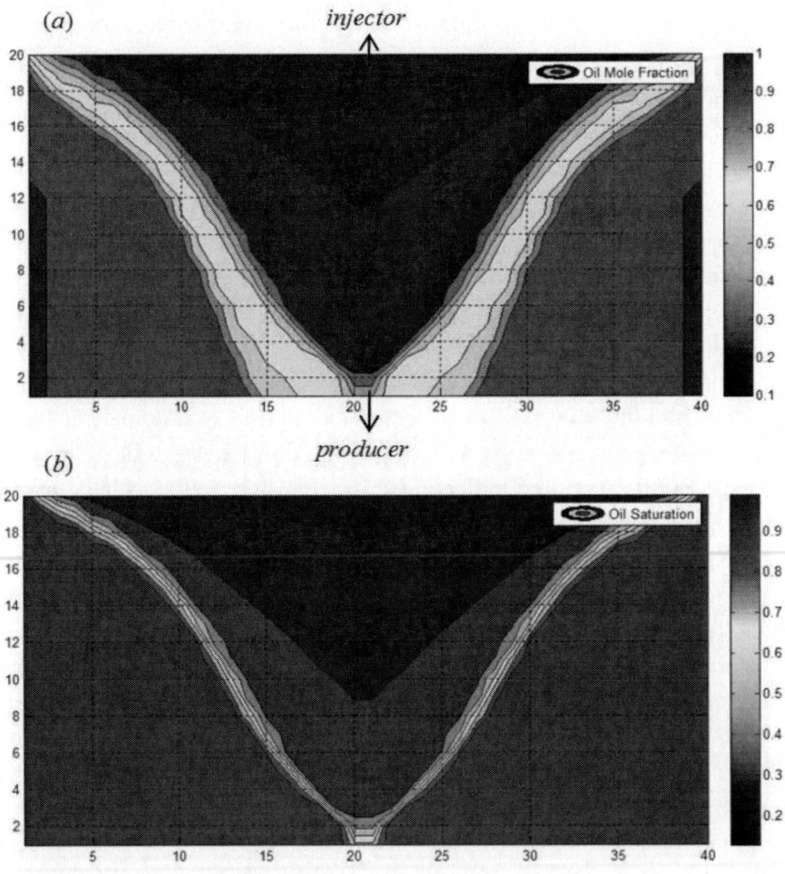

Figure 3. The bed schematics for (a) heavy oil mole fraction and (b) oil saturation at t=120 min.

Five different injection rates are applied on the present reservoir to study the effect of solvent rates on production rate, breakthrough time and solvent-oil ratio and as a result, it can estimate the optimum injection rate. It must be mentioned during the gas injection processes in homogeneous media, the desire way for passing the injected gas is mainly along the highest pressure gradient (Jiang, 2013) so that, in this case it is a straight line between two horizontal well bores. Therefore at higher pressure gradients between the injector and producer gas phase with higher mobility respect to oil phase can move faster toward the production well and occupy the neighborhood of the producer and prevents oil production effectively. Figure 4 compares the oil production rates obtained by five injection rates. It shows that the higher injection rate implies a higher production rate initially, but in a longer time, as the breakthrough times reach, these differences are reduced. The different differential pressures (which are created by the different gas injection rates) applied on the reservoir are a key of such changes in trends of production rates. Indeed, applying the higher injection rate leads to higher differential pressure between two horizontal wells and it initially makes the VAPEX process to a displacement process (Kok, Yildirim, and Akin, 2009). Since the results of model by the higher injection rate leads to the higher production rate and earlier breakthrough time at the first. Once the solvent breakthrough occurred, the grids near the production well are occupied by the vapor and effectively decrease the oil rate at the continuation of the process. Of course, agreeing with the figure, Kok et al. (2009) experimentally observed production rate of heavy for all injection rates first increased and then gradually fell down. Furthermore, Figure 5 compares the gas production rate vs. time for the five rates. It shows the effect of increasing the injection rates on the gas production. In the model with 0.00178 (cc/s) rate, it takes much time to approach in gas production, whereas the higher injection rate causes immediate gas production. The lowest rate doesn't produce tangible solvent gas because, in spite of the higher mobility ratio of gas the defined rate doesn't develop a valuable pressure difference (or differential pressure) between two wells to overcome the gravity drainage of tilted oil having higher density. It is valuable to point; initially the main driving force for oil production at the maximum rate is displacement of heavy oil, whereas at the minimum one mechanical dispersion and gravity drainage are the main mechanism throughout the oil production.

It is clear that at each grid block, higher gas saturation has a higher gas relative permeability, therefore it causes a lower mobility ratio for oil movement. So, in the VAPEX process, early breakthrough time reduced the oil production rates in the longer time. Figure 6 depicts the recovery factor at different solvent rates. As can be seen, at the beginning of the process, the highest injection rate has the most oil recovery; but totally, all injection rates have a similar oil recovery about 180 minutes after the beginning of the process. Therefore, it can be concluded after the early breakthrough, diffusion becomes the more important mechanism and oil production rate of the process approach to the same value in the different injection rates.

Generally, the net accumulative solvent employed for the VAPEX performance is implemented for filling voids as saturated gas, dissolving in the oil phase and for supplying the free or dissolved solvent as a production (Butler, Mokrys, and Das, 1995). In Figure 7, the solvent to dead oil productions are compared with the high to low solvent injection rate. This figure shows that in the case of the lowest injection rate, the amounts of SOR are limited to 0.2 *gr* solvent/*gr* dead oil during the steady oil production times, while the SOR for the most injection rate goes beyond from 2.5 *gr* solvent/*gr* dead oil. The cost of solvent is an important issue and the high SOR does not support economical aspect of the process. An appropriate plan for injection rate in large reservoirs for the VAPEX process will be combined in two high and low injection rates, so that a high solvent injection rate is applied to the reservoir to let the solvent go through the crude heavy oil and communicate between two wells with high initial rate. Afterward, a low injection rate causes the solvent to be transported in the crude oil, so that thin oil can drain strongly and eventually, a good oil recovery is obtained. Figure 8 depicts such a combined injection. As can be seen, two injection rate values 0.0178 and 0.00178 were selected to validate the suggested plan. Shortly the higher rate is subjected to the defined bed, consequently a high differential pressure is developed between the injector and producer well and the running vapor displaces the heavy oil until a communication between two wells is created. Afterward, once diffusion becomes more effective in the oil production the lower rate is applied into the well to support the swept spaces and required solvent for dissolution and diffusion in the heavy oil in the continuation of the process. It exposes a more initial rate and oil recovery with low SOR that fairly is an economical recovery. Finally, it should be noted the low injection rate should be of such value that the created differential pressure developed by vapor injection not be able to overcome the downward drainage of tilted oil derived by gravity force.

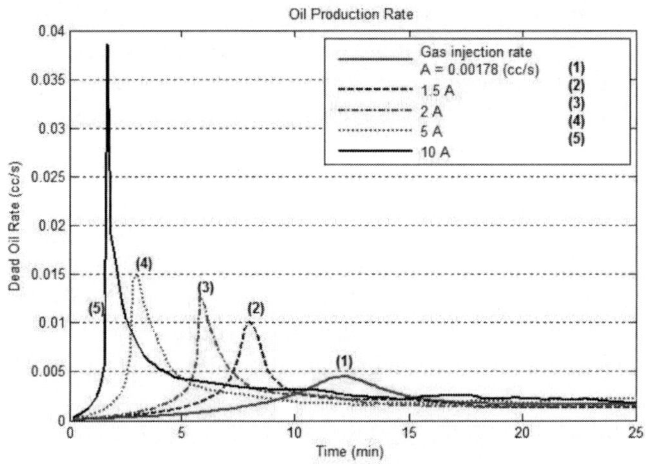

Figure 4. Production rates obtained from the model for five different injection rates.

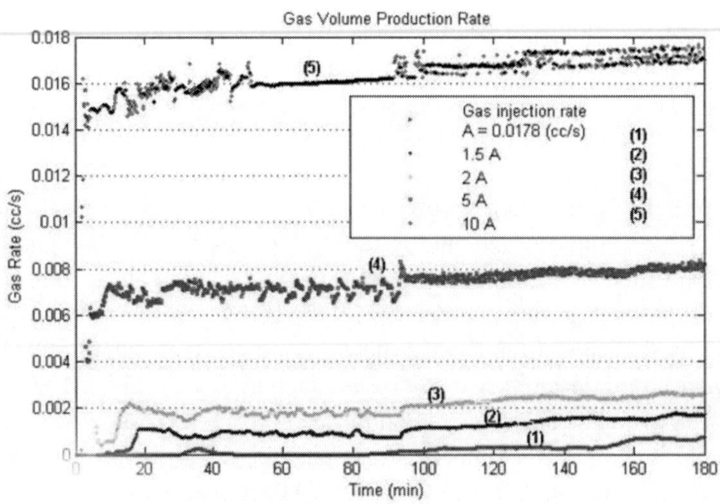

Figure 5. Effect of increasing injection rate on the early breakthrough time and gas production rate.

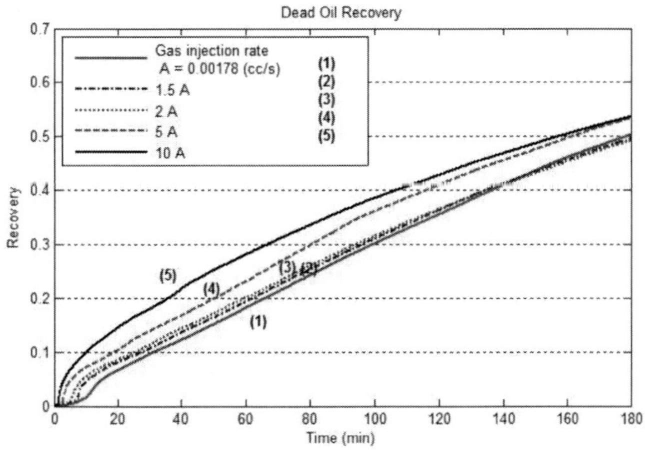

Figure 6. Effect of injection rate on the oil recovery.

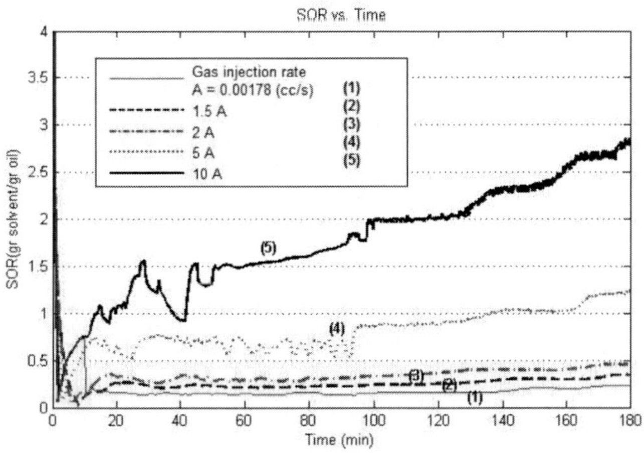

Figure 7. Solvent-dead oil ratio at the five different injection rates.

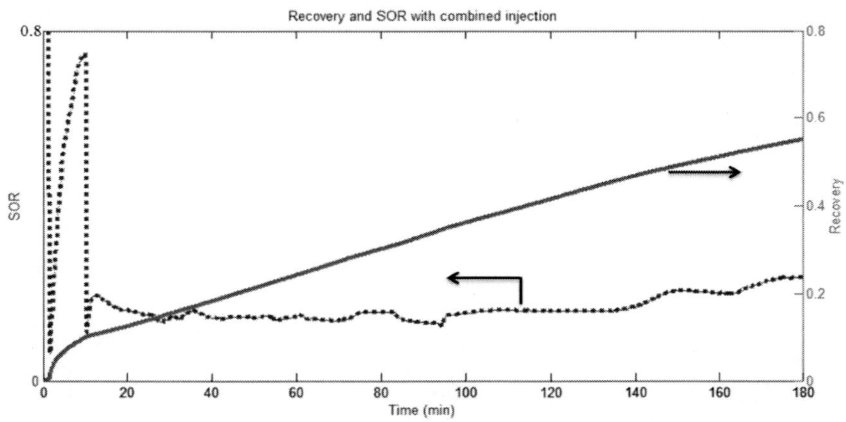

Figure 8. Recovery and SOR with injection of 0.0178 (cc/s) in the first 10 minutes and 0.00178 (cc/s) for the longer time.

Conclusion

A compositional reservoir simulator has been developed in a miscible process that could fairly estimate reservoir variables during the VAPEX process. Because in the VAPEX process, mechanical dispersion and viscosity reduction are important factor for draining of oil, therefore it was tried using reliable dispersion, diffusion and oil viscosity coefficients (Hayduk and Boustani correlations for mechanical dispersion and Suh and Lederer correlations for oil viscosity reduction) so that acceptable results could be eventuated.

After validation, several simulations were run to observe the effect of injection rate on the extraction rate in an Iranian reservoir. It was shown that a high injection rate leads to higher pressure difference between two horizontal wells and eventually at the beginning of the process it weakens the diffusion mechanism and makes the VAPEX process to a displacement one. The injected gas initially pushed the heavy oil toward producer and it increased initial production rate of oil, but in an early time, breakthrough occurred and displacement mechanism lost its ability, and then only diffusion, viscosity reduction and gravity drainage were able to produce oil effectively. On the base of this information, a proper plan was suggested and tested, so that a high solvent rate was injected at the beginning of the process until a communication between two wells was created. Afterward, once diffusion became more effective a second rate with lower value was injected to permit the solvent to be transported into the virgin zone of heavy oil. Finally, in addition to higher oil recovery, we could prevent extra vapor from wasting and we could produce more oil as economically as possible.

Appendix

Basic Differential Equations

Absolutely, total mass of the component is conserved in spite of the fact that the mass is not conserved within each phase. So, the following mass balances are satisfied at the water and hydrocarbon phases:

$$\frac{\partial(\phi[x_m \xi_o S_o + y_m \xi_g S_g])}{\partial t} + \nabla.(x_m \xi_o \boldsymbol{u}_o + y_m \xi_g \boldsymbol{u}_g) + \phi(\xi_o \mathbf{K}_{mo} \nabla x_m + \xi_g \mathbf{K}_{mg} \nabla y_m) = q_m, \quad \text{(A1)}$$

and

$$\frac{\partial(\phi \xi_w S_w)}{\partial t} + \nabla.(\xi_w \boldsymbol{u}_w) = q_w, \quad \text{(A2)}$$

where, ξ defines the molar density of each phase and $m = 1, 2, \ldots, N_c$, which N_c specifies the number of components. $\mathbf{K}_m$ introduces the mass transfer tensor of *mth* component in the fluid mixture. As long as both diffusion and dispersion present in the system, it is reasonable to use coefficients which were described at the pervious references (Perkins and Johnston, 1963; Hayduk and Minhas, 1982; Boustani and Maini, 2001).

$$\mathrm{K}_{m\alpha} = \frac{D_{m\alpha}}{F\phi} + |\alpha|.\boldsymbol{u}, \tag{A3}$$

as it can be seen, the right side of Eq. (A3) is combined of two terms to represent molecular diffusion and dispersion due to the velocity. The details of the correlations are explained in the Appendix.

In the mentioned equations, the volumetric velocity is governed by the Darcy's law:

$$\boldsymbol{u} = -\frac{k_{r\alpha}}{\mu_\alpha}\boldsymbol{K}\,(\nabla P_\alpha - \rho_\alpha g \nabla Z) \tag{A4}$$

Assuming that the hydrocarbon contents are in the phase equilibrium, other algebraic constraints are as follows:

$$\sum_{m=1}^{N_c} x_m = 1, \sum_{m=1}^{N_c} y_m = 1 \tag{A5}$$

$$K_m = y_m / x_m \tag{A6}$$

$$S_w + S_g + S_o = 1 \tag{A7}$$

The fluid pressures in each phase are defined by capillary pressures:

$$\begin{aligned} P_o &= P_{cow} + P_w, \\ P_g &= P_{cgo} + P_o \end{aligned} \tag{A8}$$

So, there are $2N_c$+9 dependent variables: x_m, y_m, $\boldsymbol{u_\alpha}$, P_α and S_α with $2N_c$+9 independent correlations besides capillary functions in order to obtain the unknowns.

In1978 Kazemi et al. (1978) demonstrated that the equilibrium correlations imply to the following equation:

$$x_m\xi_o S_o + y_m\xi_g S_g = \frac{z_m(K_m S_g\xi_g + S_o\xi_o)}{1+V(K_m-1)} = \frac{z_m(K_m S_g\xi_g + S_o\xi_o)}{1+(\frac{S_g\xi_g}{S_g\xi_g+S_o\xi_o})(K_m-1)} = \frac{z_m(K_m S_g\xi_g + S_o\xi_o)}{(\frac{S_g\xi_g+S_o\xi_o+S_g\xi_g(K_m-1)}{S_g\xi_g+S_o\xi_o})} = z_m(S_o\xi_o + S_g\xi_g) \tag{A9}$$

By substitution of Eq. (A9) into Eq. (A1) and summing up the resulting equations over all the hydrocarbon components, following flow equation is produced which is utilized for obtaining the pressure variable.

$$\frac{\partial(\phi(S_o\xi_o + S_g\xi_g))}{\partial t} = \nabla.((\xi_o\nabla\boldsymbol{u}) + (\xi_g\nabla\boldsymbol{u})) + q_o + q_g \tag{A10}$$

Next, Eq. (A4) is put into the mass equation, so:

$$\frac{\partial(\phi(S_o\xi_o + S_g\xi_g))}{\partial t} = \nabla.(T_o(\nabla P_o - \gamma_o\nabla Z) + T_g(\nabla P_g - \gamma_g\nabla Z)) + q_o + q_g, \tag{A11}$$

which the transmissibility terms are defined as;

$$T_\alpha = \frac{\xi_\alpha k_{r\alpha}}{\mu_\alpha}\boldsymbol{K} \tag{12}$$

The right hand side of Eq. (A11) describes fluid flow and injection/production in the reservoir, whereas the left hand side presents accumulation term. In order to change the left hand side of Eq. (A11) to a pressure variable form, a manipulation is applied on the accumulation term as follows:

$$\frac{\partial(\phi(S_o\xi_o + S_g\xi_g))}{\partial t} = \left(S_o\frac{\partial(\phi\xi_o)}{\partial t} + \phi\xi_o\frac{\partial S_o}{\partial t}\right) + \left(S_g\frac{\partial(\phi\xi_g)}{\partial t} + \phi\xi_g\frac{\partial S_g}{\partial t}\right) \tag{A13}$$

Applying the chain rule converts the differentiation form of porosity and molar density in respect to time to the variables dependency of pressure at constant temperature:

$$\frac{\partial(\phi\xi_o)}{\partial t} = \xi_o\frac{\partial\phi}{\partial t} + \phi\frac{\partial\xi_o}{\partial t}$$
$$\xi_o\frac{\partial\phi}{\partial t} + \phi\frac{\partial\xi_o}{\partial t} = \xi_o\frac{\partial\phi}{\partial P_o}.\frac{\partial P_o}{\partial t} + \phi\frac{\partial\xi_o}{\partial P_o}.\frac{\partial P_o}{\partial t} \tag{A14}$$

Also, the aforementioned procedure is the same for the gas phase. Now, the left hand side of Eq. (A11) is determined and the flow equation may then be written:

$$\left(S_o(\xi_o\frac{\partial\phi}{\partial P_o}.\frac{\partial P_o}{\partial t}+\phi\frac{\partial\xi_o}{\partial P_o}.\frac{\partial P_o}{\partial t})+\phi\xi_o\frac{\partial S_o}{\partial t}\right)+$$
$$\left(S_g(\xi_g\frac{\partial\phi}{\partial P_g}.\frac{\partial P_g}{\partial t}+\phi\frac{\partial\xi_g}{\partial P_g}.\frac{\partial P_g}{\partial t})+\phi\xi_g\frac{\partial S_g}{\partial t}\right)= \tag{A15}$$
$$\nabla.(T_o(\nabla P_o-\gamma_o\nabla Z)+T_g(\nabla P_g-\gamma_g\nabla Z))+q_o+q_g$$

This form of equation is more convenient and widely applicable compared to the first form, since the dependencies of fluid compressibility and porosity in respect to pressure are more available. The density and porosity can be calculated by using linear correlations to be applied at the mentioned equations.

$$\xi=\xi_0\exp(C_F(P-P_0))\approx\xi_0(1+C_F(P-P_0)),$$
$$C_F=\frac{1}{\xi}\frac{\partial\xi}{\partial P} \tag{A16}$$

$$\phi=\phi_0\exp(C_R(P-P_0))\approx\phi_0(1+C_R(P-P_0)),$$
$$C_R=\frac{1}{\phi}\frac{\partial\phi}{\partial P} \tag{A17}$$

In continuation of the work, all the required equations are discretized in two-dimensional flow in the porous media to cover all mathematical details. So, for the time step of *n+1*, the algebraic form of Eq. (A14) converts to the following finite difference form:

$$\frac{V_R}{\Delta t}\phi^n((\xi_o^nS_o^{n+1}+\xi_g^nS_g^{n+1})-(\xi_o^nS_o^n+\xi_g^nS_g^n))-$$
$$\frac{V_R}{\Delta t}((\phi^nS_o^n(\frac{\partial\xi_o}{\partial P})+S_o^n\xi_o^n(\frac{\partial\phi}{\partial P})).(P_o^{n+1}-P_o^n)+$$
$$(\phi^nS_g^n(\frac{\partial\xi_g}{\partial P})+S_g^n\xi_g^n(\frac{\partial\phi}{\partial P})).(P_g^{n+1}-P_g^n))=$$
$$\mathrm{T}^{n+1}_{o,i+\frac{1}{2},j}(P^{n+1}_{o,i+1,j}-P^{n+1}_{o,i,j})-\mathrm{T}^{n+1}_{o,i-\frac{1}{2},j}(P^{n+1}_{o,i,j}-P^{n+1}_{o,i-1,j})+$$
$$\mathrm{T}^{n+1}_{o,i,j+\frac{1}{2}}(P^{n+1}_{o,i,j+1}-P^{n+1}_{o,i,j})-\mathrm{T}^{n+1}_{o,i,j-\frac{1}{2}}(P^{n+1}_{o,i,j}-P^{n+1}_{o,i,j-1})+$$
$$\mathrm{T}^{n+1}_{g,i+\frac{1}{2},j}(P^{n+1}_{g,i+1,j}-P^{n+1}_{g,i,j})-\mathrm{T}^{n+1}_{g,i-\frac{1}{2},j}(P^{n+1}_{g,i,j}-P^{n+1}_{g,i-1,j})+$$

$$\mathrm{T}^{n+1}_{g,i,j+\frac{1}{2}}(P^{n+1}_{g,i,j+1}-P^{n+1}_{g,i,j})-\mathrm{T}^{n+1}_{g,i,j-\frac{1}{2}}(P^{n+1}_{g,i,j}-P^{n+1}_{g,i,j-1})-$$
$$(\mathrm{T}\gamma)^{n+1}_{o,i,j+\frac{1}{2}}(Z_{i+1,j}-Z_{i,j})+(\mathrm{T}\gamma)^{n+1}_{o,i,j-\frac{1}{2}}(Z_{i,j}-Z_{i-1,j})-$$
$$(\mathrm{T}\gamma)^{n+1}_{g,i,j+\frac{1}{2}}(Z_{i+1,j}-Z_{i,j})+(\mathrm{T}\gamma)^{n+1}_{g,i,j-\frac{1}{2}}(Z_{i,j}-Z_{i-1,j})+$$
$$Q^{n+1}_{o,i,j}+Q^{n+1}_{g,i,j}, \tag{A18}$$

where,

$$\mathrm{T}^{n+1}_{i+\frac{1}{2},j}=T^{n+1}_{i+\frac{1}{2},j}\left(\frac{(\Delta Z.\Delta Y)_{i,j}}{\Delta X_{i+\frac{1}{2},j}}\right)$$
$$\mathrm{T}^{n+1}_{i,j+\frac{1}{2}}=T^{n+1}_{i,j+\frac{1}{2}}\left(\frac{(\Delta X.\Delta Y)_{i,j}}{\Delta Z_{i,j+\frac{1}{2}}}\right) \tag{A19}$$
$$\mathrm{V}_{R,i,j}=(\Delta Z.\Delta Y.\Delta X)_{i,j}$$

Eq. (A18) is a system of equations, so it developed a sparse pentadiagonal coefficient matrix which must be solved by a numerical algorithm.

Overall compositions are calculated from Eq. (A1) after the pressure calculations;

$$z^{n+1}_{m}=(\mathrm{T}^{n+1}_{mo,i+\frac{1}{2},j}(P^{n+1}_{o,i+1,j}-P^{n+1}_{o,i,j})-\mathrm{T}^{n+1}_{mo,i-\frac{1}{2},j}(P^{n+1}_{o,i,j}-P^{n+1}_{o,i-1,j})+$$
$$\mathrm{T}^{n+1}_{mo,i,j+\frac{1}{2}}(P^{n+1}_{o,i,j+1}-P^{n+1}_{o,i,j})-\mathrm{T}^{n+1}_{mo,i,j-\frac{1}{2}}(P^{n+1}_{o,i,j}-P^{n+1}_{o,i,j-1})+$$
$$\mathrm{T}^{n+1}_{mg,i+\frac{1}{2},j}(P^{n+1}_{g,i+1,j}-P^{n+1}_{g,i,j})-\mathrm{T}^{n+1}_{mg,i-\frac{1}{2},j}(P^{n+1}_{g,i,j}-P^{n+1}_{g,i-1,j})+$$
$$\mathrm{T}^{n+1}_{mg,i,j+\frac{1}{2}}(P^{n+1}_{g,i,j+1}-P^{n+1}_{g,i,j})-\mathrm{T}^{n+1}_{mg,i,j-\frac{1}{2}}(P^{n+1}_{g,i,j}-P^{n+1}_{g,i,j-1})-$$
$$(\mathrm{T}_m\gamma)^{n+1}_{o,i,j+\frac{1}{2}}(Z_{i+1,j}-Z_{i,j})+(\mathrm{T}_m\gamma)^{n+1}_{o,i,j-\frac{1}{2}}(Z_{i,j}-Z_{i-1,j})-$$
$$(\mathrm{T}_m\gamma)^{n+1}_{g,i,j+\frac{1}{2}}(Z_{i+1,j}-Z_{i,j})+(\mathrm{T}_m\gamma)^{n+1}_{g,i,j-\frac{1}{2}}(Z_{i,j}-Z_{i-1,j})+$$
$$\hat{\mathrm{K}}^{n+1}_{mo,i+\frac{1}{2},j}(x_{m,i+1,j}-x_{m,i,j})-\hat{\mathrm{K}}^{n+1}_{mo,i-\frac{1}{2},j}(x_{m,i,j}-x_{m,i-1,j})+$$
$$\hat{\mathrm{K}}^{n+1}_{mo,i,j+\frac{1}{2}}(x_{m,i,j+1}-x_{m,i,j})-\hat{\mathrm{K}}^{n+1}_{mo,i,j-\frac{1}{2}}(x_{m,i,j}-x_{m,i,j-1})+$$

$$\hat{K}^{n+1}_{mg,i,j+\frac{1}{2}}(y_{m,i,j+1}-y_{m,i,j})-\hat{K}^{n+1}_{mg,i,j-\frac{1}{2}}(y_{m,i,j}-y_{m,i,j-1})+$$
$$+x_m Q^{n+1}_{o,i,j}+y_m Q^{n+1}_{g,i,j}+$$
$$\frac{V_R}{\Delta t}\phi^n(\xi^n_g S^n_g+\xi^n_o S^n_o)z^n_m)/(\frac{V_R}{\Delta t}\phi^{n+1}(\xi^{n+1}_g S^{n+1}_g+\xi^{n+1}_o S^{n+1}_o)),$$

(A20)

and for i-direction (the same procedure of j-direction is applied);

$$\hat{K}^{n+1}_{m\alpha,i+\frac{1}{2},j}=\frac{\xi_\alpha D_{m\alpha}}{F}.(\frac{(\Delta Z.\Delta Y)_{i,j}}{\Delta X}_{i+\frac{1}{2},j})+|A|.T^{n+1}_{m\alpha,i+\frac{1}{2},j}(P^{n+1}_{\alpha,i+1,j}-P^{n+1}_{\alpha,i,j})+$$

$$|B|.(T^{n+1}_{m\alpha,i,j+\frac{1}{2}}(P^{n+1}_{\alpha,i,j+1}-P^{n+1}_{\alpha,i,j})-(T\gamma)^{n+1}_{m\alpha,i,j+\frac{1}{2}}(Z_{i,j+1}-Z_{i,j})), \quad \text{(A21)}$$

$$T_{mo}=T_o x_m,\quad T_{mg}=T_g y_m$$

The hydrocarbon components are flashed at the new grid pressures and compositions and equilibrium correlations, so Eqs. (A5) and (A6) are applied to determine the compositions and the quantities of each phase. Water saturations are other parameters which are calculated from Eq. (A2),

$$S^{n+1}_w=(T^{n+1}_{w,i+\frac{1}{2},j}(P^{n+1}_{w,i+1,j}-P^{n+1}_{w,i,j})-T^{n+1}_{w,i-\frac{1}{2},j}(P^{n+1}_{w,i,j}-P^{n+1}_{w,i-1,j})+$$
$$T^{n+1}_{w,i,j+\frac{1}{2}}(P^{n+1}_{w,i,j+1}-P^{n+1}_{w,i,j})-T^{n+1}_{w,i,j-\frac{1}{2}}(P^{n+1}_{w,i,j}-P^{n+1}_{w,i,j-1})-$$
$$(T\gamma)^{n+1}_{w,i,j+\frac{1}{2}}(Z_{i+1,j}-Z_{i,j})+(T\gamma)^{n+1}_{w,i,j-\frac{1}{2}}(Z_{i,j}-Z_{i-1,j})+$$
$$Q_w+\frac{V_R}{\Delta t}(\phi^n\xi^n_w S^n_w))/(\frac{V_R}{\Delta t}(\phi^{n+1}\xi^{n+1}_w))$$

(A22)

The final series of computations are for determination of gas and oil saturations. To achieve the selected point, the following equilibrium saturation correlations are utilized,

$$L^{n+1}=S^{n+1}_o\xi^{n+1}_o/(S^{n+1}_o\xi^{n+1}_o+S^{n+1}_g\xi^{n+1}_g)$$
$$S^{n+1}_g=1-S^{n+1}_o+S^{n+1}_w$$

(A23)

Finally,

$$S_g^{n+1} = \left(\frac{(1-S_w)\xi_o(1-L)}{L\xi_g + (1-L)\xi_o}\right)^{n+1}$$
$$S_o^{n+1} = \left(\frac{(1-S_w)\xi_g L}{L\xi_g + (1-L)\xi_o}\right)^{n+1} \tag{A24}$$

Calculation Sequence

The equations developed here can be extended to the VAPEX process in two dimensions. An implicit iterative technique with small time steps is used to estimate fluids flow in the petroleum bed:

1. At the beginning of each time step, the pressure, saturations and overall composition are assumed at all grid blocks.
2. The hydrocarbon components are flashed then, all fluid and rock properties are determined under the considered conditions.
3. Each grid block pressure is obtained from Eq. (A18).
4. At the new pressures, Eq. (A20) is applied to determine the overall hydrocarbon compositions.
5. The composition and amounts of gas and oil phases are obtained from the flash calculation.
6. The water saturation at each grid block is determined from Eq. (A22).
7. The results of the previous steps are used to calculate the new hydrocarbon saturations at each grid block.

These sequences are repeated until the errors of gridpoint pressures converge to the minimum value. After that, variables are ready for the next time step. It is clear, with two hydrocarbon components, the explained procedure has the similarity to an iterative black oil computation.

NOMENCLATURE

c	Solvent volume fraction
dp	Particle diameter (m)
F	Formation resistivity factor
g	Gravity acceleration (m^2 / s)
h	Height of model (m)
k_r	Relative permeability
q	Molar flow rate per unit volume ($mole / m^3 .s$)
t	Time (s)
u	Velocity vector (m/s)
u	Horizontal velocity (m/s)
v	Vertical velocity (m/s)
x	Hydrocarbon mole fraction of liquid phase

y	Hydrocarbon mole fraction of vapor phase
z	Overall hydrocarbon mole fraction
C_F	Compressibility factor of molar density *(1/kPa)*
C_R	Compressibility factor of porosity *(1/kPa)*
D	Molecular diffusion (m^2/s)
K	Matrix permeability (m^2)
K_m	Equilibrium ratio of hydrocarbon component
L	Length of well (*m*)
P	Pressure (*kPa*)
Q	Molar flow rate ($mole^3/s$)
S	Saturation
T_α	Transmissibility
X	Horizontal direction
Z	Vertical direction

Greek Symbols

δ	heterogeneity index for porosity distribution
σ	surface tension (kg/s^2)
γ	Gradient of fluid phase (*kpa/m*)
ϕ	Porosity
μ	Viscosity (*Pa.s*)
ν	Kinematic viscosity (m^2/s)
ρ	Density (kg/m^3)
ξ	Molar density ($kmole/m^3$)
Δ	Differential operator
K	Total dispersion coefficient (m^2/s)
K	Total dispersion coefficient for X or Z-axis (m^2/s)
Ω	Cementation factor

Subscripts

α	Index for fluid phase
g	Gas
m	Index of hydrocarbon component
o	Oil
s	Solvent
w	Water

Special Definitions

$$T_\alpha = \frac{\xi_\alpha k_{r\alpha}}{\mu_\alpha} \boldsymbol{K}$$

$\nabla.$ = divergence operator

∇ = gradient operator

REFERENCES

Ahmadloo, F., Asghari, K., and Henni, A. (2014). The Interplay of Permeability, Drainage Height, Capillarity, and Solvent Type on Diffusion and Convective Dispersion in the Vapex Process: A Critical Review. *Petroleum Science and Technology,* 32 (3):345-353.

Ahmadloo, F., Asghari, K., Henni, A., and Freitag, N. (2013). Experimental Investigation of Capillarity and Drainage Height Roles in the Vapor Extraction Process. *Petroleum Science and Technology,* 31 (2):204-214.

Azad, A., and Chalaturnyk, R. J. (2012). An improved SAGD analytical simulator: Circular steam chamber geometry. *Journal of Petroleum Science and Engineering,* 82:27-37.

Azin, R., Kharrat, R., Ghotbi, C., and Vossoughi, S. (2005). Applicability of the VAPEX Process to Iranian Heavy Oil Reservoirs. *SPE Middle East Oil and Gas Show and Conference*, Kingdom of Bahrain.

Boustani, A., and Maini, B. B. (2001). The Role of Diffusion and Convective Dispersion in Vapour Extraction Process. *Journal of Canadian Petroleum Technology,* 40 (4):68-77.

Branco, C. M., and Rodriguez, F. (1996). A Semi-Implicit Formulation for Compositional Reservoir Simulation. *SPE Advanced Technology Series,* 4 (1):171-177.

Butler, R., and Mokrys, I. (1993). Recovery of Heavy Oils Using Vapourized Hydrocarbon Solvents: Further Development of the Vapex Process. *Journal of Canadian Petroleum Technology,* 32 (6).

Butler, R., and Mokrys, I. J. (1989). Solvent Analog Model of Steam-Assisted Gravity Drainage. *AOSTRA Journal of Research,* 5 (1):17-32.

Butler, R. M., and Mokrys, I. J. (1991). A New Process (VAPEX) For Recovering Heavy Oils Using Hot Water And Hydrocarbon Vapour. *Journal of Canadian Petroleum Technology,* 30 (1):97-106.

Butler, R. M., Mokrys, I. J., and Das, S. K. (1995). The Solvent Requirements for Vapex Recovery. SPE International Heavy Oil Symposium, 19-21 June 1995, Calgary, Alberta, Canada.

Das, S. K., and Butler, R. M. (1998). Mechanism of the vapor extraction process for heavy oil and bitumen. *Journal of Petroleum Science and Engineering,* 21 (1–2):43-59.

Derakhshanfar, M., and Gu, Y. (2012). An Analytical Model for Determination of the Solvent Convective Dispersion Coefficient in the Vapor Extraction Heavy Oil Recovery Process. *Transport in Porous Media*, 92 (2):495-507.

Etminan, S. R. (2007). Complex Effect of Connate Water in Vapour Extraction Process. SPE International Conferance, Anaheim, California, USA.

Ghesmat, K. (2014). In-Situ Solvent-Assisted Gravity Drainage of Bitumen: Nonlinear Numerical Analysis. *SPE Journal,* 19 (1):109-121.

Haghighat, P., Maini, B., and Abedi, J. (2013). Experimental and Numerical Study of VAPEX at Elevated Temperatures. SPE Heavy Oil Conference, Alberta, Canada.

Hayduk, W., and Minhas, B. (1982). Correlations for Prediction of Molecular Diffusivities in Liquids. *The Canadian Journal of Chemical Engineering,* 60 (2):295-299.

Hosseini Pour, A., Azin, R., and Yunan, M. (2011). Comprehensive Study to Determine Optimum Solvent Composition for VAPEX In a Heavy Oil Reservoir. SPE Enhanced Oil Recovery Conference.

James, L. A., Rezaei, N., and Chatzis, I. (2008). VAPEX, Warm VAPEX and Hybrid VAPEX - The State of Enhanced Oil Recovery for In Situ Heavy Oils in Canada. *Journal of Canadian Petroleum Technology,* 47 (4).

Jiang, T. (2013). An Improved Vapour Solvent Injection Technique for Enhanced Heavy Oil Recovery, Faculty of Graduate Studies and Research, University of Regina.

Kazemi, H., Vestal, C. R., and Shank, D. G. (1978). An Efficient Multicomponent Numerical Simulator. *Society of Petroleum Engineers Journal,* 18 (5):355-368.

Kok, M., Yildirim, Y., and Akin, S. (2009). Application of vapor extraction (VAPEX) process on carbonate reservoirs. *Energy Sources, Part A,* 31 (5):377-386.

Lederer, E. L. (1933). Zur Kenntnis der Viskositaets-Temperatur-Funktion (Study of Viscosity Temperature Function). *Proceeding of World Pet. Cong.,* 526-528.

Lin, L., Zeng, F., and Gu, Y. (2014). A circular solvent chamber model for simulating the VAPEX heavy oil recovery process. *Journal of Petroleum Science and Engineering,* 118:27-39.

Luo, P., Kladkaew, N., Gu, Y., and Saiwan, C. (2007). Detection of the Onset of Asphaltene Precipitation in a Heavy Oil-Solvent System. *Canadian International Conference*, Calgary, Alberta, Canada.

Luo, P., Wang, X., Gu, Y., Zhang, H., and Moghadam, S. (2008). Asphaltene Precipitation and Its Effects on the Vapour Extraction (VAPEX) Heavy Oil Recovery Process. International thermal operations and heavy oil symposium, Calgary, Alberta, Canada.

Mei, H., Zhang, M., Kong, X., Wu, Q., Sun, L., and Li, S. (1998). A K-Value Compositional Model for a Retrograde Condensate Reservoir. *SPE Gas Technology Symposium*, 15-18 March 1998, Calgary, Alberta, Canada.

Moghadam, S., Nobakht, M., and Gu, Y. (2009). Theoretical And Physical Modeling of a Solvent Vapour Extraction (VAPEX) Process for Heavy Oil Recovery. *Journal of Petroleum Science and Engineering,* 65 (1):93-104.

Moosavi, M. H., Naseri, A., Nasrollahi, M., Amraee, H., Nalchi, M., Masoudnia, A., and Shojaei, G. 2005. KEM Reservoir. Tehran, Iran

Nghiem, L. X., Kohse, B. F., and Sammon, P. H. (2001). Compositional Simulation of the VAPEX Process. *Journal of Canadian Petroleum Technology,* 40 (8).

Nourozieh, H., Kariznovi, M., Abedi, J., and Chen, Z. (2009). A New Approach to Simulation of the Boundary Layer in the Vapor Extraction Process. *Canadian International Petroleum Conference*, Jun 16 - 18, Calgary, Alberta.

Perkins, T. K., and Johnston, O. C. (1963). A Review of Diffusion and Dispersion in Porous Media. *Society of Petroleum Engineers Journal*, 70-84.

Pourabdollah, K., Mokhtari, B., and Kharrat, R. (2012). The VAPEX Process, from Beginning Up to Date. *Fuel*, 107 (5).

Rasti, F., Masihi, M., and Kharrat, R. (2011). The Semi-Analytical Modeling and Simulation of the VAPEX Process of "Kuh-e-Mond" Heavy Oil Reservoir. *Petroleum Science and Technology,* 29 (5):535-548.

Sabeti, M., Ehsani, M. R., Katebifar, S., and Delparish, H. (2014). Simple and Confident Approach in Modeling Miscible Flow in Petroleum Reservoirs. *The 8th International Chemical Engineering Congress & Exhibition* (IChEC 2014), Kish, Iran.

Sabeti, M., Nikookar, M., Ehsani, M. R., and Mohammadi, A. H. (2014). Numerical modelling of Solvent Injection into Heavy Oil Reservoirs: In: *Handbook on Oil Production Research,* Nova Science Publishers, Inc.

Shi, J., and Leung, J. Y. (2014). Physics-based proxy modelling of solvent transport in VAPEX process. *The Canadian Journal of Chemical Engineering,* 92 (8):1467-1480.

Shu, W. R. (1984). A Viscosity Correlation for Mixtures of Heavy Oil, Bitumen, and Petroleum Fractions. *Society of Petroleum Engineers Journal*, 24 (3):277-282.

Tam, S. (2007). VAPEX Experiments in an Annular Packing of Glass Beads and the Numerical Simulation of VAPEX using Comsol®.

Young, L. C., and Stephenson, R. E. (1983). A Generalized Compositional Approach for Reservoir Simulation. *Society of Petroleum Engineers Journal*, 23 (5):727-742.

In: Enhanced Oil Recovery
Editor: Alicia Knight
ISBN: 978-1-63463-917-0

Chapter 9

IMPACT OF VARIOUS BOUNDARY CONDITIONS ON THE SPONTANEOUS IMBIBITION MECHANISM DURING HEAVY OIL RECOVERY PROCESS THROUGH A FINITE DIFFERENCE SCHEME

***Mojgan Ebrahiminejad*[1], *Mohammad R. Ehsani*[1], *Mohammad Nikookar*[2,*], *Leili Sahranavard*[2] and *Amir H. Mohammadi*[3,4,†]**

[1]Department of Chemical Engineering, Isfahan University of Technology, Isfahan, Iran
[2]Department of Chemical Engineering, Tarbiat Modares University, Tehran, Iran
[3]Institut de Recherche en Génie Chimique et Pétrolier (IRGCP), Paris Cedex, France
[4]Thermodynamics Research Unit, School of Engineering, University of KwaZulu-Natal, Durban, South Africa

ABSTRACT

Spontaneous imbibition of water into the matrix blocks is an important mechanism for crude oil recovery from fractured reservoirs. This paper presents a numerical modeling of imbibition phenomenon when water is imbibed by capillary forces into an oil-saturated cylindrical core for studying the effect of boundary condition and shape factor on the spontaneous imbibition. In this modeling work, we assume that imbibition is a diffusion process and gravity forces are neglected. We also consider various boundary conditions. Finite difference implicit method is used to solve the spontaneous imbibition equation. Accuracy of the modeling is investigated by comparing the modeling results and the experimental data. Then, imbibition in various directions (r and z) is studied. By

* Corresponding author: M. Nikookar. Department of Chemical Engineering, Tarbiat Modares University, Tehran, Iran. E-mail: mohamad_ni@yahoo.com.

† Corresponding author: A. H. Mohammadi. Institut de Recherche en Génie Chimique et Pétrolier (IRGCP), Paris Cedex, France. Thermodynamics Research Unit, School of Engineering, University of KwaZulu-Natal, Howard College Campus, King George V Avenue, Durban 4041, South Africa. E-mail: a.h.m@irgcp.fr and amir_h_mohammadi@yahoo.com.

using the proposed model, the most effective boundary conditions for achieving the maximum oil recovery can be determined.

Keywords: Spontaneous imbibition, oil recovery, capillary diffusivity coefficient, boundary conditions, EOR

NOMENCLATURE

Ai	Surface area open to imbibition in the i-th direction
CDC	Capillary diffusivity coefficient
CY	Cylindrical core sample
D	Capillary diffusion coefficient(m^2/s)
# FC	Number of faces coated with polyester
Fs	Shape factor
i	Integer denoting cell location in x-directions
k	Integer denoting cell location in z- directions
K	Absolute permeability (m^2)
k_h	Permeability to fluid h (m^2)
k_{ro}	Relative permeability to oil
k_{rw}	Relative permeability to water
L_C	Characteristic length
m	Integer denoting cell location in r- directions.
n	Integer indicating time level
P_C	Capillary pressure (pa)
P_h	fluid h (oil or water) pressure (pa)
P_o	Oil pressure (pa)
P_w	Water pressure (pa)
R	Oil recovery
R_∞	Oil recovery at infinite time
r	Coordinate in r-direction
S	Normalized water saturation (fraction,m^3/m^3)
SAi	Distance from the open surface to the center of the matrix (m)
S_{av}	Average oil saturation at time t (fraction,m3/m3)
S_o	Oil saturation at time t in position (r,z) (fraction,m3/m3)
S_{or}	Residual oil saturation in the matrix (fraction,m3/m3)
S_w	Water saturation at time t in position (r,z) (fraction,m3/m3)
S_{wi}	Initial water saturation in the matrix (fraction,m3/m3)
t	Imbibition time(s)
u_{hx}	Fluid (oil or water) velocity in x-direction
u_{or}	Oil velocity in r-direction (m/s)
u_{oz}	Oil velocity in z-direction (m/s)
u_t	Total velocity (m/s)
u_{wr}	Water velocity in r direction (m/s)
u_{wz}	Water velocity in z direction (m/s)
V	Volume of the matrix (m^3)

x	Coordinate in x-direction
z	Coordinate in z-direction
σ	Interfacial tension (mN/m)
$\emptyset$	Matrix porosity (m^3/m^3)
μ_h	Fluid h viscosity (pa.s)
μ_o	Oil viscosity (Pa.s)
μ_w	Water viscosity (Pa.s)

INTRODUCTION

As matter of fact, a lot of researches have been done to improve recovering heavy oil and improve enhanced oil recovery methods considering increasing global oil demand, declining production of conventional oil reservoirs and also increasing oil prices in recent years. Furthermore, fractured reservoirs are more than 20 percent of the world's oil reservoirs, consequently researchers are examining fractured reservoirs.

As a result, the fluids flow in fractured reservoirs is complex as the fluid can flow in the matrix or fracture. Consequently, there are considerable differences between the fractured reservoirs and conventional reservoirs from oil producing point of view. The matrix has high storage capacity and low permeability. The second zone is fracture with low capacity and high permeability. When the matrix is saturated with non-wetting phase and wetting-phase is in the fracture, the capillary and gravity forces cause displacement of wetting-phase into the matrix network, this process is called imbibition. If the reservoir is hydrophilic and there is enough water in the fracture zone, strong capillary forces cause water as the wetting–phase to imbibe into the matrix and oil is expelled from the matrix to the fracture.

There are several mechanisms for oil recovery from natural fractured reservoirs with low permeability matrix; Spontaneous imbibition is the most important mechanism for oil recovery from such reservoirs. Ability of matrix to imbibe water spontaneously is dependent on its hydrophilic property. The rate of spontaneous imbibition is dependent on the pore structure, matrix permeability, matrix wettability and interfacial tension between the wetting and non-wetting fluids (Kinjal et al., 2012).

Imbibition can occur by co-current and counter-current flows. In the co-current imbibition, water and oil flow in the same direction; but in the counter-current imbibition, oil and water flow in the opposite directions. Co-current imbibition is faster and can be more effective than counter-current imbibition, but in the most cases, counter-current flow is the dominant mechanism.

Experimentally, imbibition can be investigated by surrounding a saturated core sample with water and measuring the oil recovery at different times (Behbahani et al., 2005). The imbibition process is an important mechanism for oil recovery from hydrophilic fractured reservoirs; so many efforts have been made to investigate the effective parameters on this process (Takahashi, 2009). Fractured reservoirs are often modeled as a combination of narrow fractured channel and the matrix.

Blair (1964) modeled numerically the counter-current imbibition process and concluded that the rate of imbibition is a function of capillary pressure, relative permeability, oil viscosity and initial water saturation (Blair, 1964). Beckner et al. (1987) modeled the imbibition process as a diffusion process. He assumed that diffusion coefficient is nonlinear and the total velocity is zero and also gravity forces are ignored (Beckner et al., 1987).

Bech et al. (1989) and Durta and Aziz (1991) used the diffusion equation for their modeling (Bech et al., 1989; Durta and Aziz 1991). The nonlinear diffusion equation is solved numerically by Chen et al. (1995) using a semi-implicit approach (Chen et al., 1995). Baker and Wilson (1997) modeled the fracture and matrix system in the radial coordinate. They compared their results with the experimental data, and concluded that imbibition process is a function of both relative permeability and capillary pressure (Baker and Wilson, 1997).

Several analytical models have been presented for prediction of oil recovery by imbibition process. Arnofsky et al. proposed an empirical model using the following equation:

$$R = R_\infty (1 - e^{-\alpha t}) \tag{1}$$

where, R is the oil recovery, $R_\infty = (1 - S_{wi} - S_{or}).\emptyset_m$ is the oil recovery at infinite time, S_{wi} is the initial water saturation in the matrix, S_{or} is the residual oil saturation in the matrix, Φ_m is the matrix porosity, α is a constant and equal to the required time for achievement of 63% oil recovery, and t is time (Takahashi, 2009).

The Aronofsky's model was improved by Kazemi et al. (1992) using multi-exponential function as below (Kazemi et al., 1992):

$$R = R_\infty(1 - R_1 e^{-a_1 t} - R_2 e^{-a_2 t} - \cdots) \tag{2}$$

$$\mathrm{R}_1 + \mathrm{R}_2 + \cdots = 1 \tag{3}$$

where, R_1, R_2, ... and α_1, α_2,... are empirical constants. Two or three terms given in Equation (2) are sufficient for the convergence of the empirical results. Durta and Aziz (1991) have proposed similar exponential equations (Durta and Aziz, 1991). Oil recovery in the imbibition process can also be predicted as a function of dimensionless times, the following equations are proposed for the dimensionless time by Mattax and Kyte (1962):

$$t_d = t\sqrt{\frac{K}{\emptyset}\frac{\sigma}{\mu_w L_c^2}} \tag{4}$$

$$R = R_\infty(1 - e^{-at_d}) \tag{5}$$

where, t is time, K is absolute permeability, $\emptyset$ is porosity, σ is interfacial tension, μ_w is water viscosity and L_c is a characteristic length (Mattax and Kyte, 1962).The characteristic length is modified by Kazemi et al. (1992) as below

$$L_s = \frac{1}{\sqrt{F_s}} = \sqrt{\frac{V_{matrix}}{\sum_{i=1}^{n}\frac{A_i}{S_{Ai}}}} \tag{6}$$

where, F_s is the shape factor, A_i is the surface area in the i-th direction, S_{Ai} is the distance from the surface to the center of the matrix block and V_{matrix} is the matrix volume (Takahashi, 2009; Standnes, 2010; Yildiz et al., 2006).

The dimensionless time is modified by using the geometric mean of water and oil viscosities instead of water viscosity by Ma et al. (1995-1997) (Takahashi, 2009). Kashchiev and Firoozabadi (2003) showed that capillary diffusivity coefficient (CDC) is a function of water saturation and CDC curve versus S is a bell-shaped curve with a maximum approximately at S=0.5 (Kashchiev and Firoozabadi, 2003). The value of CDC approaches nearly zero at high and small values of normalized water saturation.

For median values of the normalized water saturation, using a constant CDC is a better assumption, so it is expected that the maximum difference between the experimental and numerical results is in the saturation values close to unity (Standnes, 2006).

In this study, a two phase, two-dimensional model is presented for study of spontaneous imbibition process in the fractured reservoirs. In this model, we assume that the imbibition is a diffusion process and capillary diffusivity coefficient (CDC) is constant and is independent of water saturation and different boundary conditions are also considered. As a result, the aim of present study is to investigate various boundary conditions.

Therefore, CDC can be assumed constant with changing the permeable surfaces. Without considering early and late times, the assumption is almost correct which has been examined in cylindrical cores by Standnes (2005); (Standnes, 2004; Standnes, 2006). The two-phase mass balance equations for water and oil using Darcy's equation and other auxiliary equations are solved simultaneously by finite difference method. The accuracy of this model is studied by comparing the results with some experimental data.

Then, this model is used for studying the effect of some important parameters on imbibition performance in the fractured reservoir.

MATHEMATICAL MODELING

A mathematical formulation of the spontaneously imbibition process is presented here. Assumptions used in deriving the governing equations are as below:

1 Gravity terms are neglected
2 Water and oil are incompressible fluids
3 Core is incompressible
4 Capillary pressure is the only driving force
5 The total velocity is zero

The model consists of two main equations containing of conservation mass equation for water and oil. Conservation mass equation for each phase (oil and water) in r and z directions in a cylindrical core (Figure 1) is given as below:

$$\emptyset\frac{\partial(S_w)}{\partial t}+\frac{1}{r}\frac{\partial(u_{wr}.r)}{\partial r}+\frac{\partial(u_{wz})}{\partial z}=0 \tag{7}$$

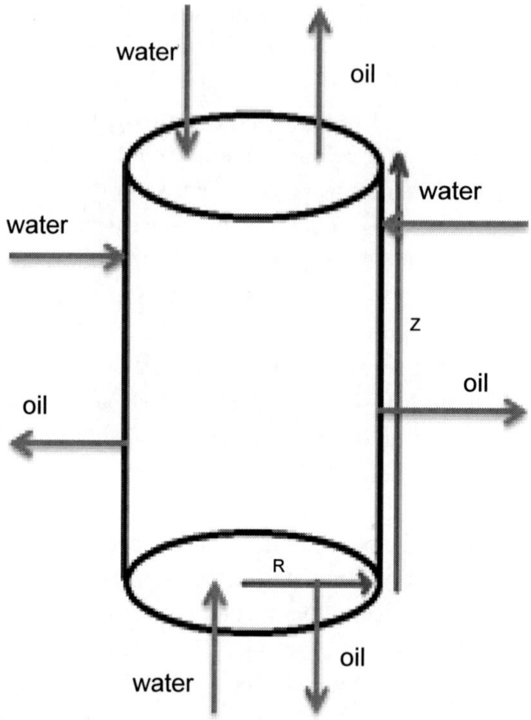

Figure 1. Core geometry for the spontaneous imbibition when all faces are open.

$$\emptyset\frac{\partial(S_o)}{\partial t}+\frac{1}{r}\frac{\partial(u_{or}.r)}{\partial r}+\frac{\partial(u_{oz})}{\partial z}=0 \tag{8}$$

Water and oil velocity can be obtained using Darcy's law as follow:

$$u_{hx}=\frac{-kh}{\mu h}\frac{\delta ph}{\delta x} \tag{9}$$

The total velocity for counter-current imbibition process is zero:

$$u_t=0 \tag{10}$$

$$u_{or}+u_{wr}=0,\ u_{oz}+u_{wz}=0 \tag{11}$$

By inserting Darcy's equation in the equation (11):

$$\frac{k_o}{\mu_o}\frac{\partial P_o}{\partial r}+\frac{k_w}{\mu_w}\frac{\partial P_w}{\partial r}=0 \tag{12}$$

Some auxiliary equations are used to reduce variables, such as the correlation of fluid saturations in the reservoir as follow:

$$S_o + S_w = 1 \tag{13}$$

Other auxiliary equations are the correlation between capillary pressure and water saturation that can be used to replace the capillary pressure by saturation equations as follows:

$$P_{cw} = P_o - P_w = f(S_w) \tag{14}$$

where,

$$\frac{\partial P_o}{\partial x} = \frac{\partial P_c}{\partial r} + \frac{\partial P_w}{\partial r} \tag{15}$$

By inserting Equation (13) into Equation (11):

$$\frac{\partial P_w}{\partial r} = -\frac{\frac{k_{or}}{\mu_o}}{\frac{k_{or}}{\mu_o} + \frac{k_{wr}}{\mu_w}} \cdot \frac{\partial P_c}{\partial r} \tag{16}$$

By substituting Equation (16) into the Darcy's equation, the oil and water velocities are obtained as function of capillary pressure.

$$u_{wr} = K.\frac{\frac{k_{or}}{\mu_o} \cdot \frac{k_{wr}}{\mu_w}}{\frac{k_{or}}{\mu_o} + \frac{k_{wr}}{\mu_w}} \cdot \frac{\partial P_c}{\partial S_w} \cdot \frac{\partial S_w}{\partial r} \tag{17}$$

$$u_{wz} = K.\frac{\frac{k_{or}}{\mu_o} \cdot \frac{k_{wr}}{\mu_w}}{\frac{k_{or}}{\mu_o} + \frac{k_{wr}}{\mu_w}} \cdot \frac{\partial P_c}{\partial S_w} \cdot \frac{\partial S_w}{\partial z} \tag{18}$$

Finally, by inserting Equation (17) and Equation (18) into Equation (7), the governing equation for the counter-current imbibition will be obtained as follows:

$$\emptyset \frac{\partial}{\partial t}(S_w) + \frac{1}{r}\frac{\partial}{\partial r}\left(\frac{K.k_{or}}{\mu_o} \cdot \frac{1}{1 + k_{ro} \cdot \mu_w \cdot \frac{1}{k_{rw} \cdot \mu_o}} \cdot \frac{\partial P_c}{\partial S_w} \cdot \frac{\partial S_w}{\partial r} . r\right) + \frac{\partial}{\partial z}\left(\frac{K.k_{or}}{\mu_o} \cdot \frac{1}{1 + k_{ro} \cdot \mu_w \cdot \frac{1}{k_{rw} \cdot \mu_o}} \cdot \frac{\partial P_c}{\partial S_w} \cdot \frac{\partial S_w}{\partial z}\right) = 0 \tag{19}$$

The capillary diffusivity coefficient (CDC) is defined as:

$$D(S_W) = -\frac{Kk_{ro}}{\mu_o}\frac{1}{1+\frac{k_{ro}\,\mu_w}{k_{rw}\,\mu_o}}\frac{dP_C}{dS_w} \quad (20)$$

D is a non-linear function of water saturation. We define the normalized water saturation as below:

$$S = f(r,z,t) = \frac{S_w - S_{wi}}{1 - S_{or} - S_{wi}} \quad (21)$$

S is the ratio of recovered oil to the total recoverable oil, thus S is a dimensionless parameter; if S is used in above equations, final equations will not be sensitive to initial water saturation and S will always change from 0 to 1. By inserting the parameter S into the Equation (7), Equation (19) converts into the following correlation:

$$\emptyset\frac{\partial}{\partial t}(S) + \frac{1}{r}\frac{\partial}{\partial r}\left(D.\frac{\partial S}{\partial r}.r\right) + \frac{\partial}{\partial z}\left(D.\frac{\partial S}{\partial z}\right) = 0 \quad (22)$$

Equation (22) will be solved and S(r,z,t) will be calculated by using numerical method.

SOLUTION METHOD

This equation is discretized using finite difference method. The first order spatial differential equations are presented as below:

$$\frac{\partial f}{\partial x} \approx \frac{f_{i+1} - f_{i-1}}{2\Delta x} \quad (23)$$

For time differential equation, forward finite difference form is used as follow:

$$\frac{\partial f}{\partial t} \approx \frac{f^{n+1} - f^{n}}{\Delta t} = \frac{\delta f}{\Delta t} \quad (24)$$

By using the above correlations, equation (22) is discretized as below:

$$\emptyset.\frac{S(m,k)^{n+1} - S(m,k)^{n}}{\Delta t} + \frac{D_{m+\frac{1}{2},k}.(S(m+1,k)^{n+1} - S(m,k)^{n+1})}{h_r^2}$$
$$+\frac{D_{m-1/2,k}.(S(m-1,k)^{n+1} - S(m,k)^{n+1})}{h_r^2}$$
$$+\frac{D_{m,k+1/2}.(S(m,k+1)^{n+1} - S(m,k)^{n+1})}{h_z^2}+$$

$$\frac{D_{m,k-\frac{1}{2}} \cdot (S(m,k-1)^{n+1} - S(m,k)^{n+1})}{h_z^2} +$$

$$\frac{1}{(m-1)h_r} . D . \frac{S(m+1,k)^{n+1} - S(m-1,k)^{n+1}}{2h_r} = 0 \quad (25)$$

$$D . \frac{S(m+1,k)^{n+1} - S(m-1,k)^{n+1}}{2h_r} \quad (26)$$

Equation (26) is valid for the internal blocks. At the permeable surfaces in contact with water, boundary conditions are set to be $S_w = 1 - S_{or}$, therefore $S=\frac{(1-S_{or})-S_{wi}}{1-S_{or}-S_{wi}}=1$. The blocks of impermeable surfaces are unknown, thus the mass balance must be written for impermeable surfaces. Initial conditions are equal to initial water saturation hence $S_w = S_{wi}$ and $S=\frac{S_{wi}-S_{wi}}{1-S_{or}-S_{wi}}=0$.

Results and Discussion

In this work, 2-D finite difference method is used and finally, counter-current imbibition is simulated using MATLAB program. If the core is fully saturated with oil and S_{wi}=0, S_w is a parameter that will indicate imbibed water saturation to each control volume of core at different times. The volume of imbibed water is equal to recovered oil from core. In this simulation, the fluid properties, core data and experimental data of Standnes are used (Tables 1 and 2) (Standnes, 2004; Standnes, 2006).

The numerical results are compared with the experimental data of oil recovery for the cylindrical samples, while the flow is in the radial direction in Figures 2-5 to examine the validity of the proposed method (Standnes, 2004; Standnes, 2006).

These figures show a little difference between numerical and experimental results. This issue is resulted by our assumption (considering constant CDC). After validation; the proposed model is used for studying the effect of core size and boundary conditions on the oil recovery.

Table 1. Core data form imbibition test (Standnes, 2004; Standnes, 2006)

Test	Sample shape boundary condition	Diameter and radius(m)	Height(m)	Porosity %	Capillary diffusion coefficient(m2/s)
3	CY-2FC	0.0199 0.00995	0.05	43	4.5 x 10^{-8}
4	CY-2FC	0.0339 0.01695	0.0496	42.6	3 x 10^{-8}
5	CY-2FC	0.06 0.03	0.032	42.4	4.x 10^{-8}
6	CY-2FC	0.1 0.05	0.0397	43.2	2.8 x 10^{-8}

Table 2. Fluid properties (Standnes, 2004; Standnes, 2006)

Liquid phase	Density at 20°C (g/cm3)	Viscosity at 20 C (mPa. s)	IFT using decane as oil phase at 20 C (mN/m)
Distilled water	0.9982	1.0	46.0
n-Decane	0.731	0.95	-

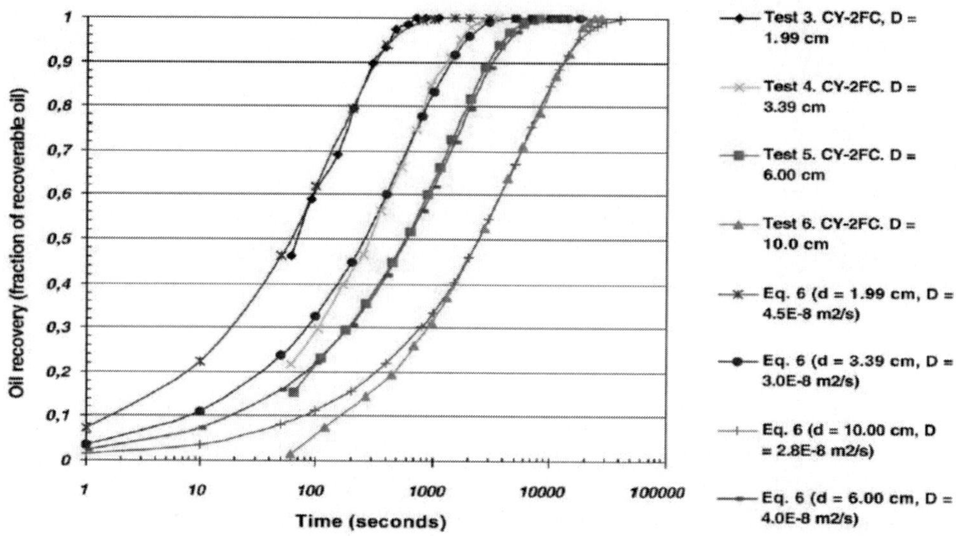

Figure 2. Comparison of experimental and simulated oil recovery curves for cylindrical cores having the top and bottom faces closed for fluid exchange (logarithmic time scale) (Standnes, 2006).

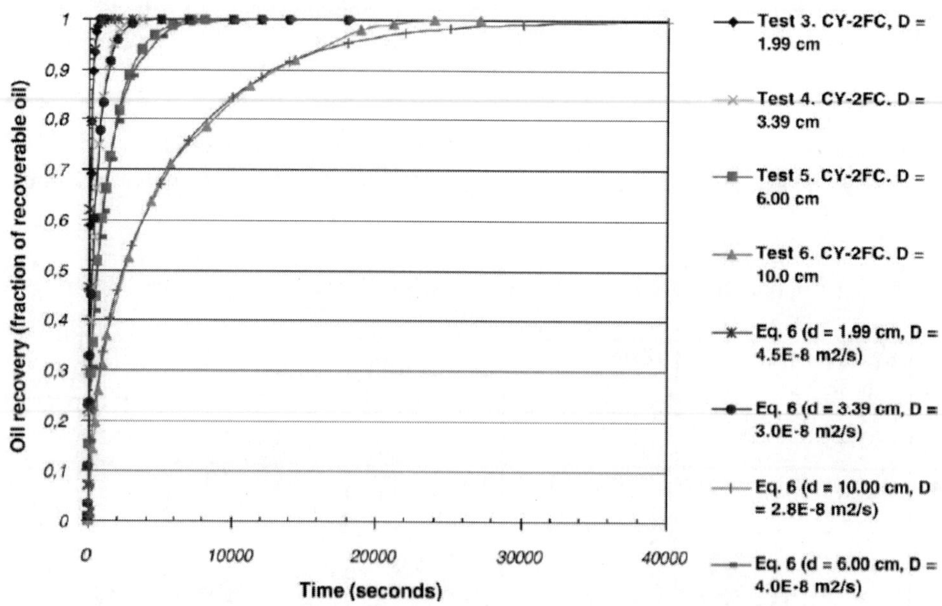

Figure 3. Comparison of experimental and simulated oil recovery curves for cylindrical cores having the top and bottom faces closed for fluid exchange (linear time scale) (Standnes, 2006).

Figures 4 and 5 indicate only the results of numerical method, when flow is only in r direction. As can be seen, with increasing core diameter, the imbibition rate has declined and the time of maximum production (S=1) has also increased.

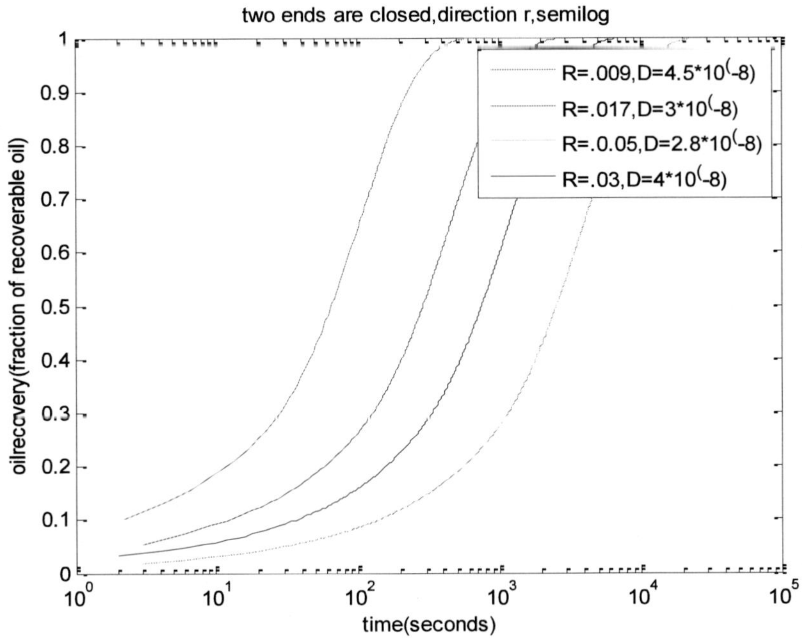

Figure 4. Oil recovery curves predicted by model for cylindrical cores having the top and bottom faces closed for fluid exchange (logarithmic time scale).

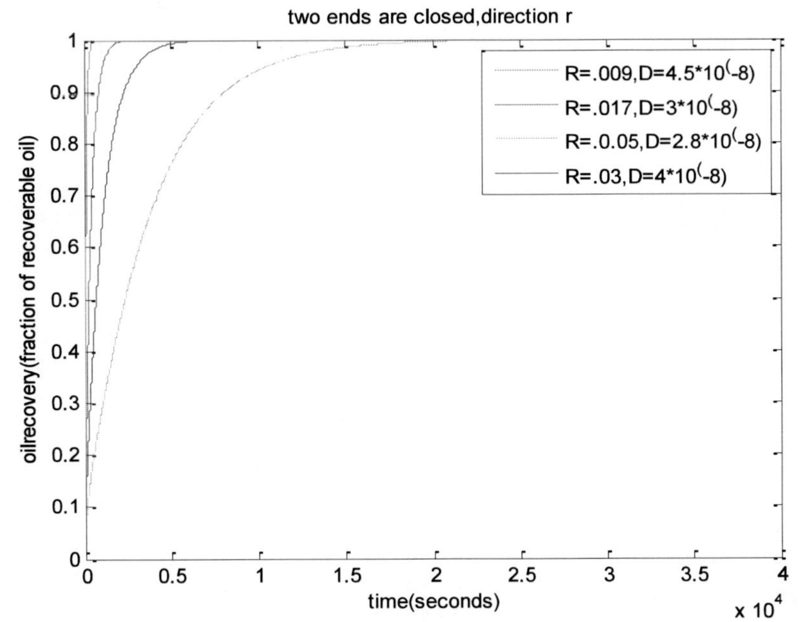

Figure 5. Oil recovery curves predicted by model for cylindrical cores having the top and bottom faces closed for fluid exchange (linear time scale).

Figure 6 shows that if the fluids flow only in the z direction, the cores have the top and bottom faces opened for transmitting fluids. According to Figure 6, with increasing core height, the imbibition rate has decreased. In Figure 6, the blue and red lines are for the cores with the same height and various radiuses. In this case, with decreasing the core radius, imbibition rate has increased.

Figure 7 shows the oil recovery for cores with only their top or bottom face open for transmitting fluids. In this case, the imbibition rate has significantly declined compared to the previous case. Thus, if the core has the top and bottom faces open to the flow, the imbibition rate will increase for several times. According to the Figures 4 and 5, for cores with small radius (0.09-0.017), the recovery rate in r direction is more than in z direction, but for cores with higher diameter and height, the recovery rate in z direction is more. Hence, we cannot conclude that in which direction, the recovery will increase and it is dependent on the radius and height of each sample.

Figure 8 shows oil recovery for cores with their top or bottom faces closed and other faces open for transmitting fluid and figures 9 and 10 show oil recoveries for cores with all faces open for exchanging fluid. According to these figures, generally the imbibition rate increases with increasing the permeable surfaces in contact with water, but increasing rate for each core is different.

Figure 11 shows oil recovery for a core with d=10 cm at different boundary conditions and shows the effect of each permeable face on the oil recovery. Therefore, the most effective face for fluid exchanging can be determined and consequently, the maximum oil recovery will be achieved within the minimum time.

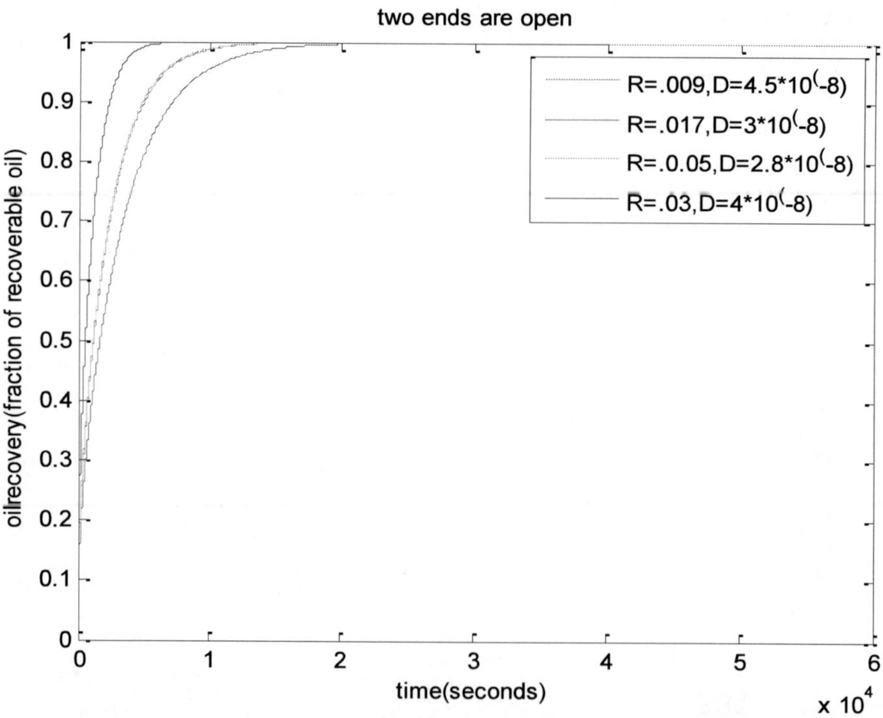

Figure 6. Oil recovery curves predicted by model for cylindrical cores having the top and bottom faces opened for fluid exchange (linear time scale).

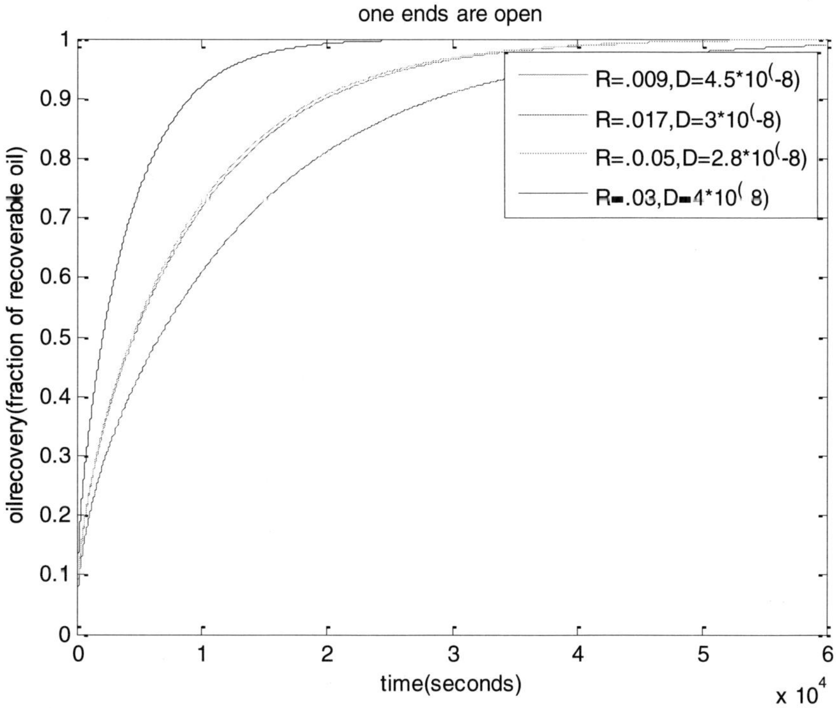

Figure 7. Oil recovery curves predicted by model for cores that have only the top or bottom face opened for transmitting fluid.

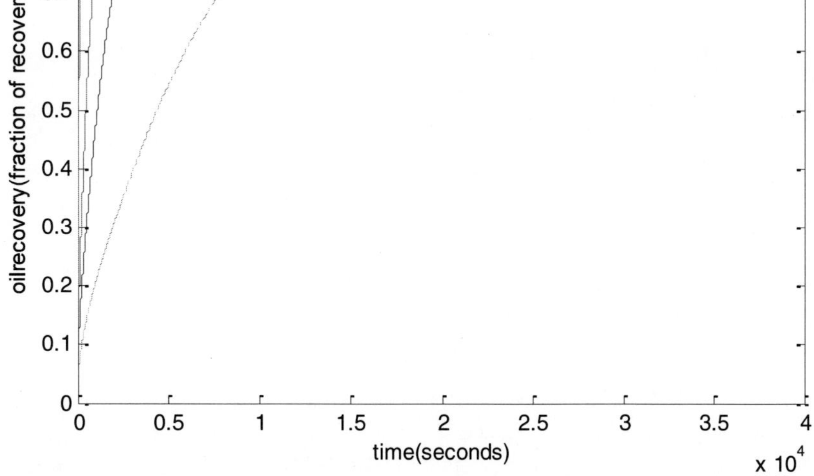

Figure 8. Oil recovery curves predicted by model for cores that have the top or bottom face closed and other faces are open for transmitting fluid.

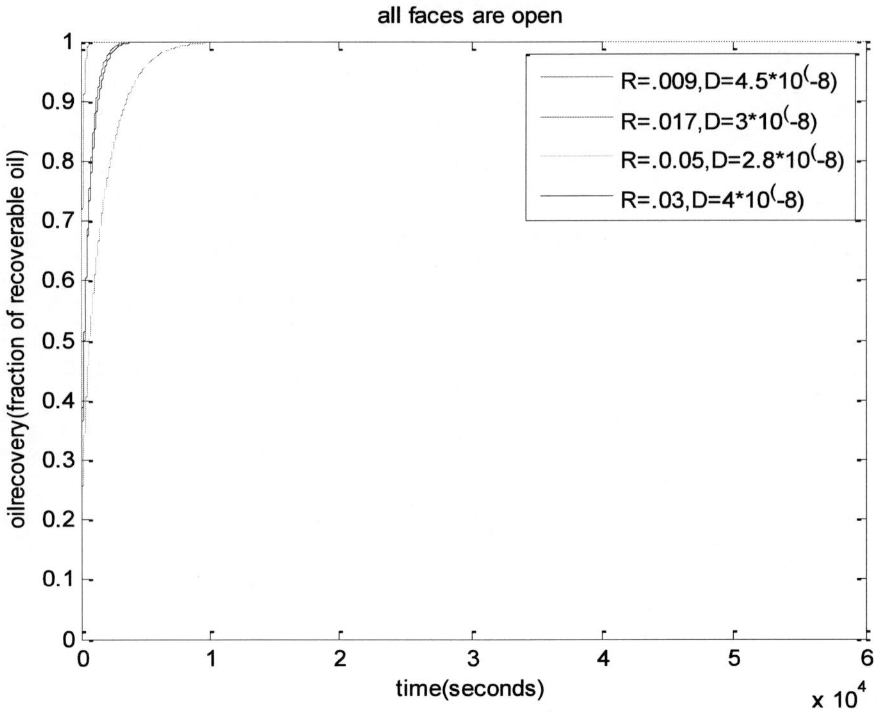

Figure 9. Oil recovery curves predicted by model for cores that all faces are open for exchanging fluid (linear time scale).

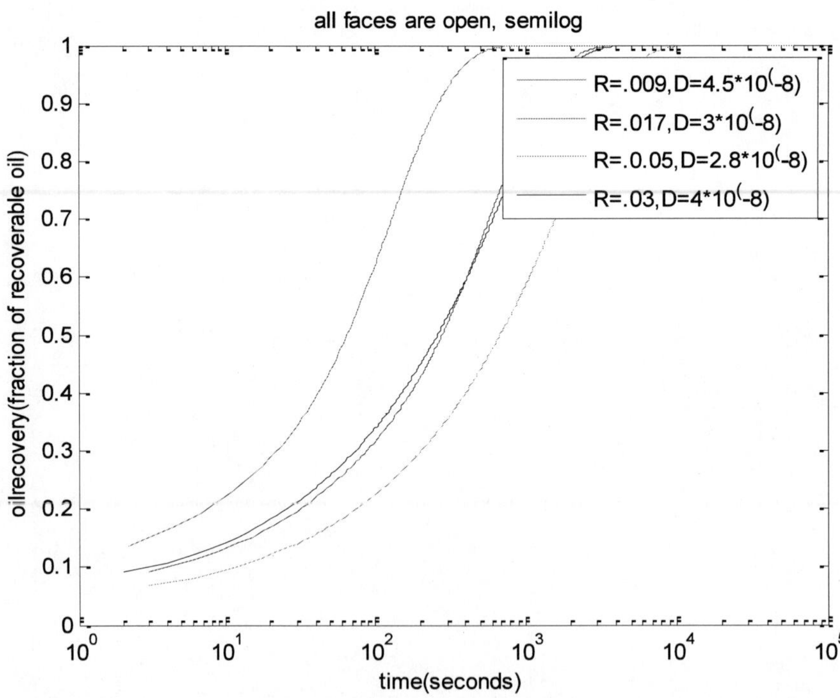

Figure 10. Oil recovery curves predicted by model for cores that all faces are open for exchanging fluid (logarithmic time scale).

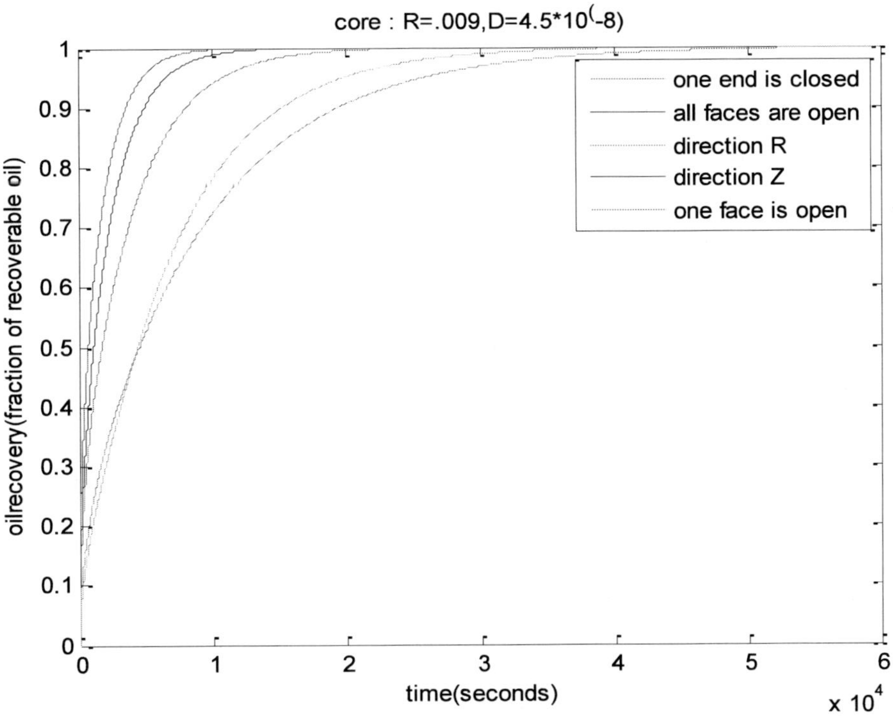

Figure 11. Oil recovery fraction predicted by model for a core with d=10 cm at different boundary conditions.

Figure 12 shows normalized water saturation in a core with d=10 cm within a specific time. The fluids flow in r and z directions and all faces are opened for exchanging fluid. There is symmetry in the z direction, because the gravity is neglected. This figure shows that with increasing distance from the permeable faces, water saturation in the core decreases. Water saturation is a function of S_{wi} and wettability.

CONCLUSION

1 In the cores with small radius (0.09-0.017), the recovery rate in r direction is more than in z direction, but for cores with higher diameter and height, the recovery rate in z direction is more. Thus, we cannot conclude that in which direction, the recovery will increase and it is dependent on the radius and height of each sample.
2 Generally, the imbibition rate increases with increasing the permeable surfaces in contact with water
3 The problems related to the measurement of recovered oil at early time is high; also assumption of constant CDC is not correct at early time; therefore, comparison between experimental and numerical results will be difficult at early time of imbibition process.

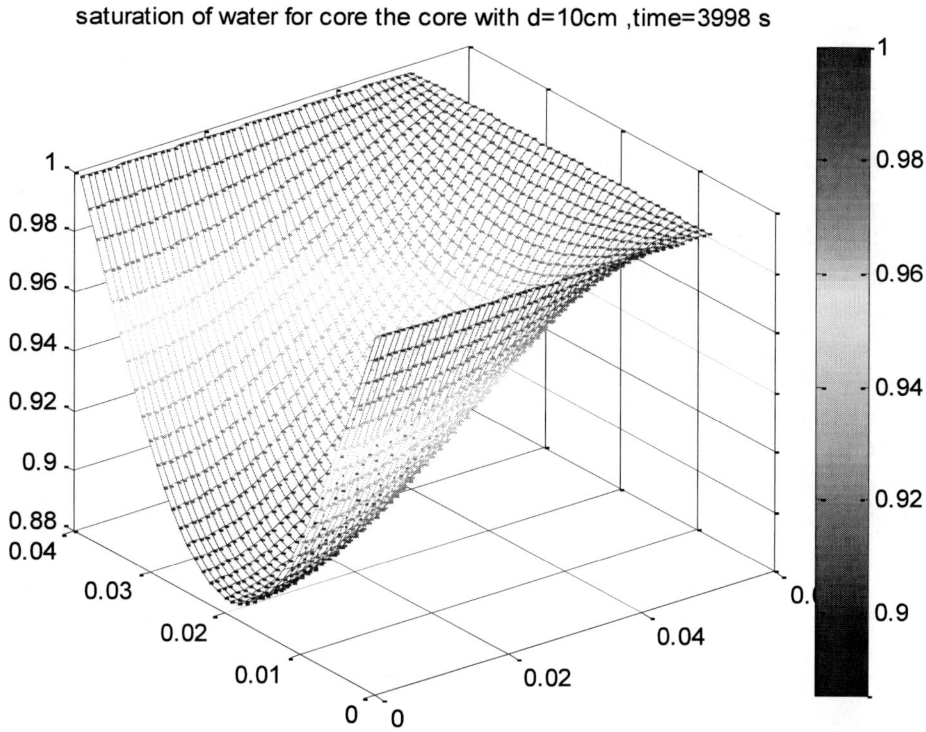

Figure 12. Normalized water saturation profile from a core with d=10 cm for a specific time.

REFERENCES

Baker R., G. Wilson, *"Numerical Simulation of Laboratory Scale Imbibition Experiment" Internal Report*, Epic Consultant Services Ltd, (1997).

Bech N., O. K. Jenson, B. Nielsen, "Modeling of Gravity-Imbibition and Gravity-Drainage Processes: Analytic and Numerical Solutions", *SPE Reservoir Engineering 18428-PA*, Volume 6. Number 1 (1991).

Beckner B. L., K. Ishimoto, S. Yamaguchi, A. Firoozabadi, K. Azis, *"Imbibition-Dominated Matrix-Fracture Fluid Transfer in Dual Porosity Simulators," presented at the SPE Annual Technical Conference and Exhibition, Dallas*. Texas, 27-30 September, 1987.

Behbahani H., G. Donato, M. Blunt, "Simulation of Counter-Current Imbibition in Water-wet Fractured Reservoirs", *Journal of Petroleum Science and Engineering*. 50, (2005), 21-39.

Blair P. M., Humble Oil and Refining Co, "Calculation of Oil Displacement by Countercurrent Water Imbibition", *SPE Journal 873-PA*, Volume 4. Number 3, (1964).

Chen J., M. A. Miller, K. Sepehrnoori, *"Theoretical Investigation of Countercurrent Imbibition in Fractured Reservoir Matrix Blocks", presented at the Symposium on Reservoir Simulation, San Antonio*. Texas, 12-15 February, 1995.

Durta T. V., K. Aziz: "A New Double-Porosity Reservoir Model for Oil/Water Flow Problems," *SPE Reservoir Engineering 21248-PA*, Volume 7. Number 4 (1991).

Kashchiev D., A. Firoozabadi, "Analytical Solution for 1-D Countercurrent Imbibition in Water-Wet Media", *SPE Journal 87333-PA*, Volume 8. Number 4 (2003).

Kazemi H., J. Gilman, A. Elsharkawy, "Analytical and Numerical Solution of Oil Recovery from Fractured Reservoirs with Empirical Transfer Function (includes associated papers 25528 and 25818)", *SPE Reservoir Engineering 19849-PA*, Volume 7. Number 2, (1992).

Kinjal R. P., N. M. Manoj, R. P. Twinkle, "A Mathematical Model of Imbibition Phenomenon in Heterogeneous Porous Media During Secondary Oil Recovery Process", *Applied Mathematical Modelling* (2012).

Mattax C., J. Kyte, "Imbibition Oil Recovery from Fractured, Water-Drive Reservoirs", Reservoirs", *SPE Journal 187-PA,* Volume 2. Number 2, (1962).

Standnes D. C., "Spontaneous imbibition of water into cylindrical cores with high aspect ratio: Numerical and experimental results", *Journal of Petroleum Science and Engineering*. 50, (2006), 151-160.

Standnes D. C., "Experimental study of the impact of boundary conditions on oil recovery by co-current and counter-current spontaneous imbibition", *Energy Fuels* 18, (2004), 271-282.

Standnes D. C., "Scaling Group for Spontaneous Imbibition Including Gravity", *Energy Fuels*. 24, (2010), 2980-2984.

Takahashi S., *"Water Imbibition, Electrical Surface Forces, and Wettability of Low Permeability Fractured Porous Media", dissertation submitted to the department of energy resources engineering and the committee on graduate studies of Stanford university, (2009).*

Yildiz H. O., M. Gokmen, Y. Cesur, "Effect of shape factor, characteristic length, and boundary conditions on spontaneous imbibitions", *Journal of Petroleum Science and Engineering*. 53, (2006), 158-170.

In: Enhanced Oil Recovery
Editor: Alicia Knight
ISBN: 978-1-63463-917-0

Chapter 10

REDUCTION OF HEAVY OIL VISCOSITY USING NANOPARTICLES IN ENHANCED OIL RECOVERY PROCESS

S. Afzal,[1] Mohammad R. Ehsani,[1] Mohammad Nikookar,[2*] A. R. Khodabandeh-Shahraki,[2] E. Roayaei[2] and Amir. H. Mohammadi[3,4,*]

[1]Department of Chemical Engineering, Isfahan University of Technology, Isfahan, Iran
[2]IOR Research Institute, National Iranian Oil Company (NIOC), Tehran, Iran
[3]Institut de Recherche en Génie Chimique et Pétrolier (IRGCP), Paris Cedex, France
[4]Thermodynamics Research Unit, School of Engineering, University of KwaZulu-Natal, Howard College Campus, Durban, South Africa

ABSTRACT

The reduction of heavy oil viscosity is the most important issue in enhanced heavy oil recovery. The viscosity reduction should be achieved in a quick economical way. In this communication, the effect of different concentrations of various metal oxide including Fe_2O_3, NiO, CuO, ZnO, Al_2O_3, TiO_2 and WO_3 at various temperatures is studied on the heavy oil viscosity. The experimental tests show that some of these nanoparticles decrease the heavy oil viscosity less than 50% in certain concentrations at different temperatures. The reason of this viscosity reduction is that some of the nanoparticles act as a catalyst. Our study shows that NiO, Fe_2O_3 are the most effective nanoparticles for viscosity reduction.

Keywords: heavy oil, viscosity, nanoparticle, metal oxide, EOR

---

* Corresponding Authors: M. Nikookar: E-mail: mohamad_ni@yahoo.com & A.H. Mohammadi, E-mail: a.h.m@irgcp.fr & amir_h_mohammadi@yahoo.com.

NOMENCLATURES

A	Constant of Equation (2) (cp)
B	Constant of Equation (2) (1/K)
r	Correlation coefficient
T	Temperature (K)
μ	Total viscosity (cp)
μ_0	Viscosity of dispersion medium (cp)
Φ	Volume fraction of solid spheres

1. INTRODUCTION

High viscosity of heavy oil causes low primary production from heavy oil reservoirs. Thermal recovery methods are the best for increasing production from such kind of reservoirs, because thermal methods reduce the viscosity of heavy oil and increase its mobility and as a result increase their recovery [1]. However, the yield of thermal recovery is low.

Nanotechnology has already contributed significantly to the technological advances in industry, such as biomedical, materials and manufacturing industries and more recently in the petroleum industry. Nanotechnology has the potential to introduce revolutionary changes in several areas of oil and gas industry, for example exploration, production, enhanced oil recovery and refining [2, 3].

It has been shown that reduction of heavy oil viscosity is not only due to high temperature effect but also a series of chemical reactions. These reactions known as aqua-thermolysis change the physical properties [4]. Clark et al. (1990) found that when aqueous metal is added into the process, the heavy oil viscosity is further reduced in comparison with steam-only experiments [5]. Cacciola et al. (1993) investigated the catalytic effect of oil soluble naphthene on aqua-thermolysis, the experimental result showed that Ni, Pt, *etc.* have benign activity, selectivity and stability for dehydrogenation of cyclohexane and methylcyclohexane [6]. Fan et al. (2004) investigated the catalytic effects of minerals on aqua- thermolysis of heavy oils. They found that heavy oil of Liaohe oilfield could undergo aqua-thermolysis in the steam stimulation process, and reduced the viscosity and molecular weight of heavy oils. Their results showed that the reservoir rock had the catalytic roles on aqua-thermolysis. When added 10 wt% reservoir rock into the reaction system, they observed that the heavy oil viscosity reduced from 88.5 to 55.8 Pa.s, in 36.9wt% reservoir rock, and the average molecular reduced from 587 to 475, in 19.1% [7]. Fan et al. (2007) used ionic liquid to upgrade heavy oil. They found that ionic liquid, [(Et)NH][AlCl], could decrease viscosity, average molecular weight, and asphaltene content of the heavy oil. Their results showed that the bonds which were formed between ionic liquids and organic sulfur in heavy oil, are very weak due to weakens of C-S bonds [8]. Li et al. (2007) studied the effect of nano-nickel catalyst in viscosity reduction of Liahoextra-heavy oil by aqua-thermolysis reaction. They observed that this issue reduced viscosity, resin and asphaltene content of extra-heavy oil effectively [9]. Hascakir et al. (2008) investigated the effect of micron sized iron particles on heavy oil viscosity without steam treatment. They reported a noticeable viscosity reduction

[10]. Chen et al. (2009) studied viscosity reduction of nano-keggin-$K_3PMo_{12}O_{40}$ in catalytic aqua-thermolysis of heavy oil. They found that it changes in oxygen-containing groups mainly during the catalytic aqua- thermolysis [11]. Hamedi Shokrlu and Babadagli (2010) studied the effect of nano-sized metals on viscosity reduction of heavy oil/bitumen during thermal applications. They used different metal types such as iron, nickel and copper with different sizes. Their results showed that nano sized particles had a remarkable effect on heat transfer through heavy oil [4].

Enistein (1906) presented an equation for calculating of viscosity of a dilute liquid suspension of solid spheres (μ) as follows [5]:

$$\mu = \mu_0 (1+2.5\Phi) \quad (1)$$

where Φ is the volume fraction of solid spheres in the suspension and μ_0 is the viscosity of dispersion medium. As given by equation (1), the effect of interaction between particles and the crude oil may not be significant to overcome the increase of viscosity due to physical effects [5].

Aqua-thermolysis is a chemical reaction that occurs at high temperature and pressure between water, oil, and reservoir rock, and improves the physical properties of heavy oil. Viscosity reduction happens due to breakage of big molecules, like asphaltene by breaking its weak C-S bonds. The energy required to break these bonds can be provided by exothermic chemical reaction between metal particles and oil phase.

As the particle size decreases in nano-scale, it behaves based on its atoms and molecules behavior and shows different properties [12]. Also, because of being small size of pores and throats in the porous medium, using nanoparticles in oil reservoirs has more benefits.

Furthermore, the role of heat transfer in thermal recovery of heavy oil is very important, especially when nano-particles addition to this process improves heat transfer and then increases the oil recovery.

The purpose of this work is to investigate the effect of nanoparticles on heavy oil viscosity. A set of experiments, including measuring viscosity at various concentrations of nanoparticles and different temperatures and concentrations of nanoparticles when mixing with heavy oil sample, were conducted. Also, studied are the effects of the nanoparticle types on viscosity reduction.

2. Experimental

2.1. Materials

In this study, we used various Nano-metal oxide such as NiO, TiO_2, Al_2O_3, ZnO, WO_3, Fe_2O_3, and CuO. We selected these metal oxides for the following reasons:

- TiO_2, ZnO, Fe_2O_3 and NiO were used as catalysts in cracking reaction.
- Fe_2O_3 and CuO were used for their heat properties.
- Al_2O_3 was used for changing the tension and surface tension.
- WO_3 was used as catalyst in heterogen reaction.

Table 1. Specification of the used nanoparticles

Particle type	Formula	Form	Purity (wt%)	APS	SSA	Appearance
Nickel oxide	NiO	nano powder	99.5%	<30 nm	>50 m^2/g	Green black
Titanium oxide	TiO_2	nano powder	99%	10 nm	150 m^2/g	Anatase
Iron (III) oxide	Fe_2O_3	nano powder	99%	<30 nm	<30 m^2/g	Red powder
Tangstan oxide	WO_3	nano powder	99%	<50 nm	<20 m^2/g	Light yellow
Alumina (α)	Al_2O_3	nano powder	99%	~80 nm	<10 m^2/g	White powder
Copper (II) oxide	CuO	nano powder	99%	<50 nm	>80 m^2/g	Black powder
Zinc oxide	ZnO	nano powder	99%	<50 nm	>60 m^2/g	-

Table 1 shows the specifications of the nanoparticles. Heavy oil with API of 21 was used in all the experiments. These nanoparticles were purchased from Notrino Company.

2.2. Equipment

A DBR electromagnetic viscometer was used for viscosity measurement. This apparatus determines the viscosity using electromagnetic waves by importing a constant force on a piston inside the measurement chamber. Motion of the piston is impeded by viscous flow in the annulus between the piston and the measurement chamber wall. This chamber is covered by a jacket and supplied with liquid bath. This system is supplied with the Cambridge ViscoPro 2000 electronic package and can measure viscosity of oil at various temperatures continuously [13]. Details of this apparatus is shown in Figure 1.

An ultrasonic lab devices UP200S (Hielscher Ultrasonics, 200W, 24 kHz) was used for mixing different percentages of nanoparticles in our heavy oil sample. This device is suited for all general ultrasonic applications in small and medium scales.

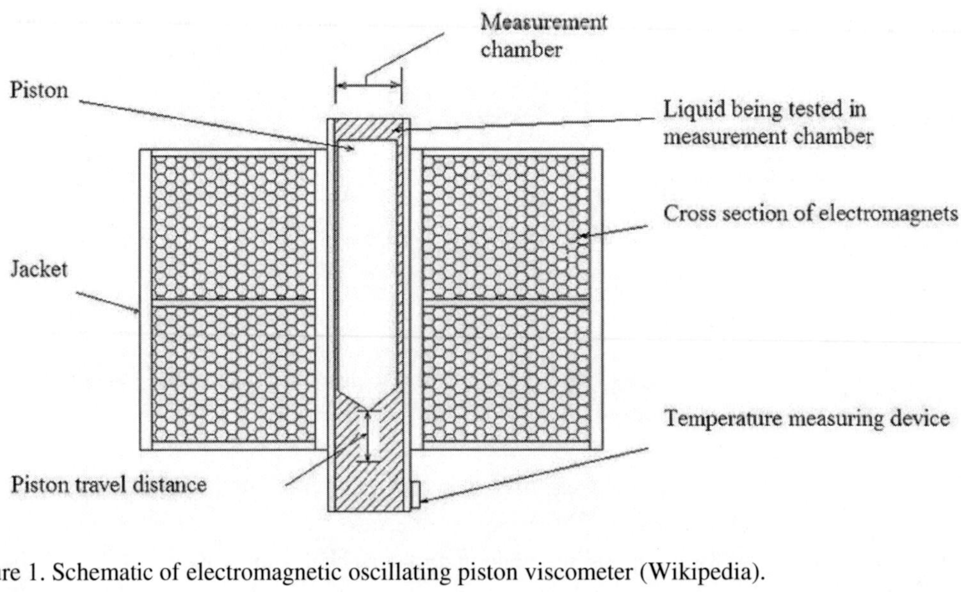

Figure 1. Schematic of electromagnetic oscillating piston viscometer (Wikipedia).

Table 2. Viscosity measurements at different temperatures using different types of nanoparticles with different concentrations

Nano type	Concentration of particles wt%	Viscosity (cp) at 40 °C	Viscosity (cp) at 50 °C	Viscosity (cp) at 80 °C	Viscosity (cp) at 100 °C
none	0	169.74	105.3	40.45	29.59
Fe_2O_3	0.2	104.04	59.23	18.22	9.011
	0.5	114.1	62.77	18.51	8.98
	1	170.82	86.72	20.85	11.55
	2	234.00	71.61	26.44	16.59
NiO	0.2	184.06	121.61	45.15	25.62
	0.5	167.10	93.65	24.89	11.41
	1	198.20	125.78	28.58	13.08
	2	280.6	138.00	30.45	15.92
WO_3	0.2	140.35	81.03	27.11	17.13
	0.5	154.33	104.391	37.85	24.52
	1	203.04	116.44	67.12	17.02
	2	265.61	165.99	58.41	37.26
TiO_2	0.2	212.49	123.52	50.55	28.02
	0.5	64.86	49.11	35.12	29.59
	1	257.20	149.32	64.78	47.27
	2	396.12	275.17	126.10	94.70
CuO	0.2	187.54	106.20	41.64	25.88
	0.5	202.08	119.57	46.29	30.17
	1	203.89	120.65	43.69	27.15
	2	240.02	141.12	55.02	32.42
Al_2O_3	0.2	175.40	103.24	36.93	26.59
	0.5	193.84	112.95	42.63	29.72
	1	199.03	121.19	48.61	34.58
	2	199.96	144.21	54.32	36.59
ZnO	0.2	245.27	147.96	52.52	33.10
	0.5	282.81	174.45	53.53	33.55
	1	291.88	175.65	60.99	35.81
	2	377.45	332.07	91.65	58.49

2.3. Experimental Procedure

Experiments for viscosity determination were implemented so that our mixtures (heavy oil and nano-particles) were prepared first, a definite amount of nano-particle was added to the heavy oil sample, then the mixtures were exposed to 10 minutes of ultrasonic with 100% power and 1 cycle in each second for well mixing using an ultrasonic lab device UP200S. In the next stage, the mixture was imported in DBR electro-magnetic viscometer and then, its temperature was adjusted at the desired temperature; after some minutes, viscosity of the mixture can be read on the device.

Four concentrations including 0.2, 0.5, 1, and 2 wt% of each nanoparticle were used and viscosity of each mixture were measured at four temperatures of 40, 50, 80, 100 °C.

Table 3. The percentage of viscosity deviation at different temperatures using different types of nanoparticles with different concentrations

Nano type	Concentration of particles wt%	percent of viscosity deviation at 40 °C	percent of viscosity deviation at 50 °C	percent of viscosity deviation at 80 °C	percent of viscosity deviation at 100 °C
Fe_2O_3	0.2	-38.71	-43.75	-54.96	-64.78
	0.5	-32.78	-40.39	-54.24	-64.88
	1	0.64	-17.65	-48.45	-54.85
	2	-37.86	-32.00	-34.36	-35.28
NiO	0.2	8.43	15.49	11.62	0.10
	0.5	-1.56	-11.06	-38.47	-55.39
	1	16.77	19.44	-29.34	-48.87
	2	65.31	31.05	-24.72	-37.78
WO_3	0.2	-17.31	-23.01	-32.98	-33.05
	0.5	-9.08	-0.86	-6.43	-4.16
	1	19.62	10.58	65.94	-33.49
	2	56.48	57.64	44.40	45.58
TiO_2	0.2	25.19	17.30	24.97	9.48
	0.5	-61.79	-53.35	-13.18	15.65
	1	51.52	38.96	60.16	84.73
	2	133.37	161.32	211.74	270.08
CuO	0.2	10.49	0.85	2.95	1.13
	0.5	19.05	13.55	14.45	17.91
	1	20.12	14.58	8.00	6.11
	2	41.41	34.01	36.02	26.67
Al_2O_3	0.2	3.33	-1.95	-8.70	3.91
	0.5	14.20	7.29	5.38	16.15
	1	17.25	15.09	20.17	35.12
	2	17.81	36.95	34.29	42.98
ZnO	0.2	44.50	40.51	29.84	29.33
	0.5	66.61	65.66	32.35	31.11
	1	71.95	66.81	50.79	39.93
	2	122.37	215.35	126.57	128.57

3. Results and Discussion

The results of the viscosity measurement are presented in Table 2 and Figures 2 and 3. The measurements for each sample were repeated at least three times to confirm the results. Values given in Table 2 are the average amounts of measurements. The deviation percent $((\mu-\mu_0)/\mu_0\times 100\%)$ is shown in Table 3.

In this work, we investigate the effects of temperature, nanoparticle types and their concentrations. The results and discussion are presented as follow:

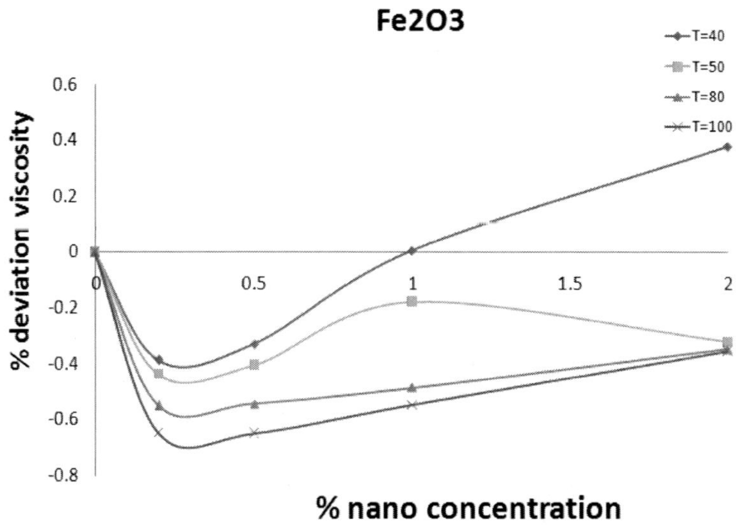

a

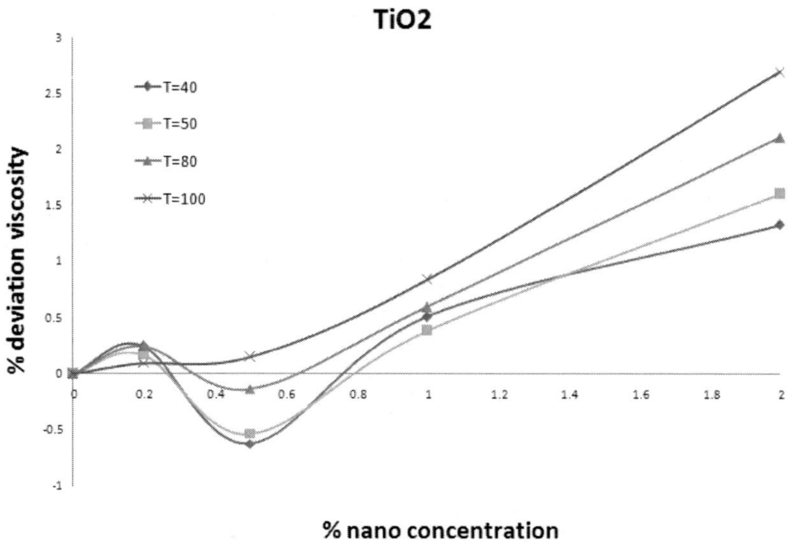

b

Figure 2. (Continued)

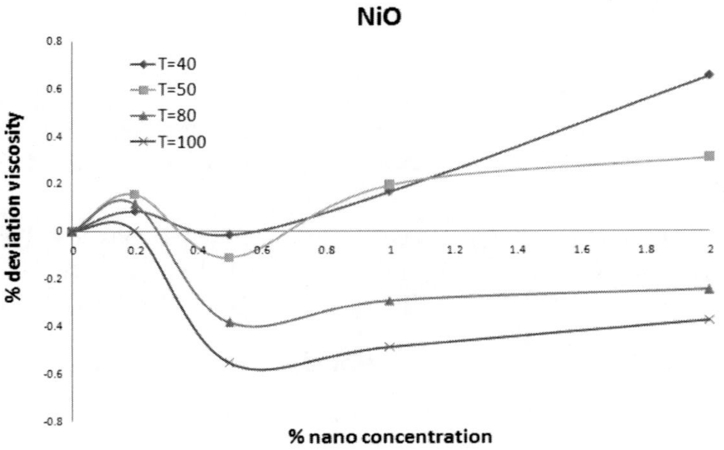
NiO
T=40
T=50
T=80
T=100
% deviation viscosity
% nano concentration

c

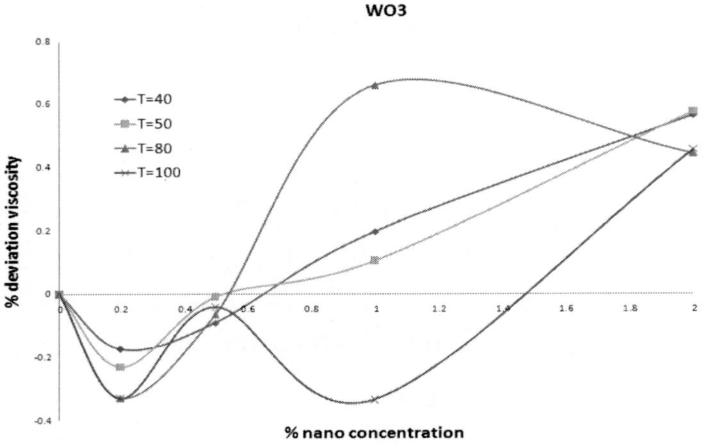
WO3
T=40
T=50
T=80
T=100
% deviation viscosity
% nano concentration

d

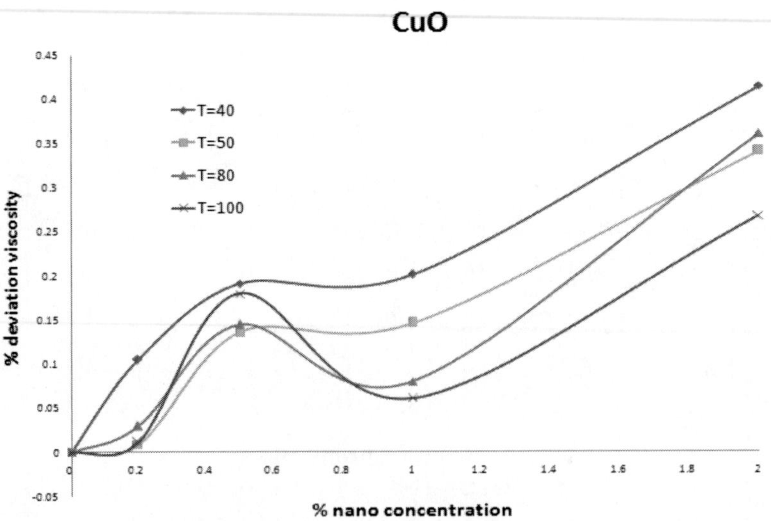
CuO
T=40
T=50
T=80
T=100
% deviation viscosity
% nano concentration

e

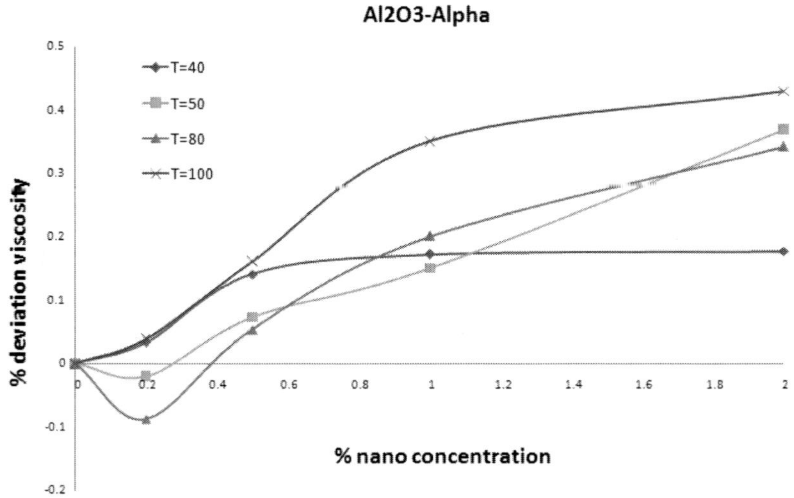

f

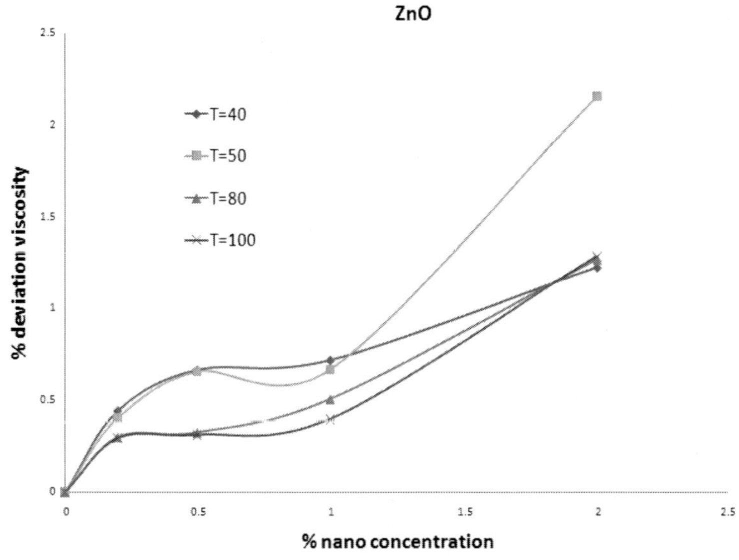

g

Figure 2. Effect of the nanoparticles types on the percentage of the viscosity deviation versus concentration behavior at 40, 50, 80 and 100 ºC (a- Fe_2O_3, b- TiO_2, c- NiO, d- WO_3, e- CuO, f- Al_2O_3-α, g- ZnO).

3.1. Effect of Nanoparticle Types and Their Concentrations

The overall effect of nanoparticles on the sample viscosity leads to an optimum concentration of the particles, in which the viscosity reduction effect is maximum. To understand the effect of the nanoparticle types on the viscosity reduction, the sample viscosity versus concentration of different nanoparticles at different temperatures are compared in Figure 2.

Viscosities of heavy oil samples at different temperatures and various percentages of nanoparticles are shown in Figures 2 and 3. As indicated by the results, Fe_2O_3, WO_3, TiO_2 and NiO decrease viscosity of heavy oil more than other nanoparticles. The most reduction of viscosity happens at the lowest concentration of nanoparticles. This issue happens because these nanoparticles act as catalysts in cracking reaction. Therefore, heavy oil samples crack to the lighter components. Our results show that the cracking reaction happens at low temperature, when using nanoparticles such as Fe_2O_3, WO_3, TiO_2 and NiO. In studied researches, Fe, Ni and Ti act as catalyst cracking of C-S, C=C, C ≡ C bonds of heavy components of heavy oil. Our results show this fact. This effect is explained by referring to the catalytic effect of metal particles on breaking of the C-S bonds in the aqua-thermolysis process [5].

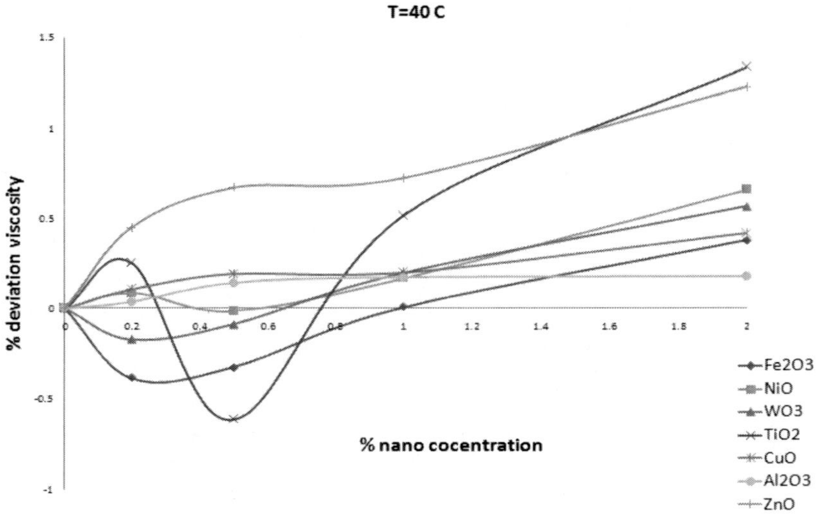

a

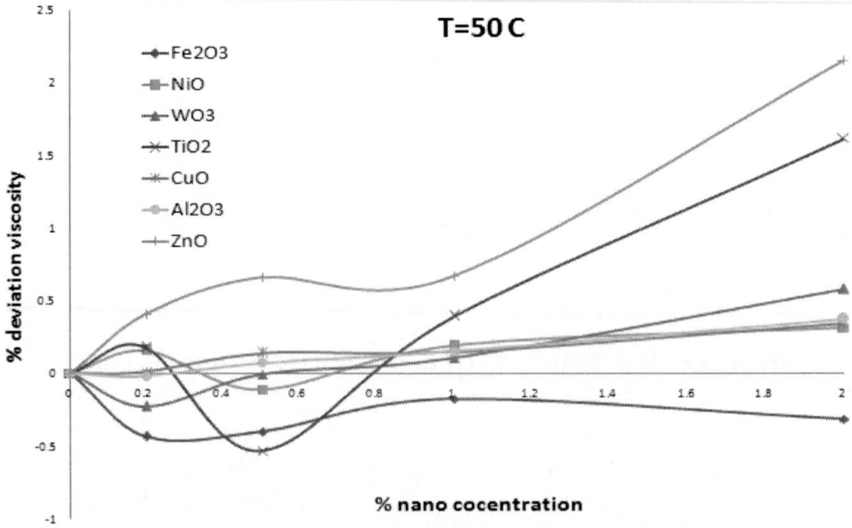

b

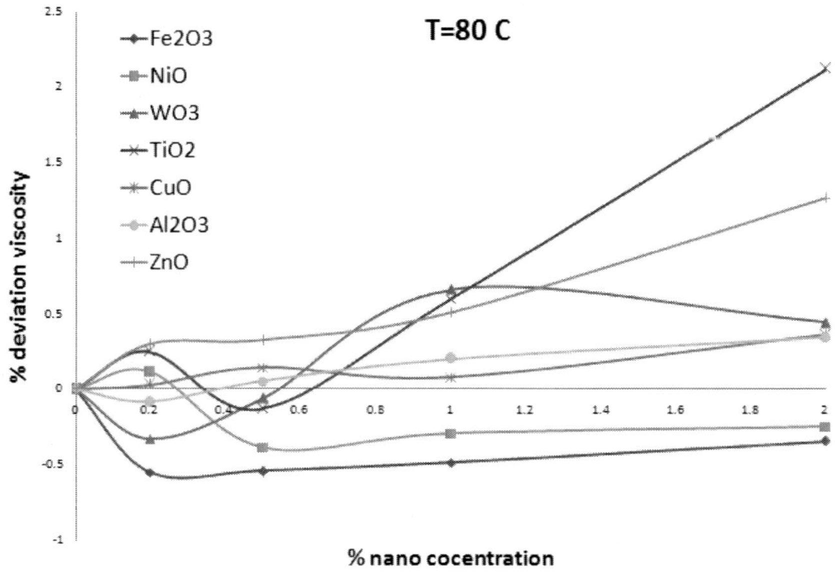

c

Figure 3. Effect of nanoparticle types on the percentage of viscosity deviation versus concentration behavior (a- Fe_2O_3, b- TiO_2, c- NiO, d- WO_3, e- CuO, f- Al_2O_3-α, g- ZnO).

This result is similar to the obtained result of Hamidi Shokrlu and Babadagli [4] and Hascakir et al. [10]. This issue is due to larger surface areas of nanoparticles that cause an increase in the contact area of the particles with the oil phase [5]. Also, when nanoparticles are added to the oil; they are placed between the oil layers and lead to movement of fluid layers on each other easily [14].

Some types of nanoparticles such as Al_2O_3 and CuO at any concentration, and some others such as Fe_2O_3, WO_3 and TiO_2, at high concentrations, increase the crude oil viscosity. At these concentrations, interaction between oil and nanoparticles increase resistance against the movement.

3.2. Effect of Temperatures

Figure 2 shows the behavior of each type of nanoparticle at definite temperature. This figure indicates that at low temperatures, percentage of viscosity deviation is more than at high temperatures.

Figure 4 shows the temperature effect on viscosity without attention to nanoparticles types causes the same behavior, but addition of nanoparticles changes the value of viscosity. This figure shows that sample viscosity versus temperature have exponentially behavior by the following equation:

$$\mu = A \exp (B/T) \quad (2)$$

Table 4. The exponential equation (2) fitting parameters and correlation coefficients

Nano type	Concentration of particles wt%	A(cp)	B(1/K)	r^2
none	0	3.69×10^{-6}	5450	0.981
Fe_2O_3	0.2	1.154×10^{-5}	5037	0.983
	0.5	5.786×10^{-6}	5295	0.985
	1	1.225×10^{-5}	5101	0.982
	2	1.051×10^{-4}	4564	0.991
NiO	0.2	9.229×10^{-4}	3822	0.977
	0.5	1.023×10^{-5}	5210	0.991
	1	9.540×10^{-6}	5282	0.995
	2	8.72×10^{-6}	5373	0.979
WO_3	0.2	1.921×10^{-5}	4886	0.971
	0.5	9.717×10^{-5}	4432	0.986
	1	2.498×10^{-5}	5631	0.982
	2	2.450×10^{-4}	4256	0.983
TiO_2	0.2	5.235×10^{-6}	5372	0.987
	0.5	1.573×10^{-3}	3260	0.974
	1	3.574×10^{-5}	4815	0.963
	2	3.311×10^{-4}	4273	0.986
CuO	0.2	8.112×10^{-4}	3844	0.989
	0.5	1.015×10^{-3}	3810	0.986
	1	5.080×10^{-4}	4031	0.989
	2	8.99×10^{-4}	3898	0.992
Al_2O_3	0.2	5.780×10^{-4}	4141	0.991
	0.5	6.174×10^{-4}	4157	0.994
	1	1.382×10^{-3}	3915	0.993
	2	6.711×10^{-3}	3056	0.991
ZnO	0.2	7.98×10^{-4}	3949	0.987
	0.5	3.31×10^{-4}	4278	0.993
	1	6.27×10^{-4}	4079	0.996
	2	5.67×10^{-4}	4276	0.988

T is temperature (K), A = 3.69×10^{-6}, B = 5450 and r^2 (the correlation coefficient) = 0.981. Table 4 presents these parameters for each type of nanoparticle. This equation shows that values of *B* are in the same range, but values of *A* are different. However, value of parameter *A* for NiO, Fe_2O_3, WO_3 and TiO_2 is maximum at low concentrations. Then, low concentrations (0.2 and 0.5) as the optimum concentrations are reported.

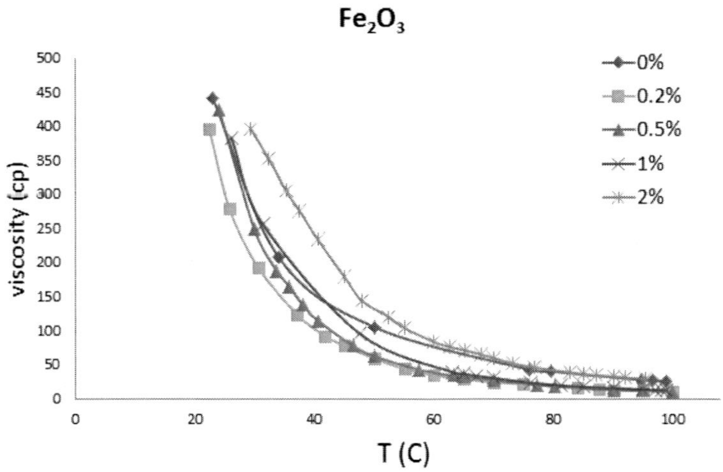

a

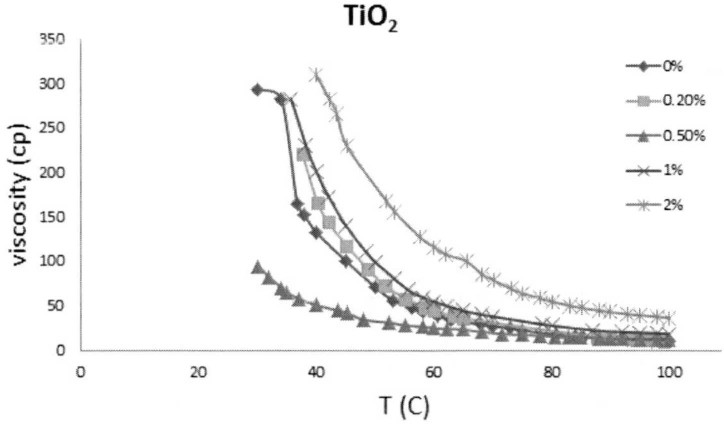

b

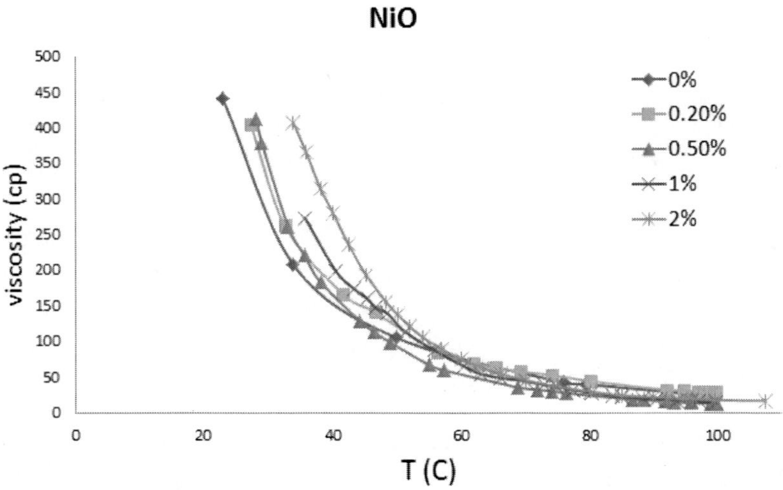

c

Figure 4. (Continued)

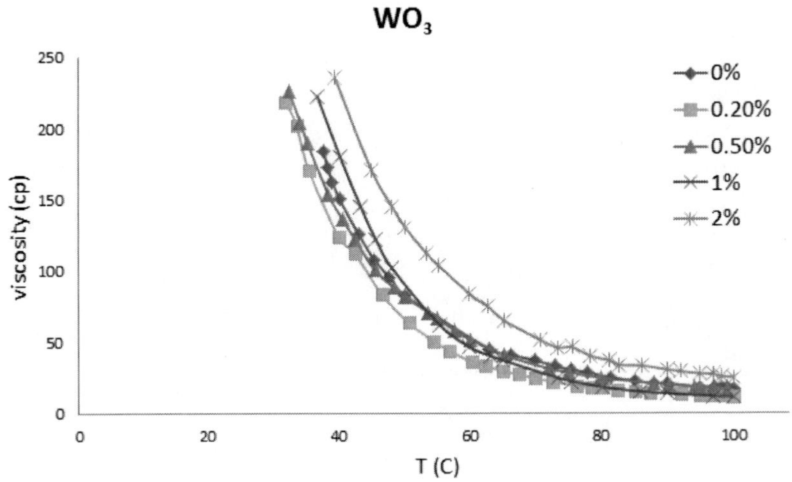

WO3
250
200
150
100
50
0
viscosity (cp)
0
20
40
60
80
100
T (C)
0%
0.20%
0.50%
1%
2%

d

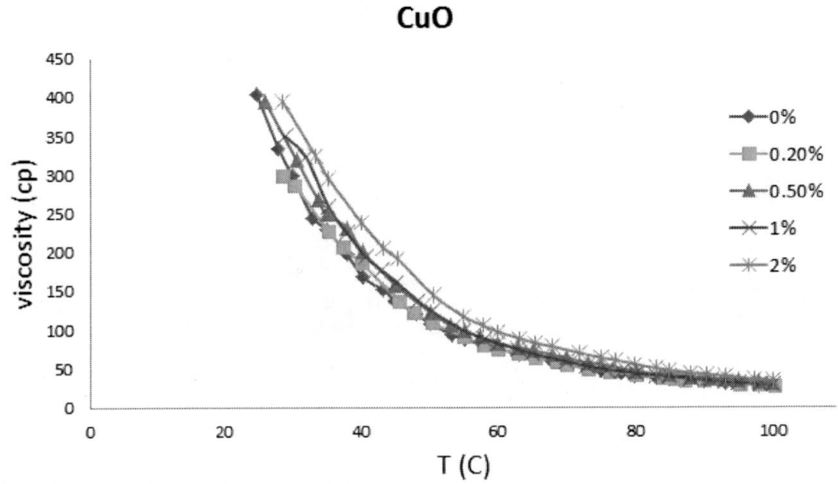

CuO
450
400
350
300
250
200
150
100
50
0
viscosity (cp)
0
20
40
60
80
100
T (C)
0%
0.20%
0.50%
1%
2%

e

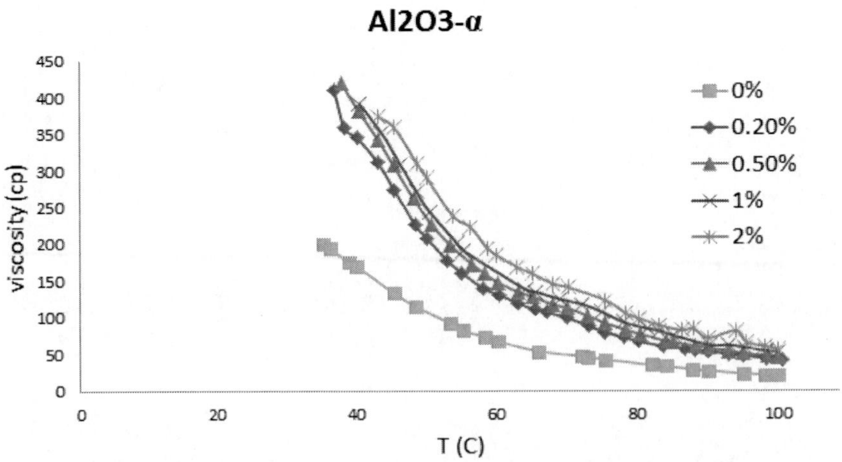

Al2O3-α
450
400
350
300
250
200
150
100
50
0
viscosity (cp)
0
20
40
60
80
100
T (C)
0%
0.20%
0.50%
1%
2%

f

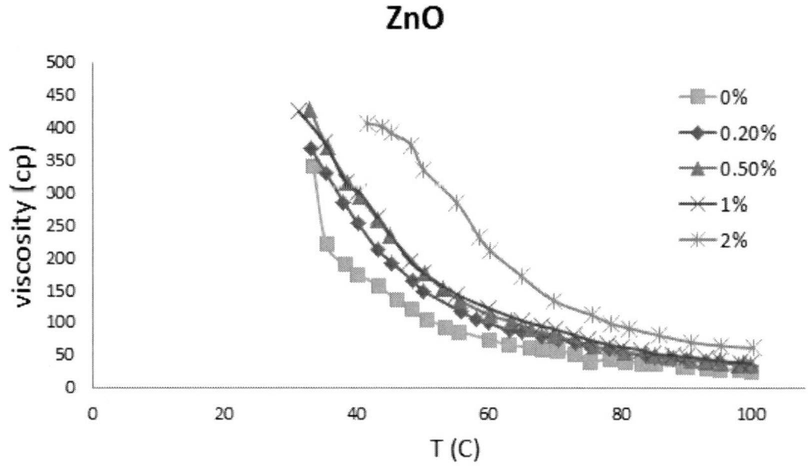

g

Figure 4. Viscosity behavior of the oil sample mixed with different types of nanoparticles with different concentration versus temperature (a- Fe_2O_3, b- TiO_2, c- NiO, d- WO_3, e- CuO, f- Al_2O_3-α, g- ZnO).

3.3. Effect of Temperatures and Nanoparticles

At higher temperatures, the effect of NiO and Fe_2O_3 in all concentrations, causes decreasing of oil viscosity but in low concentrations (0.2 and 0.5 %) viscosity decreases more than any concentrations. Tables 2 and 3 and Figures 2 - 4 show that ZnO and Al_2O_3 do not have any effect on viscosity reduction. Therefore, NiO, Fe_2O_3, and WO_3 are the best nanoparticles at low concentrations (0.2 and 0.5 %) and high temperatures for decreasing oil viscosity.

CONCLUSION

In this study, the effect of nanoparticles on the heavy oil viscosity was investigated experimentally. The obtained viscosity data show the effect of nanoparticle types, with different concentrations and temperatures. The following conclusions are obtained after analysis of the set of experiments conducted in this study:

1. Heavy oil viscosity could be decreased by some of metal oxide nanoparticles even at different temperature. The catalytic reaction and heat properties may be the reason of this reduction. Some of these nanoparticles types act as cracking catalyst and cause cracking the C-S, C=C, C ≡ C bonds of heavy components of heavy oil and increase the heavy oil viscosity.
2. The percentage of viscosity reduction and optimum concentration of nanoparticles strongly depend on the type of nano-metal oxide. Different types of nanoparticles reduce the heavy oil viscosity via a different series of exothermic catalytic reactions.
3. Increasing of heavy oil viscosity in high concentration of nanoparticle occurs at different temperatures for all the types of nanoparticles.

4. The optimum nanoparticle selected types are Fe_2O_3, NiO and WO_3. The optimum concentrations are 0.2 and 0.5 % and maximum reduction of heavy oil viscosity happens at high temperatures. However, effect of nanoparticles on heavy oil viscosity reduction appears to be more in thermal process recovery for high temperatures.

REFERENCES

[1] Lashanizadegan A., Ayatollahi Sh., Homayoni M., (2008), Simultaneous Heat and Fluid Flow in Porous Media: Case Study: Steam Injection for Tertiary Oil Recovery, *Chem. Eng. Comm.* 195: 521–535.

[2] Shah R.D., (2009), Application of Nanoparticle Saturated Injectant Gases for EOR of Heavy Oils, SPE 129539- STU, *Presented at the SPE International Student Paper Contest at the SPE Annual Technical Conference and Exhibition,* Louisiana, USA. http://en.wikipedia.org/wiki/Viscometer.

[3] Omole O., Olieh M.N., Osinowo T., (1999), Thermal visbreaking of heavy oil, *Fuel,* 78 (12): 1489-1496.

[4] Hamedi Shokrlu Y., Babadagli T., (2010), *Effect of nano sized metals on viscosity reduction of heavy oil/bitumen during thermal applications, CSUG/SPE 137540, presented at the Canadian Unconventional Resources & International Petroleum Conference,* Calgary, Alberta, Canada.

[5] Clark P.D., Clarke R.A., Hyne J.B., Lesage K.L., (1990), Studies on the effect of metal species on oil sands undergoing steam treatments, *AOSTRA J Res,* 6(1): 53-64.

[6] Cacciola G., Aristov Y. I., Restuccia G., Parmon V.N., (1993), Influence of hydrogen-permeable membranes upon the efficiency of the high-temperature chemical heat pumps based on cyclohexane dehydrogenation-benzene hydrogenation reactions, *International Journal of Hydrogen Energy,* 18(8): 673–680.

[7] Fan H., Zhang Y., Lin Y., (2004), The catalytic effects of minerals on aqua- thermolysis of heavy oils, *Fuel,* 83 (14-15): 2035-2039.

[8] Fan H., Li Z.B., Liang T., (2007), Experimental study on using ionic liquids to upgrade heavy oil, *Journal of Fuel Chemistry and Technology,* 35(1):32–35.

[9] Li W., Zhu J., Qi J., (2007), Application of nano- nickel catalyst in the viscosity reduction of Liaohe heavy oil by aquathermolysis, *Journal of Fuel Chemistry and Technology,* 35 (2): 176-180.

[10] Hascakir B., Babadagli T., Akin S., (2008), Experimental and numerical modeling of heavy oil recovery by electrical heating, *SPE 117669, SPE/PS/CHOA International Thermal Operations and Heavy Oil Symposium,* Calgary Canada,.

[11] Chen Y., Wang Y., Lu J., Wu C., (2009), The viscosity reduction of nano-keggin-K3PMo12O40 in catalytic aquathermolysis of heavy oil, *Fuel,* 88: 1426–1434.

[12] Yokoyoma T., Masuda H., Suzuki M., Ehara K., Nogi K., Fuji M., Fukui T., et al., (2007), Basic properties and measuring methods of nanoparticles, *Nanoparticle technology handbook,* ed. M. Hosokawa, K., Nogi, M. Natio, T. Yokoyama, Chap.1, 5-48, Elsevier.

[13] Operating and maintenance manual of DBR Sand-Alone EM viscometer system, DB Robinson design and manufacturing Ltd. Edminston, Alberta, Canada, (2000). http://www.dbra.com.

[14] Ettefaghi E., Ahmadi H., Rashidi A., Mohtasebi S., Alaei M., (2013), *International Journal of Industrial Chemistry,* 4:28.

In: Enhanced Oil Recovery
Editor: Alicia Knight
ISBN: 978-1-63463-917-0

Chapter 11

SCREENING OF POTENTIAL APPLICATION OF EOR PROCESSES IN A NATURALLY FRACTURED OIL RESERVOIR

Arash Kamari[1], Mohammad Nikookar[2], Abdolhossein Hemmati-Sarapardeh[3], Leili Sahranavard[2] and Amir H. Mohammadi[1,4*]

[1]School of Engineering, University of KwaZulu-Natal, Howard College Campus, King George V Avenue, Durban, South Africa
[2]IOR Research Institute, National Iranian Oil Company (NIOC) - R&T, Tehran, Iran.
[3]Department of Chemical and Petroleum Engineering, Amirkabir University of Technology, Tehran, Iran
[4]Institut de Recherche en Génie Chimique et Pétrolier (IRGCP), Paris Cedex, France

ABSTRACT

High potential of enhanced oil recovery (EOR) methods to progressively expand hydrocarbon exploitation makes the decisions on the applicability of certain recovery techniques for Iranian oil reservoirs, which have reached the late stage of their life. Due to high operational costs and high technical sensitivity of these methods particularly in naturally fractured reservoirs because of complexity, natural heterogeneity and non-uniformity, studies are required for selection of the most appropriate EOR method. In this communication, a comprehensive EOR screening for 12 EOR methods by criteria definition based on numerous reported successful EOR projects and presented criteria tables by some previous researchers have been carried out in an Iranian oil reservoir. In addition, for showing the importance of EOR screening approach during selection process of EOR methods, a sensitivity analysis was conducted. The results revealed that polymer, CO_2 miscible and immiscible flooding can be considered technically the most appropriate methods for applying in the Iranian naturally fractured carbonate reservoirs.

* Corresponding Authors: A.H. Mohammadi: E-mail: a.h.m@irgcp.fr & amir_h_mohammadi@yahoo.com

Keywords: enhanced oil recovery (EOR); screening; fractured reservoirs; sensitivity analysis

NOMENCLATURE

EOR	enhanced oil recovery
IFT	interfacial tension
CSS	cyclic steam injection
CSF	continues steam flooding
SAGD	steam assisted gravity drainage
ISC	in situ combustion
WOR	water oil ratio
MMP	minimum miscibility pressure
NREC	not recommended
HC	hydrocarbon
N_2	nitrogen
CO_2	carbon dioxide
S_w	water saturation
Φ	porosity

1. INTRODUCTION

Because of oil price increase and reduction of natural production of oil reservoirs over the past few years, recovery methods have become attractive. Oil recovery techniques in petroleum engineering involve primary, secondary and tertiary methods, which the third case is named as Enhanced Oil Recovery (EOR). Enhanced oil recovery methods involve the implementation of external substances to change physical and chemical interactions in hydrocarbon reservoirs in a manner that promotes desired recovery conditions (Al Adsani and Bai, 2011). Enhanced oil recovery techniques have drawn great attention in the oil and natural gas industry because of their high potential for recovering more oil from depleted reservoirs with conventional production technique (Gharbi, 2005). Development of numerous oil reservoirs in Iran entering to the final stage of their life makes such reservoirs an excellent selection for study of enhanced oil recovery methods.

While performing enhanced recovery methods in a specific reservoir, it should be sufficiently accurate to select an appropriate EOR method with respect to influencing parameters of rock and fluid properties as these methods are so expensive and sometimes the expenses are much higher than the revenue, hence the significance of screening approach is understood. Selection and implementation of proper EOR technique for a specific reservoir, which is nowadays named as EOR screening, can save time and cost, and also reduces the risk. For this purpose, there are different methods for the screening project such as determination of parameters in a specified range, using tables and graphs, fuzzy logic, artificial intelligence, and artificial neural network methods.

The problem of EOR screening has been carried out in several studies (Al-Mutairi and Kokal, 2011; Al Adsani and Bai, 2010; Alvardo et al., 2002; Barzegari et al., 2010; Dickson et al., 2010; Gharbi, 2000; Guerillot, 1988; Parkinson et al., 1994; Shokir et al., 2002, *etc*). In one of the first studies on screening of EOR methods, screening criteria was briefly presented in a series of tables and also, various kinds of EOR methods were presented by simple graphs (Taber and Martin, 1983). Then, EOR methods were classified based on field data and EOR mechanisms (Taber et al. 1997a, b) and were updated in 2011 (Al-Adsani and Bai, 2011).

Lee et al. (2011) developed an artificial neural network in order to select an appropriate EOR method with the given reservoir properties. The input layer of this network was composed of the key reservoir parameters, while the output layer was composed of the five EOR methods to be evaluated. The results indicated that the developed artificial neural network can be used to select the most appropriate EOR process according to the reservoir rock and fluid characteristics in a time and cost effective way. Gharbi (2000) applied an artificial intelligence technique to assist the selection and design of EOR processes. The results showed a good performance for developed intelligent method in selection of EOR method. Abbas and Song (2011) designed a capable intelligent system for modification of a parameter. The results indicated that application of emerging expert-system technology to select EOR method is very important and useful. Ibatullin et al. (2002) expressed that available analytical technologies theoretically make it possible to solve the problem of selection of optimum EOR methods. Bennion et al. (2000) described issues which need to be taken into consideration from a reservoir perspective when considering development of a gas storage reservoir.

In this study, problem of selecting appropriate EOR method for a south western Iranian naturally fractured oil reservoir is solved. Most important data of the reservoir have been selected and analyzed. A comprehensive study for selection of the proper EOR methods, among 12 EOR methods, was done and also effect of each parameter on the selection was discussed. Finally, to investigate the effects of permeability, oil density (API°) and water drive on the selected EOR methods for the mentioned reservoir, a sensitivity analysis was employed.

2. THE UNDER-SURVEY RESERVOIR

To study EOR screening methods, a naturally fractured reservoir in south western of Iran, whose properties are summarized in Table 1, is considered as a case study. The oil reservoir is an oil-wet carbonate reservoir located at a depth of 2999 ft. The reservoir, which belongs to a main Iranian formation, has a total net pay thickness of 882.54 ft. Also, this reservoir has a maximum gross thickness of 1961.21 ft and contains 446 million bbl original volume of oil in place.

Oil viscosity, reservoir temperature and °API are 0.39cp, 132.8°F and 45.8, respectively. The reservoir water has high salinity (38000 ppm). The initial pressure and bubble point pressure are 2160.9 and 1969.8 psi, respectively, which this pressure condition gives an under-saturated reservoir. The water drive in the reservoir is not active and according to petrophysical evaluation, the formation is a poor reservoir as it has a porosity of 5.7% and water saturation (Sw) around 40%.

Table 1. The properties of under-survey reservoir

Parameter	Value	Units
Depth:	2999	ft
Formation Type:	Carbonate	
Clay Content:	Low	
Reservoir Temperature:	132.8	°F
Water Hardness:	800	ppm
Water Salinity:	38000	ppm
Initial Pressure:	2160.9	psi
Bubble Point Pressure, BPP:	1969.8	psi
Current Pressure:	1911	psi
Solution Gas/Oil Ratio:	126	scf/bbl
Dead Oil Density @ Surface:	45.8	°API
Oil Viscosity:	0.39	cp
Fracturing:	Yes	
Horizontal Permeability:	50	mD
Active water Drive:	No	
Bottom Water:	None	
Gas Cap:	Local	
Water Oil Ratio:	9	ratio
Producing Gas/Oil Ratio:	141.42	scf/bbl
Gross Pay Thickness (formation):	1961.21	ft
Net Pay Thickness (formation):	882.54	ft
Porosity:	0.057	fraction
Connate Water Saturation:	0.4	fraction
Oil Formation Volume Factor:	1.38	
Original Oil in Place:	446456096	bbl

The ultimate recovery factor in this reservoir has been reported 17.2 % which makes it a good candidate for implementation of EOR methods. Therefore, the aim of present study is a comprehensive feasibility analysis to find the most appropriate EOR methods in this reservoir.

3. EOR Methods and Screening Criteria

As previously mentioned, oil recovery methods involve the primary, secondary and tertiary methods. In the primary method, oil is produced by natural energy of reservoir. The secondary recovery is used when natural drive energy is depleted or too small for economic oil recovery; energy must be added to the reservoir to permit additional oil recovery (Stosur et al., 2003a, b). The tertiary recovery is used when a reservoir fails to produce enough oil by the secondary recovery. EOR processes refer to the methods for increase of the oil recovery efficiency from an oil reservoir after the primary and secondary stages by using fluids injection (Zerafat et al., 2011). The aforementioned concepts are well illustrated in Figure 1.

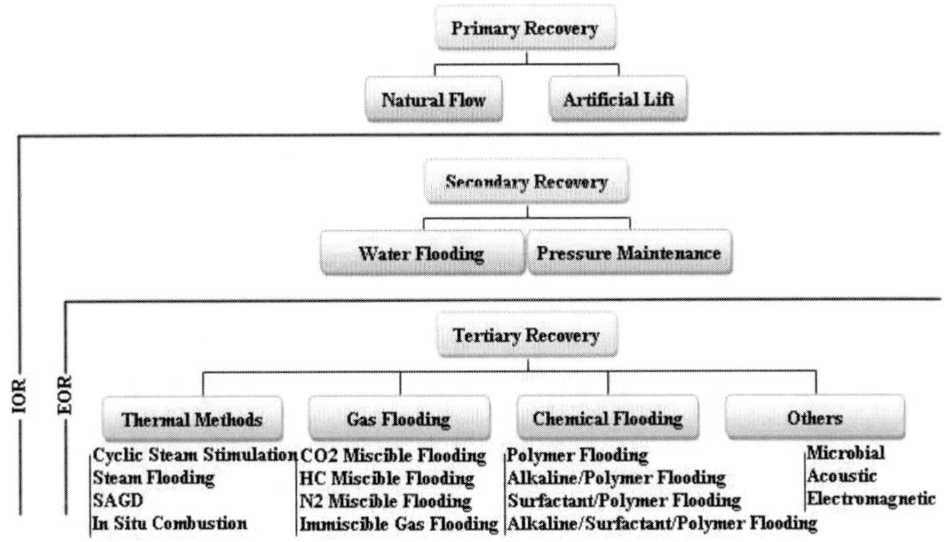

Figure 1. Terms of oil recovery: the initial design from Stosur et al. (2003a, b).

Basically, there are three main enhanced oil recovery mechanisms, which are divided into three phases.

These mechanisms are solvent extraction through miscibility by miscible gas injection methods, reducing oil viscosity by thermal methods and decreasing interfacial tension (IFT) between oil and displacing fluid by chemical methods (Taber et al.1997a). There are three main groups of EOR methods which are applied based on these mechanisms. In this study, each main group involves four EOR techniques, which based on the above-mentioned mechanisms can be classified as follows:

1. Chemical Flooding Methods:
   - Polymer Flooding
   - Alkaline/Polymer Flooding
   - Surfactant/Polymer Flooding
   - Alkaline/Surfactant/Polymer Flooding
2. Gas Flooding Methods:
   - Carbon Dioxide Miscible Flooding
   - Hydrocarbon Miscible Flooding
   - Nitrogen Miscible Flooding
   - Immiscible Gas Flooding
3. Thermal Methods:
   - Cyclic Steam Stimulation
   - Steam Flooding
   - Steam Assisted Gravity Drainage
   - In-Situ Combustion

Before implementation of the EOR methods to any reservoir, an exact and complete investigation of reservoir properties is necessary. In addition, a geological study should be conducted, because of geological factors perhaps causes EOR methods fail due to unexpected reservoir problems (Alkafeef and Zair, 2007).

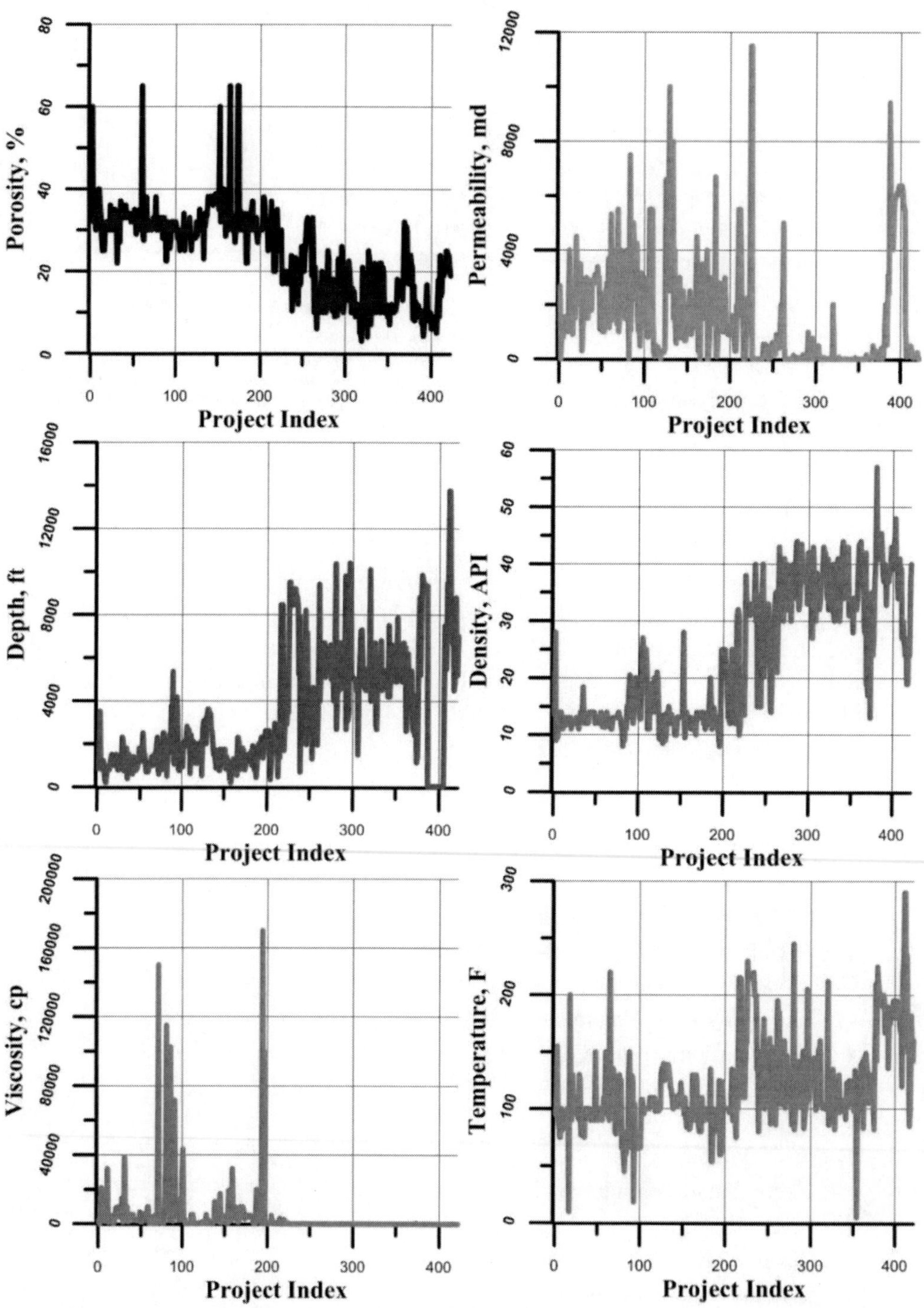

Figure 2. Distributions of the effective parameters for successful producing EOR projects. Data index: Steam flooding (1-207), hot water injection (208-215), in situ combustion (216-236), chemical flooding (237-263), CO_2 miscible (264-367), CO_2 immiscible (268-375), hydrocarbon miscible (376-422).

Totally, there are several parameters including production, petrophysical, crude oil chemistry, produced water chemistry and field data. But, the most effective parameters were classified into two main groups: rock parameters (porosity, permeability, initial oil saturation and depth) and fluid parameters (gravity, viscosity and temperature).

A convenient way to show EOR methods is their arrangement by the most important rock and fluid parameters such as porosity, permeability, depth, gravity, temperature and viscosity as shown in Figure 2.

This figure shows oil gravity range that is the most important parameter (Taber et al., 1997a), for thermal EOR techniques entail comparatively lower compared to chemical and gas flooding. Moreover, porosity and permeability values for the chemical and gas flooding methods are lower compared to the thermal methods. From this figure, it can be concluded that the thermal methods are more suitable for the reservoirs with low temperature. The gas flooding methods must be applied in deep reservoirs and the crude oil with high °API unlike the thermal methods which are implemented for improving the recovery in shallow reservoirs and the crude oil with low API, while the chemical methods are applied for the crude oil with mid-density in the mid-depth reservoirs.

Figure 3 illustrates correlation between depth and API for the successful EOR methods reported by worldwide EOR survey, Oil & Gas Journal (1998, 2006, and 2008). Criteria of each parameter for various EOR methods will be discussed in next sections by details.

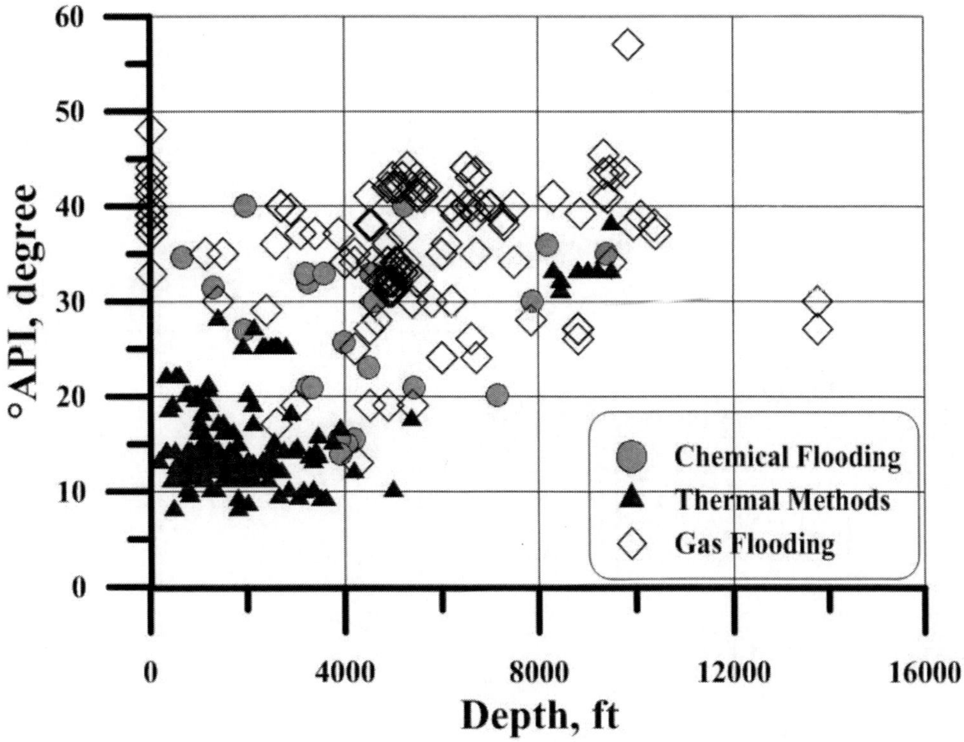

Figure 3. Reservoir depth vs. oil density of producing EOR projects.

4. Screening Procedure

To provide reasonable range of rock and fluid parameters for criteria definition of EOR methods that are either the most important or still have some promise, we used a series of worldwide EOR surveys (worldwide EOR survey, Oil & Gas Journal, 1998, 2006, 2008) including field name, start date, number of production and injection wells, formation type, porosity, area, permeability, depth, start and end saturation, gravity, viscosity, temperature, total and enhanced production, etc.

Table 2. Screening results of polymer flooding for the under-survey reservoir

Result	Screening Criteria	Reservoir Value	Units
Pass.	Reservoir Temperature < 176	132.8	°F
Pass.	Oil Visc. < 150	0.39	cp
Pass.	Oil Density (Surface) > 15	45.8	°API
Pass.	Horizontal Permeability > 50	50	mD
Pass.	Active Water Drive = No	No	Yes, No
Pass.	Bottom Water =< Local	None	Select
Pass.	Gas Cap = < Local	Local	Select
Pass.	Water Hardness < 1000	800	ppm
Pass.	Water Salinity < 100000	38000	ppm
Pass.	WOR < 10	9	ratio
Pass.	Current Oil Saturation >= 0.5	0.5	ratio
Result:	PASSED		

Table 3. Screening results of alkaline/polymer flooding method for the under-survey reservoir

Result	Screening Criteria	Reservoir Value	Units
NREC.	Formation Type = Sandstone	Carbonate	select
Pass.	Reservoir Temperature < 158	132.8	°F
NREC.	Oil Density (Surface) < 35	45.8	°API
Pass.	Oil Visc. < 150	0.39	cp
Pass.	Horizontal Permeability > 50	50	mD
Pass.	Active Water Drive = No	No	Yes, No
Pass.	Bottom Water = < Local	None	Select
Pass.	Gas Cap = < Local	Local	Select
Pass.	Clay Content = < Low	Low	Select
Pass.	Water Hardness < 1000	800	ppm
Pass.	Water Salinity < 50000	38000	ppm
Pass.	Current Oil Saturation >= 0.5	0.5	ratio
Result:	NOT RECOMMENDED		

In addition, some screening tables presented by a number of researchers are used in this study (Abbas and Song, 2011; Al Adsani and Bai, 2011; Dickson et al., 2010; Dusseault,

2006; Greaves and AL-Shamali, 1995; Shafiei et al., 2007; Taber et al. 1997a, b, *etc*) in order to select the most important screening parameters, and also to determine the range of these parameters for each EOR method. After such determination, the most important and critical screening criteria for 12 EOR methods are listed in 12 tables including the reservoir values in addition to the criteria.

Table 4. Screening results of surfactant/polymer flooding method for the under-survey reservoir

Result	Screening Criteria	Reservoir Value	Units
Pass.	Reservoir Temperature < 158	132.8	°F
Pass.	Oil Visc. < 150	0.39	cp
Pass.	Horizontal Permeability > 50	50	mD
Pass.	Active Water Drive = No	No	Yes, No
Pass.	Bottom Water = < Local	None	Select
Pass.	Gas Cap – Local	Local	Select
Pass.	Clay Content = < Low	Low	Select
Pass.	Water Hardness < 1000	800	ppm
NREC.	Water Salinity < 35000	38000	ppm
Pass.	Current Oil Saturation > =0.5	0.5	ratio
Result:	NOT RECOMMENDED		

Table 5. Screening results of alkaline/surfactant/polymer flooding method for the under-survey reservoir

Result	Screening Criteria	Reservoir Value	Units
NREC.	Formation Type = Sandstone	Carbonate	select
Pass.	Reservoir Temperature < 158	132.8	°F
NREC.	Oil Density (Surface) < 35	45.8	°API
Pass.	Oil Visc< 150	0.39	cp
Pass.	Horizontal Permeability > 50	50	mD
Pass.	Active Water Drive = No	No	Yes, No
Pass.	Bottom Water = < Local	None	Select
Pass.	Gas Cap = < Local	Local	Select
Pass.	Clay Content = Low	Low	Select
NREC.	Water Salinity < 35000	38000	ppm
Pass.	Current Oil Saturation >= 0.5	0.5	ratio
Result:	NOT RECOMMENDED		

Afterward, to select a proper EOR method for the under-survey reservoir, two conditions are considered as "Passed" and "Not Recommended" (NREC). If the method is able to cover all defined screening criteria, it will be passed through screening filter and can be selected as a suitable EOR method in the under-study reservoir. Otherwise, if the method cannot cover even one criterion, therefore it fails and is not recommended as an enhanced oil recovery method in the reservoir. After that, the most suitable methods are selected and in order to investigate the effect of permeability, density and presence of water drive on the selection process in this case study, a sensitivity analysis is applied. Finally, two sets of passed and not recommended EOR techniques are listed for application in the under-survey reservoir.

5. Results and Discussion

5.1. Screening of Chemical Flooding Methods

For screening of chemical flooding methods in our reservoir case study, we used four methods of polymer flooding, alkaline/polymer flooding, surfactant/polymer flooding and alkaline/surfactant/polymer flooding. Table 2 lists selected screening criteria for polymer flooding, the reservoir values and their units. The recommended temperature criterion for polymer flooding is lower than 176 °F. The reservoir temperature value is reported 132.8 °F, which passes the recommended criteria. Oil viscosity, water hardness, water salinity and water-oil ratio (WOR) must be less than 150 cp, 1000 ppm, 100000 ppm and 10, respectively. The screening results show that the reservoir values are in agreement with these criteria. The horizontal permeability of the reservoir is 50 mD which is passed through screening filter. Chemical flooding method is not a good candidate for the reservoir with an active water drive, because of the low remaining oil saturation followed by water flooding (Alkafeef et al., 2007). The under-survey reservoir has no active water drive; therefore all chemical methods pass this criterion. Totally, polymer flooding passed all defined criteria, and may be a good candidate for enhanced oil recovery method in the Iranian reservoirs.

For applying alkaline/polymer flooding, the formation type must be sandstone and also API is recommended less than 35. The under-survey reservoir is a naturally fractured carbonate reservoir with oil density of 45.8 API which is not in agreement with the recommended criteria for alkaline/polymer method. Table 3 summarizes criteria of the most important parameters for enhanced oil recovery by alkaline/polymer method. According to the established procedure in this study, an EOR method must pass all its defined criteria; otherwise, the method will fail during the selection process of EOR methods. Therefore, alkaline/polymer flooding may not be selected for the under-survey reservoir.

Table 4 lists screening criteria for surfactant/polymer flooding method. All the effective reservoir parameters on selection of surfactant/polymer flooding are in agreement with screening criteria except water salinity. High salinity solutions are relatively ineffective during reduction of IFT in the added solvent to polymer flooding and totally chemical flooding methods due to decreasing viscosity with increasing salinity (Taber et al.1997a). High water salinity of the under-survey reservoir causes surfactant/polymer flooding not to be selected as a proper method.

Like alkaline/polymer flooding, alkaline/surfactant/polymer flooding is recommended in sandstone formation type. In addition, API and water salinity parameters cannot pass the defined screening criteria for alkaline/surfactant/polymer flooding and subsequently it is not an appropriate method. Table 5 summarizes the results of screening for alkaline/surfactant/polymer flooding method.

5.2. Screening of Gas Flooding Methods

To evaluate the gas flooding methods in the under-survey reservoir, four methods have been considered including CO_2 miscible, nitrogen miscible, hydrocarbon miscible and immiscible flooding. Table 6 lists the most important screening criteria and related reservoir values for CO_2 miscible flooding method. Critical parameters such as viscosity, density, temperature and depth can be passed through screening filter. For applying CO_2 miscible flooding, depth, temperature and density are suggested more than 1968.5 ft, 86°F and 22.3°API, respectively.

Table 6. Screening results of carbon dioxide miscible gas flooding for the under-survey reservoir

Result	Screening Criteria	Reservoir Value	Units
Pass.	Depth > 1968.5	2999	ft
Pass.	Reservoir Temperature > 86	132.8	°F
Pass.	Oil Density (Surface) > 22.3	45.8	°API
Pass.	Oil Visc. < 10	0.39	cp
Pass.	Gas Cap = < Local	Local	Select
Pass.	Current Oil Saturation > =0.25	0.5	ratio
Result:	PASSED		

Table 7. Screening results of hydrocarbon gas miscible flooding method for the under-survey reservoir

Result	Screening Criteria	Reservoir Value	Units
NREC.	Depth > 3937.008	2999	ft
Pass.	Oil Density (Surface) > 23.995	45.8	°API
Pass.	Oil Visc. < 5	0.39	cp
Pass.	Gas Cap = < Local	Local	Select
Pass.	Current Oil Saturation > =0.3	0.5	ratio
Result:	NOT RECOMMENDED		

In addition to the aforementioned parameters, optimal determination of CO_2 minimum miscibility pressure (MMP) plays a key role in CO_2 method because of expensive provision of the required CO_2, which is discussed later in the economical screening. It should be noted here that in naturally fractured reservoirs, when there is not an active water drive, CO_2 MMP must be less than the fracturing pressure.

The screening results for CO_2 method illustrates that the related reservoir values are in agreement with suggested criteria; therefore, CO_2 miscible flooding method is selected as an adequate method.

The critical screening conditions and related reservoir values for hydrocarbon (HC) and nitrogen (N_2) miscible gas flooding methods are in Tables 7 and 8, respectively. The main factor to be considered in miscible flooding methods of HC and N_2 is the essential pressure to achieve optimal miscibility, which is depth dependent (Taber, 1988). Therefore, these methods have been implemented in deep and very deep reservoirs with depth of higher than 3937 and 5900 ft for HC and N_2 miscible methods, respectively. The under-survey reservoir depth is 2988 ft that according to this criterion, HC and N_2 miscible methods may not be selected for such reservoir. Table 9 lists the most important criteria for selection of immiscible gas flooding. The reservoir values pass the criteria that suggest a proper EOR method for our reservoir case study.

Table 8. Screening results of nitrogen gas miscible flooding method for the under-survey reservoir

Result	Screening Criteria	Reservoir Value	Units
NREC.	Depth > 5905.512	2999	ft
Pass.	Oil Density (Surface) > 35	45.8	°API
Pass.	Oil Visc. < 2	0.39	cp
Pass.	Gas Cap = < Local	Local	Select
Pass.	Current Oil Saturation > =0.35	0.5	ratio
Result:	NOT RECOMMENDED		

Table 9. Screening results of immiscible gas flooding method for the under-survey reservoir

Result	Screening Criteria	Reservoir Value	Units
Pass.	Depth > 656	2999	ft
Pass.	Oil Density (Surface) > 12.888	45.8	°API
Pass.	Visc. < 600	0.39	cp
Pass.	Active Water Drive = No	No	Yes / No
Pass.	Bottom Water = < Local	None	Select
Pass.	Gas Cap = < Local	Local	Select
Pass.	Current Oil Saturation > =0.5	0.5	fraction
Result:	PASSED		

5.3. Screening of Thermal Recovery Methods

To investigate the applicability of thermal recovery methods in the under-survey reservoir, four methods have been considered including cyclic steam stimulation (CSS), continuous steam flooding (CSF), steam assisted gravity drainage (SAGD) and in-situ combustion (ISC). In thermal recovery, to be produced at profitable rates, the reservoir must have a high permeability (Taber et al. 1997a). Table 10 lists the critical criteria for CSS method and related reservoir values. This table suggests that permeability and porosity distributions must be higher than 200 mD and 18 %, respectively, which indicates that the related reservoir values are not in agreement with the suggested criteria.

In general, thermal recovery techniques have been implemented for heavy-oil reservoirs that cannot be produced in any other way because of their very viscous oil (Taber et al. 1997a). Therefore, viscosity reduction plays a key role in any thermal enhanced oil recovery processes (Hemmati-Sarapardeh et al., 2013a, b). The under-survey reservoir is a light oil reservoir with the viscosity of 0.39 cp, which is not in agreement with the suggested viscosity criteria.

As previously seen in Figures 2 and 3, thermal recovery methods have been used for reservoirs with shallow depth; therefore, it is obvious that shallow reservoirs have less pressure compared to deep reservoirs. Consequently, the current reservoir pressure may not pass through the screening filter. Based on the above comments, it is concluded that CSS may not be an efficient and appropriate EOR method for the under-survey reservoir.

Table 10. Screening results of CSS method for the under-survey reservoir

Result	Screening Criteria	Reservoir Value	Units
Pass.	Depth < 4593	2999	ft
NREC.	Current Pressure < 1501	1911	psi
NREC.	8<Oil Density (Surface) <35	45.8	°API
NREC.	100 < Visc. < 20,000	0.39	psi
NREC.	Horizontal Permeability > 200	50	mD
Pass.	Gas Cap = None	Local	Select
Pass.	Clay Content =< Low	Low	Select
Pass.	Net Pay Thickness > 197	882.5	ft
NREC.	Porosity > 0.180	0.057	Fraction
Result:	NOT RECOMMENDED		

Table 11. Screening results of CSF method for the under-survey reservoir

Result	Screening Criteria	Reservoir Value	Units
Pass.	Depth < 4593	2999	ft
NREC.	Current Pressure < 1501	1911	psi
NREC.	8<Oil Density (Surface) < 20	45.8	°API
NREC.	50 <Oil Visc. < 50,000	0.39	psi
NREC.	Horizontal Permeability > 200	50	mD
Pass.	Gas Cap =< Local	Local	Select
Pass.	Clay Content < =Low	Low	Select
NREC.	Fracturing = No	Yes	Yes / No
Pass.	Water Oil Ratio < 10	9	ratio
Pass.	Net Pay Thickness > 197	882.5	ft
NREC.	Porosity > 0.200	0.057	Fraction
Result:	NOT RECOMMENDED		

Table 11 summarizes the suggested screening criteria and related reservoir values for CSF method. All mentioned criteria for CSS method are used for steam flooding method except formation type. In steam flooding method, due to continuously steam injection to reservoir, steam escapes through fractures and as a result, heat loss occurs. The current pressure, permeability, porosity, formation type, viscosity, etc. are not in agreement with the suggested criteria. Therefore, CSF method may not be selected as a proper EOR method for the under-survey reservoir like CSS method.

The recommended criteria and related reservoir values for SAGD method are in Table 12. Due to heat cost and losses in reservoirs with net to gross pay zone ratio more than about 0.75, SAGD method and totally thermal methods are suggested (Dusseault, 2006; Greaves and AL-Shamali, 1995; Shafiei et al., 2007). However, net to gross pay zone ratio of the under-survey reservoir is not in agreement with the recommended criterion in addition to other criteria such as permeability, porosity, viscosity, *etc.* Therefore, SAGD method fails for applying in the under-survey reservoir.

In-situ combustion (ISC) is classified as a commercial heavy-oil technology, though it is used only at very limited conditions (Shafiei et al., 2007). However, the suggested critical screening criteria and reservoir values for ISC method are summarized in Table 13. The reservoir oil viscosity, density, fracturing and porosity parameters cannot be passed through the screening filter. Therefore, this method failed for enhanced oil recovery in the under-survey reservoir as same as other thermal methods.

Table 12. Screening results of SAGD method for the under-survey reservoir

Result	Screening Criteria	Reservoir Value	Units
Pass.	Depth < 4593	2999	ft
NREC.	Current Pressure < 1501	1911	psi
NREC.	7<Oil Density (Surface) <12	45.8	°API
NREC.	Oil Visc. > 2000	0.390	psi
NREC.	Horizontal Permeability > 1000	50	mD
Pass.	Gas Cap = < Local	Local	Select
Pass.	Clay Content =< Low	Low	Select
NREC.	Fracturing = No	Yes	Select
Pass.	Water Oil Ratio < 10	9	ratio
Pass.	Net Pay Thickness > 49.2	882.5	ft
NREC.	Porosity > 0.260	0.057	Fraction
NREC.	Net Pay / Gross Pay > 0.7	0.45	Ratio
Result:	NOT RECOMMENDED		

Table 13. Screening results of ISC method for the under-survey reservoir

Result	Screening Criteria	Reservoir Value	Units
Pass.	492 < Depth < 5906	2999	ft
NREC.	2 < Oil Visc. < 5000	0.390	cp
NREC.	10<Oil Density (Surface) < 20	45.8	°API
Pass.	Horizontal Permeability > 50	50	mD
Pass.	Bottom Water = < Local	None	Select
Pass.	Gas Cap = <Local	Local	Select
NREC.	Fracturing = No	Yes	Yes/No
Pass.	Net Pay Thickness > 9.84	882.5	ft
NREC.	Porosity > 0.18	0.057	fraction
Result:	NOT RECOMMENDED		

5.4. Sensitivity Analysis

The importance of EOR screening approach becomes apparent when a parameter is placed in the border between two methods. This means that the range of only a parameter can change the selection of a method for a specific reservoir. To illustrate our purpose, we conducted the sensitivity analysis for the three selected proper EOR methods. In the first sensitivity analysis, to observe the effect of water drive on selection of the three EOR methods, we assumed that water drive is active. The results in Table 14 demonstrate that water drive is a critical parameter for polymer and immiscible flooding methods and only CO_2 miscible flooding has been selected in presence of water drive.

Table 14. Summary of sensitivity analysis for application of polymer, CO_2 miscible and immiscible gas flooding methods

Active Water Drive	Permeability=200	Permeability=25	°API=48	°API=10
-	Polymer	-	Polymer	-
-	Immiscible Gas	Immiscible Gas	Immiscible Gas	-
CO_2 Miscible	CO_2 Miscible	CO_2 Miscible	CO_2 Miscible	-

In the next sensitivity analysis, we used rock and fluid parameters of permeability and oil density. 25 and 200 mD for permeability and 48 and 10 for API are considered. The results show that with permeability of 200 mD and density of 48 API, the three EOR methods still are still selected for the under-survey reservoir. The permeability of 25 mD causes the polymer flooding method fails and also density of 10 API makes the fail conditions for both CO_2 and immiscible gas flooding methods as well as polymer flooding method. Here, it should be mentioned that both CO_2 and immiscible gas flooding methods can be used in any permeability range as far as permeability is not a critical parameter for above methods (Taber et al.1997a). The results of all sensitivity analysis are summarized in Table 14. This table confirms the key role of EOR screening approach in proper selection of an EOR method; also it shows that the value of only one parameter can change selection process.

CONCLUSION

In this study, to identify the effective parameters criteria, a large database including more than 400 successful EOR projects has been gathered in addition to several presented EOR criteria tables by the researchers. Then, a comprehensive study for selection of the proper EOR methods was performed. To pursue our objective, 12 EOR techniques were screened and the three methods of polymer flooding, CO_2 miscible flooding and immiscible gas flooding have been selected as appropriate EOR methods for applying in an Iranian naturally fractured reservoir. Some of the important conclusions derived from this study are as follows:

1. Totally, chemical methods have been used in mid-viscosities, mid-depths and mid-densities; gas flooding methods are applied in light oils, low viscosities and deep

reservoirs; and thermal methods have been applied in shallow reservoirs, viscous and heavy oils.

2. High water salinity for the added solvents to polymer is ineffective due to decreasing their viscosities.
3. The sandstone formation type for the added alkaline solvent method is a critical parameter; thus they cannot be applied for the naturally fractured carbonate reservoirs.
4. Reservoir depth is a critical parameter for all gas flooding methods, because light oils are found in deep reservoirs with gas flooding methods suggested for them.
5. Presence of fracture causes steam escapes through fractures and heat loss occurs; therefore, thermal methods generally are not suitable for the naturally fractured reservoirs.

ACKNOWLEDGMENT

IOR Research Institute; NIOC R&T is acknowledged for providing the required data and supporting this study.

REFERENCES

Abbas, E., Song, Ch.L., 2011. Artificial Intelligence Selection with Capability of Editing a New Parameter for EOR Screening criteria. *J. Eng. Sci. Tech.* 6, 628-638.

Al Adasani, A., and Bai, B, 2010. Recent Developments and Updated Screening Criteria of Enhanced Oil Recovery Techniques. In: *the CPS/SPE International Oil & Gas Conference and Exhibition,* Beijing, China.

Al Adsani, A., Bai, B., 2011, Analysis of EOR Projects and Update Screening Criteria. *J. Pet. Sci. Eng.* 79, 10–24.

Alkafeef, S.F., Zaid, A.M., 2007. Review of and outlook for enhanced oil recovery techniques Kuwait oil reservoirs. In: the International Petroleum Technology Conference, Dubai, UAE.

Al-Mutairi, S.M., Kokal, S.L, 2011. EOR potencial in the Middle East: current and future trends. In: SPE Europe/ EAGE Annual Conference and Exhibition, Vienna, Austria.

Alvardo, V., Ranson, A., Hernandez, K, Manrique, E., Matheus, J., Liscano, T., Prosperi, N., 2002. Selection of EOR/IOR Oporunities Based on Machine Learning. In: *the 13th European Petroleum Conference*, Aberdeen, UK.

Barzegari, D., Ayatollahi, Sh., Zerafat, M.M., Roosta, A.A., 2010, EOR Screening Using Artificial Intelligence Bayesian Network. In: 14th International Oil, Gas & Petrochemical Congress, Tehran, Iran.

Bennion, D. B., Thomas, F. B., Ma, T., Imer, D., 2000. Detailed Protocol for the Screening and Selection of Gas Storage Reservoirs. In: *the 2000 SPE/CERI Gas Technology Symposium held in Calgary,* Alberta, Canada.

Dickson, J. L., Leahy-Dios,A., Wylie, P.L., 2010, Development of Improved Hydrocarbon Recovery Screening Methodologies. In: *SPE Improved Oil Recovery Symposium held in Tulsa*, Oklahoma, USA.

Dusseault, M.B., 2006. Sequencing Technologies to Maximize Recovery. In: the Petroleum Society's 7th Canadian International Petroleum Conference (57th *Annual Technical Meeting*), Calgary, Canada.

Gharbi, R. B.C., 2000, An Expert System for Selecting and designing EOR Processes. *J. Pet. Sci. Eng.* 27, 33–47.

Gharbi, R., 2005, Application of an Expert System to Optimize Reservoir Performance. *J. Pet. Sci. Eng.* 49, 261–273.

Greaves, M. and AL-Shamali, O., 1995. Wet In-Situ Combustion Process Using Horizontal Wells. In: *The Sixth UNITAR International Conference on Heavy Crude and Tar Sands,* Houston, Texas, USA.

Guerillot, D.R., 1988, EOR Screening With an Expert System. In: *The Symposium on Petroleum Industry Applications of Microcomputers,* San Jose, California, USA.

Hemmati-Sarapardeh, A., Khishvand, M., Naseri, A. Mohammadi, A.H., 2013a, Toward Reservoir Oil Viscosity Correlation. *J. Chem. Eng. Sci.* 90, 53-68.

Hemmati-Sarapardeh, A., Hashemi Kiasari, H., Alizadeh, N., Mighani, S., Kamari, A., 2013b. Application of Fast-SAGD in Naturally Fractured Heavy Oil Reservoirs: A Case Study. In: *The 18th Middle East Oil & Gas Show and Conference,* Bahrain.

Ibatullin, R.R., Ibragimov, N.G., Khisamov, R.S., Podymov, E.D., Shutov, A.A. 2002. Application and Method Based on Artificial Intelligence for Selection of Structures and Screening of Technologies for Enhanced Oil Recovery. In: *SPE/DOE Improved Oil Recovery Symposium*, Tulsa, Oklahoma, USA.

Lee, J.Y., Shin, H.J., Lim, J.S., 2011, Selection and Evaluation of Enhanced Oil Recovery Method Using Artificial Neural Network. *Geosyst. Eng.*, 14, 157-164.

Moritis, G., 1998, Worldwide EOR Survey, Oil & Gas Journal, V. 96, Issue 16, pp to59. URL:http://www.ogj.com/articles/print/volume-96/issue-16/in-this-issue/general-interest/1998-worldwide-eor-survey.html

Moritis, G., 2006, Worldwide EOR Survey, Oil & Gas Journal, V. 104, Issue 15, pp to57. URL:http://www.ogj.com/articles/print/volume-104/issue-15/special-report/2006-worldwide-eor-survey.html

Moritis, G., 2008, Worldwide EOR Survey: Oil & Gas Journal, V. 106, Issue 15, pp to 59. URL:http://www.ogj.com/articles/print/volume-106/issue-15/drilling-production/special-report-2008-worldwide-eor-survey.html

Parada, C.H., Ertekin, T., 2012, A new Screening Tool for improved Oil Recovery Methods Using Artificial Neural Network. In: SPE Western Regional Meeting held in Bakersfield, California, USA.

Parkinson, W.J., Luger, G.F., Bretz, R.E., Osowski, J., 1994, Using an Expert System to Explore Enhanced Oil Recovery Methods. *Comput. Elect. Eng.* 20, 181-197.

Shokir, E.M., Goda, H.M., Sayyouh, M.H., and Fattah, Kh.A., 2002. Selection and Evaluation EOR Method Using Artificial Intelligence. In: the 26rd Annual International Technical Conference and Exhibition, Abuja, Nigeria.

Shafiei, A., Dusseault, M.B., Memarian, H., Samimi Sadeh, B., 2007. Production Technology Selection for Iranian Naturally Fractured Heavy Oil Reservoirs. Presented at the

Petroleum Society's 8th Canadian International Petroleum Conference (58th Annual Technical Meeting), Calgary, Alberta, Canada, 12 – 14 June.

Stosur, G.J., Hite, J.R., Carnahan, N.F., Miller, K., 2003a. A alphabet soup of IOR, EOR and AOR: effective communication requires a definition of terms. Presented at the SPE International Improved Oil Recovery Conference in Asia Pacific, Kuala Lumpur, October.

Stosur, G.J., 2003. EOR: Past, Present and What the Next 25 Years May Bring. Presented at the SPE IOR Conference in Asia Pacific, Kuala Lumpur, Malaysia, October 20-21.

Taber, J.J., 1988. The use of flue gas for enhanced recovery of oil. Presented at EOR by Gas Injection Symposium, Copenhagen, Denmark, September.

Taber, J.J., Martin, F.D., 1983. Technical Screening Guides for the Enhanced Recovery of Oil. In: *Proceeding of the 58th Annual Technical Conference and Exhibition*, San Francisco, California, USA.

Taber, J.J., Martin, F.D., Seright, R.S., 1997a, EOR Screening Criteria Revisited - Part1: Introduction to Screening Criteria and Enhanced Recovery Field Projects. In: *the SPE/DOE Improved Oil Recovery Symposium, Tulsa*, Oklahoma, U.S.A.

Taber, J.J., Martin, F.D., Seright, R.S., 1997b. EOR Screening Criteria Revisited - Part2: Application and Impact of Oil Prices. In: *Proceeding of the SPE/DOE Improved Oil Recovery Symposium*, Tulsa, Oklahoma, USA.

Zerafat, M.M., Ayatollahi, Sh., Mehranbod, N., and Barzegari, D., 2011. Bayesian Network Analysis as a Tool for Efficient EOR Screening. In: *the SPE Enhanced Oil Recovery Conference,* Kuala Lumpur, Malaysia.

In: Enhanced Oil Recovery
Editor: Alicia Knight

ISBN: 978-1-63463-917-0

Chapter 12

STUDY OF THE PERFORMANCE OF CYCLIC STEAM STIMULATION (CSS) OIL RECOVERY METHOD IN A NATURALLY FRACTURED CARBONATE RESERVOIR

***Arash Kamari*[1], *Mohammad Nikookar*[2,•] *and Amir H. Mohammadi*[1,3,*]**

[1]Thermodynamics Research Unit, School of Engineering,
University of KwaZulu-Natal, Howard College Campus,
Durban, South Africa
[2]IOR Research Institute, National Iranian Oil Company (NIOC), Tehran, Iran
[3]Institut de Recherche en Génie Chimique et Pétrolier (IRGCP),
Paris Cedex, France

ABSTRACT

In this communication, we simulate the cyclic steam stimulation (CSS) oil recovery method in the fractured carbonate reservoirs containing heavy-oil and also analyze the influences and contribution of different reservoir properties on ultimate oil recovery and cumulative Steam Oil Ratio (cSOR). Kuh-e Mond field, located in the south western of Iran, is considered as a specific case. The results indicate three different stages of oil production in CSS process in Kuh-e Mond NFR. First, oil production immediately increases due to fast-heating of oil by abrupt injection of steam; second, due to more heating by steam escape through fractures drainage of matrix blocks begins and a reduction in oil rate occurs; third, steam penetrates into upper fractures and therefore a pulse in oil production rate seems. The conclusions indicate that CSS performance within Kuh-e Mond reservoir is impacted by the fracture geometrical properties and how the fluid flow through matrix blocks of Kuh-e Mond reservoir in which permeability, porosity and rock wettability of matrix.

• E-mail: mohamad_ni@yahoo.com.
* E-mail: a.h.m@irgcp.fr; e-mail: amir_h_mohammadi@yahoo.com.

Keywords: cyclic steam stimulation (CSS), naturally fractured reservoir (NFR), fracture, matrix, recovery factor (RF), cumulative steam oil ratio (cSOR)

1. INTRODUCTION

About 25- 30% of the world supply of oil are contained in Naturally Fractured Reservoirs (NFRs) (Saidi, 1988). Many of the NFRs resources including heavy and/or extra heavy oil in the world carry consequential potential but recovery of oil in the matrix in those resources is still big challenge for petroleum reservoir engineers (Babadagli and Al-Bemani, 2007). According to the latest researches, there are more than 80 billion bbls of heavy oil in the Iranian reservoirs (Razavi and Kharrat, 2009). Hence, Iranian heavy oil NFRs can only be recovered by thermal enhanced oil recovery (EOR) processes including cyclic and/or continuous steam injection, in-situ combustion, SAGD, *etc*. It should be mentioned that the thermal EOR processes are best suited for reservoirs containing tar sands (≤10° API) and heavy oils (10-20 °API). As a consequence, the major mechanisms involve a large decreasing in viscosity (Kamari et al., 2014; Hemmati Sarapardeh et al., 2013), and therefore mobility ratio by increasing the reservoir temperature as thermal EOR processes supply heat to the reservoir, and vaporize some of the oil.

Cyclic Steam Stimulation (CSS) or huff-and-puff and continues steam flooding or steam drive are two main forms of steam injection EOR method (Butler, 1997). Cyclic steam stimulation EOR technique for thermal recovery of heavy oil reservoirs has received attention since early 1960's (Escobar et al., 2000). CSS is a "single well" process (Owens and Suter, 1965), and includes three stages, as shown in Figure 1. First, steam is injected for about one month. Then, the CSS well is shut in for a few days for distribution of heat in the reservoir, in other words for soaking. In the latest stage of CSS process, the well is put on production as oil rate increases quickly to a high rate, and stays at that level for a short time, and decreases over several months. Recovery factors for CSS process are low (10-40% OOIP), however, this method is especially attractive because it has quick payout (Thomas, 2007). Figure 1 shows the typical CSS process by a horizontal well in naturally fractured reservoir.

Fractured networks and matrix blocks are two parts of a naturally fractured rock. Consequently, NFRs can be considered as the excessive case of heterogeneous oil/gas reservoirs. Naturally induced fractures are channels through which fluid can flow towards a production well. These fractures may have a negative or positive impact on ultimate recovery of oil reservoir (van Golf-Racht, 1982). Fast thermal conduction is one of the most important positive influences, which lets heat to sweep areas of the reservoir that are not contacted by steam (Bagci, 2004). Beside, heat loss or steam escape through fractures of upper matrix blocks can be a negative impact. Furthermore, matrix properties such as permeability and porosity have significant role on oil recovery and express the relationship between pores. Therefore, study the pore geometry as an effective and fundamental parameter in controlling the fluid flow in porous medium is necessary.

The aim of this work is to study of the influences of fracture/matrix properties on CSS efficiency in Iranian heavy oil NFR (Kuh-e Mond). To pursue our objective, we firstly describe the geological background of Kuh-e Mond reservoir, and then develop a rectangular model to simulating CSS process.

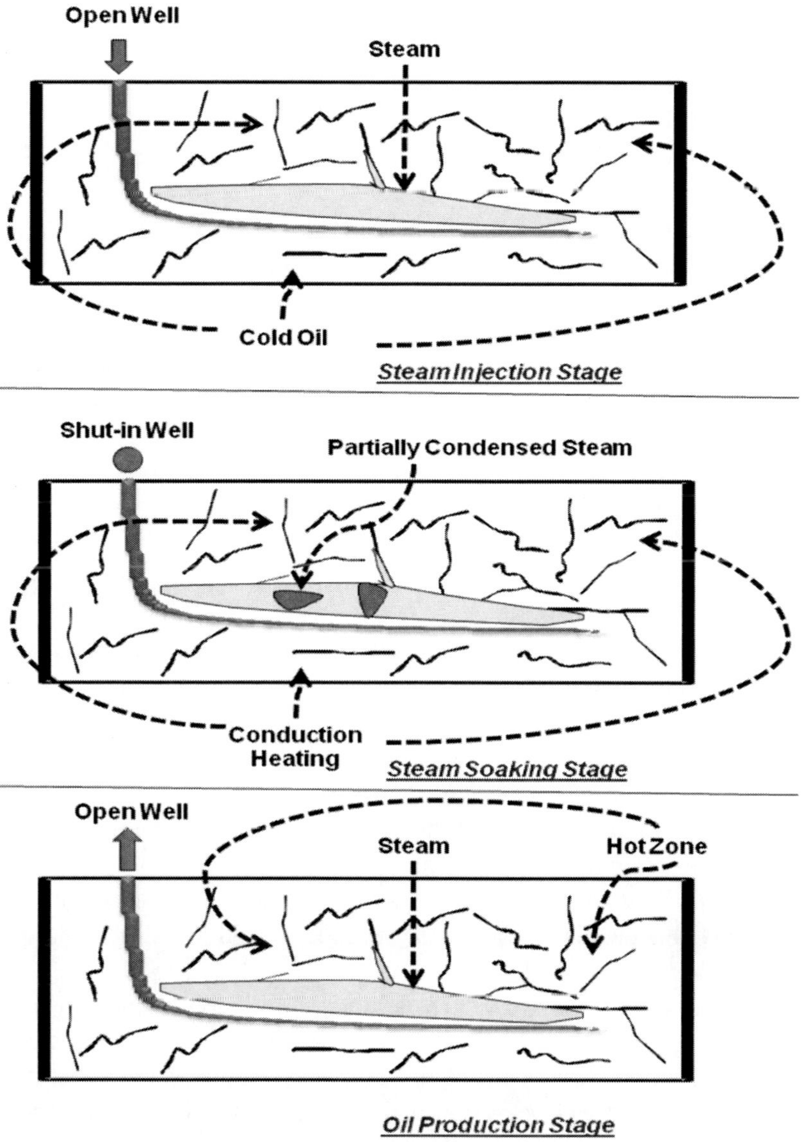

Figure 1. Typical cyclic steam stimulation in NFRs (steam zone is presented in yellow).

After that, effect of parameters such as matrix and fracture permeability, matrix and fracture porosity, fracture spacing and matrix wettability were observed. A Computer Modeling Group (CMG), thermal advanced reservoir simulator (STARS) have been utilized for the simulation investigations.

2. Literature Survey

Some attempts have been made in the literature to examine and simulate the CSS method and to more understand the sensitivity of the process to various operational and reservoir parameters, especially in the case of sandstone systems (Alajmi and King, 2011; Watkins et al., 1987; Liu et al., 2006; Beattie et al. 1991; Gunadi, 1999; Rodrigues, 1999). However, a few studies have been implemented on investigation of naturally fractured formations (Razavi and Kharrat, 2009; Reis, 1990; Kamari, 2013) despite the fact that they are large reservoirs containing one third of worldwide heavy oil and bitumen resources.

Tamim and Farough Ali (1998) developed a new analytical model for prediction of CSS performance. Their model was proposed based on a fracture heating computation, coupled with fluid flow. The conclusions demonstrated that the developed mathematical model can apply as a valuable adjunct to physical modeling or numerical simulation. Gozde et al. (1993) proposed a new mathematical model for heavy oil reservoirs including pressure drive in order to examine the cyclic steam method. To verify the developed model, they compared the results with available field data. Moreover, the obtained results were compared with another analytical model, which the proposed model was much better equipped to history match a wide variety of heavy oil reservoirs. Revana and Erdogan (2007) developed a heavy oil reservoir model including single well. Furthermore, they coupled their model with spreadsheet-based economical model. Escobar et al. (2000) presented a novel methodology to optimize CSS method for horizontal and vertical wells. Their new method implemented characterization of oil production using numerical simulation, validation, tuning and comparison model, and maximization of net present value. In other words, the new methodology was successfully compared and verified with previously published results in the literature. Hascakir and Kovscek (2010) simulated the CSS process into a low permeable, hydraulically fractured and diatomaceous reservoir which includes heavy oil. Wettability alteration was rationalized by relative permeability. Temperature-dependent relative permeability was examined and the results demonstrated that liquid-gas relative permeability affects the process performance most sensitively. Yitang et al. (2005) introduced an efficient way to increase the influence of proposing waxy oil low-permeability reservoir by cyclic stimulation method. Their results showed that high temperature steam injection into the reservoir can lead to the thermal expansion of the underground fluid, decrease the viscosity of waxy oil and distillate the crude oil.

Up to now, a comprehensive study is not well done to quantify the effects of fracture/matrix properties on performance of CSS process in naturally fractured reservoir. However, there are some efforts to fundamental understanding of how fracture geometrical characteristics control the efficiency of oil recovery in other types of enhanced oil recovery technique. Kamari et al. (2012) performed a series of laboratory works in strongly oil-wet glass micro-models involving different fracture geometries. The results demonstrated that by increasing the fracture's length, the breakthrough time reduces. In contrast, by increasing the orientation angle associated with flow direction, breakthrough time increases. Fatemi et al. (2011a) assessed the influence of fractures geometrical properties on the performance of the toe-to-heel air injection recovery method using experimental works. Some evaluated geometrical parameters were fracture location, fracture density, and fracture orientation. The obtained results demonstrated that the followed recovery method is more appropriate, efficient and applicable on highly networked fractured reservoirs. Farzaneh et al. (2011) performed solvent injection methods on one-quarter five-spot fractured micromodels at a fixed flow rate condition that initially saturated with the heavy oil. Azin et al. (2008) investigated the impact of fracture spacing geometrical property on the performance of VAPEX process. Hence, they utilized a compositional tool to simulate VAPEX process in a number of fractured systems including single and multiple blocks. Fatemi et al. (2011b) studied various fracture geometrical properties such as discontinuity, extension, orientation, dispersion, networking, and location during Steam Assisted Gravity Drainage (SAGD) recovery process. One of the most important results obtained by Fatemi et al. (2011b) was improving sweep efficiency and oil recovery in the presence of vertical fractures. Bagci (2004) performed a laboratory study to find the influence of well configurations and fractures on the SAGD recovery method. The obtained results demonstrated that vertical fractures increased the efficiency of SAGD process. Fattahi et al. (2012) studied the impact of reservoir heterogeneity of fracture spacing, matrix permeability and fracture permeability on SAGD process in NFR reservoirs. Furthermore, they investigated solutions to overcome negative influences of the aforementioned geometrical properties on SAGD process.

3. Model Development

In this work, the employed simulator was CMG-STARS. PVT properties were calculated and matched using CMG-Winprop software. Due to high run time because of large area of Kuh-e Mond, a rectangular sector of the field with dimension of 378 × 1400 × 96 ft was selected as base case model. To exclude grid sensitivity of the results, the grid sensitivity analysis has been done and 36 × 1 × 12 cells in ***i***, ***j***, ***k*** coordinate were selected as reservoir grid numbers. Average porosity, permeability and irreducible water saturation of water-wet matrix were set to 19.5%, 50 md and 15%, respectively. Formation thermal conductivity and rock heat capacity of matrix were set to 24 Btu/ft3. °F and 30 Btu/Day.ft.°F. The matrix properties are summarized in Table 1.

Modeling the fracture networks is one of the challenges faced by reservoir engineers and geoscientists when study NFRs (Zellou and Ouenes, 2001). Therefore, to simulate naturally fractured reservoirs, the dual porosity model has been employed in this study. In the dual porosity models, there is no direct communication between interblock matrices, *i.e.*, neighboring blocks are connected just through fracture flow. In other words, only communication way is fractures as the heat or fluid inside matrix can be transferred only to fracture. The fractures with permeability of 2000 md and 30 ft dimension in ***i***, ***j*** and ***k*** direction made a network fracture with porosity of 0.006 fully saturated by the crude oil. The oil thermal conductivity in the fractures was set to 2 Btu/ft3. °F. The fracture parameters are shown in Table 2.

X_{2+} (gaseous phase) C_{2+} (oil phase), and C_{7+} (oil phase) are three pseudo-components known of crude oil. The Peng-Robinson equation of state (Peng and Robinson, 1976) was utilized to match fluid properties. The fluid data are summarized in Table 3. The oil viscosity at reservoir temperature (140 °F) is about 2000 cp and its changes versus temperature are shown in Figure 2. The method which was utilized to find the relative permeability and capillary pressure for the Kuh-e Mond rock was Stone-2 model. The matrix oil-water relative permeability and capillary pressure are shown in Figures 3 and 4, respectively. Moreover, fracture oil-water relative permeability curves are indicated in Figure 5. Distribution of oil, water and gas saturations of matrix and fracture systems are shown in Figure 6 and Figure 7, respectively.

In order to study the effect of fractures/matrix geometrical properties on the CSS performance, a base case should be defined.

Table 1. The matrix properties

Parameters	Unit	Value
Permeability	md	50
Porosity	%	19.5
Residual Oil Saturation	%	40
Irreducible Water Saturation	%	15
Formation Thermal Conductivity	Btu/Day .ft. °F	24
Rock Heat Capacity	Btu/ft3. °F	30

Table 2. The fracture properties

Parameters	Unit	Value
Permeability	md	2000
Porosity	%	0.006
Oil Saturation	%	100
Fracture Spacing	ft	10
Oil Thermal Conductivity	Btu/Day .ft. °F	2

Table 3. Mole fraction and molecular weight of the crude oil used

Component	Mw	Composition (model fraction)
X_2^+	17	0.1124
C_2^+	74	0.1854
C_7^+	411	0.7022

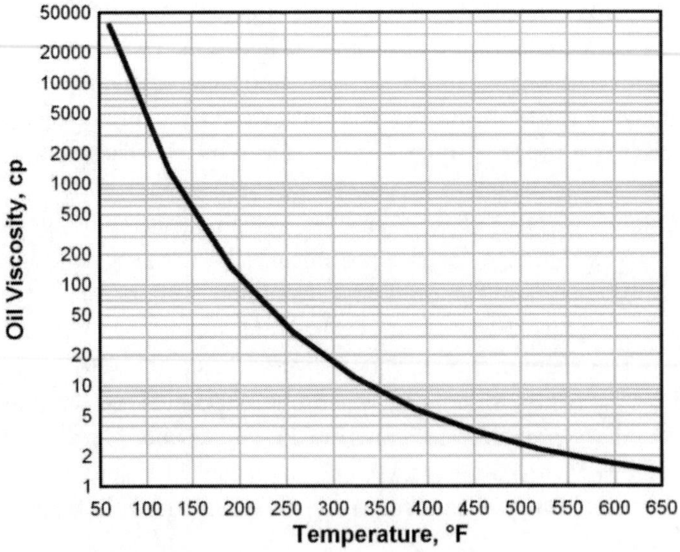

Figure 2. Oil viscosity versus temperature.

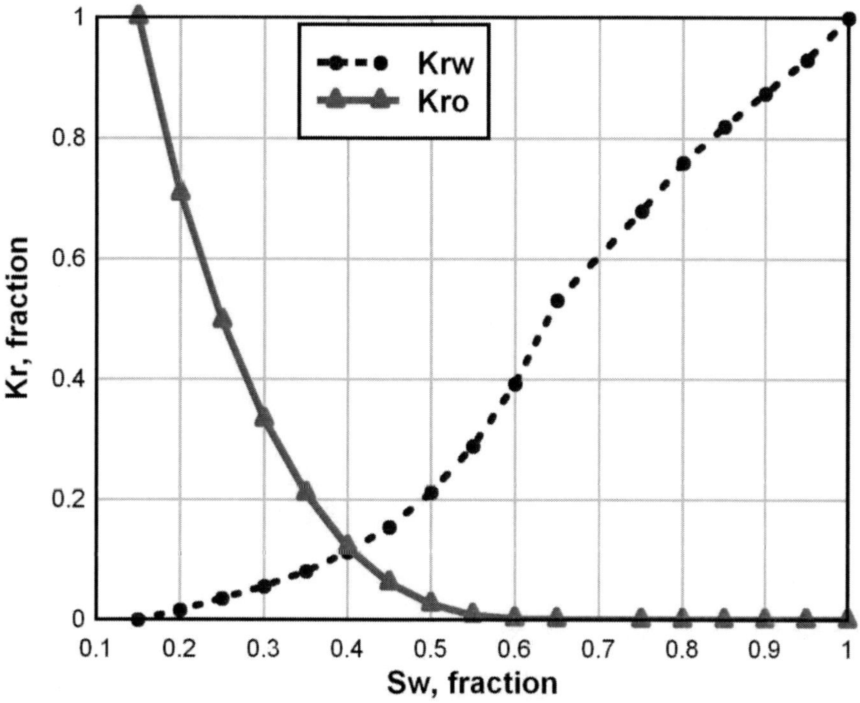

Figure 3. Oil-water relative permeability of Kuh-e Mond matrix.

Table 4. Reservoir and operational parameters used in the proposed model

Parameters	Unit	Value
Viscosity	cp	2000
Reservoir Pressure	psi	1200
Reservoir Temperature	°F	140
Reference Depth	ft	1030
Oil in Place	Ft^3	$7.5810\times10^{+06}$
Injector/Producer Well	#	3
Maximum Injection Pressure	psi	1700
Maximum Injection Temperature	°F	500
Steam Quality	-	0.9
Bottom Hole Pressure at Producer	psi	640
Steam Injection Period	day	30
Steam Soaking Period	day	15
Production Period	day	150
Total Operation Time	year	9

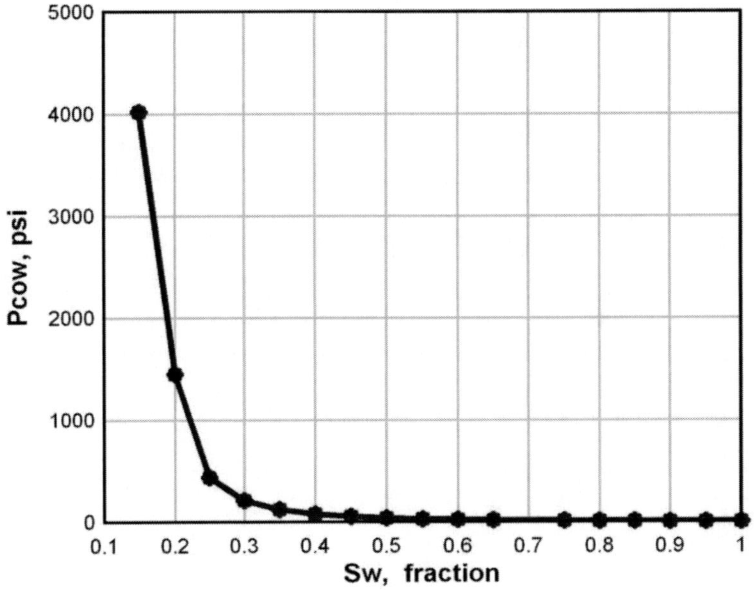

Figure 4. Capillary pressure curve of Kuh-e Mond matrix.

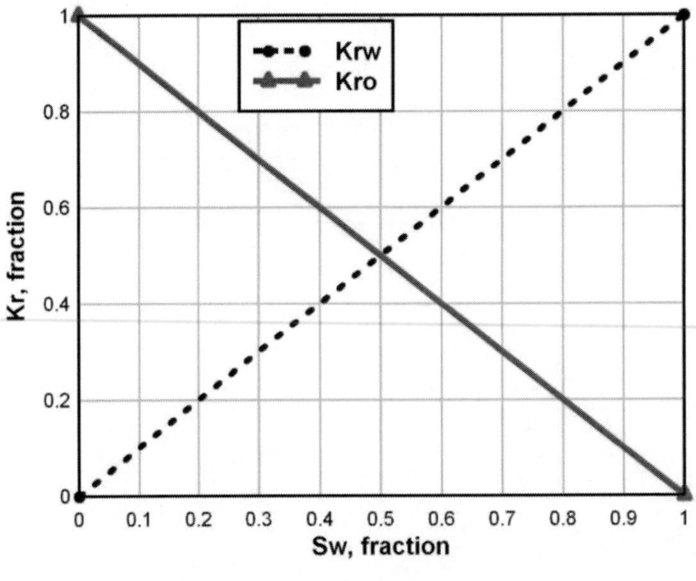

Figure 5. Oil-water relative permeability of Kuh-e Mond fracture.

Therefore, a series of reservoir and operational parameters were introduced to the basic model for making the base case model. The Bottom Hole Pressure (BHP) for production is 640 psi, for injection is 1700 psi and the steam temperature is 500 °F. The steam quality and injection rate are set over 0.9 and 1000 bbl/day, respectively. The runs were conducted with 30 days steam injection, 15 days soaking period and 150 days production period. Table 4 lists the inputs variables and operational parameters used in this model.

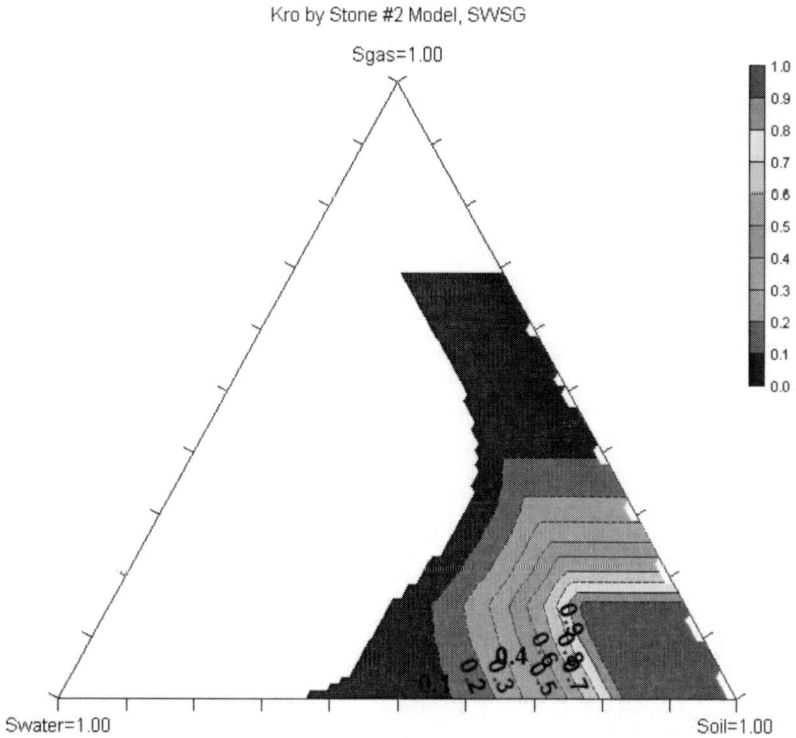

Figure 6. Distribution of oil, water and gas saturations of Kuh-e Mond matrix.

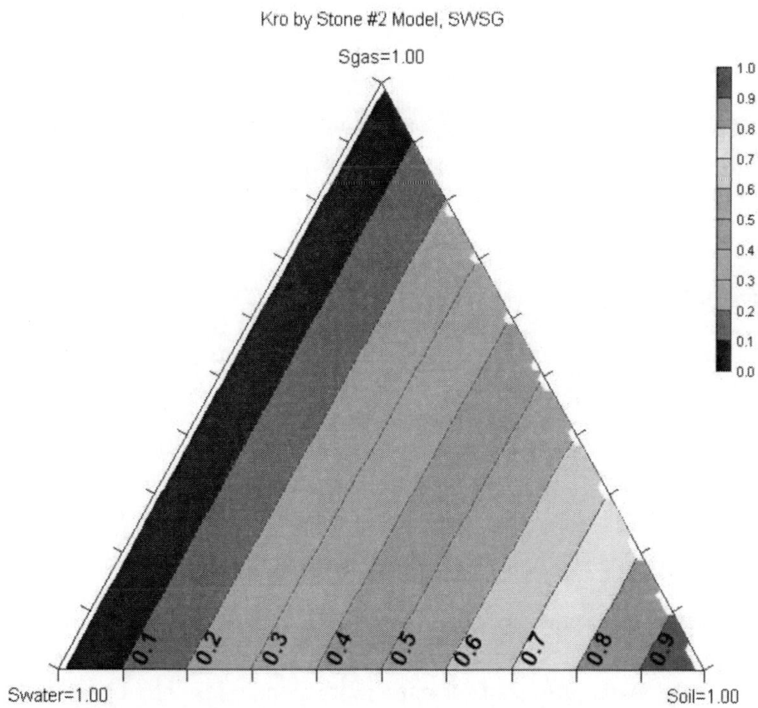

Figure 7. Distribution of oil, water and gas saturations of Kuh-e Mond fractures.

4. Results and Discussion

4.1. Base Case Analysis

After the first run, cumulative Steam Oil Ratio (cSOR) and Recovery Factor (RF) have been reported 0.926 and 26.71 %, respectively. Higher injection pressure and rate are needed to achieve a better recovery. Moreover, time periods have a significant impact on oil production and recovery factor during cyclic steam injection that optimization is necessary.

Figure 8 shows yearly oil production rate and cumulative oil rate in the base case model, and also Figure 9 indicates yearly water injection rate and cumulative water injection. These figures confirm that the oil production amounts are more than water injection amounts. As can be seen in Figure 8, the production profile can be divided into three main stages. At first stage, oil production immediately increases due to fast-heating of oil by abrupt injection of steam, reduction of oil viscosity, and somewhat steam invasion. Moreover, in this stage the oil of Near Well Region (NWR) becomes adequately hot and movable to production due to long preheating period (Kiasari et al., 2010).

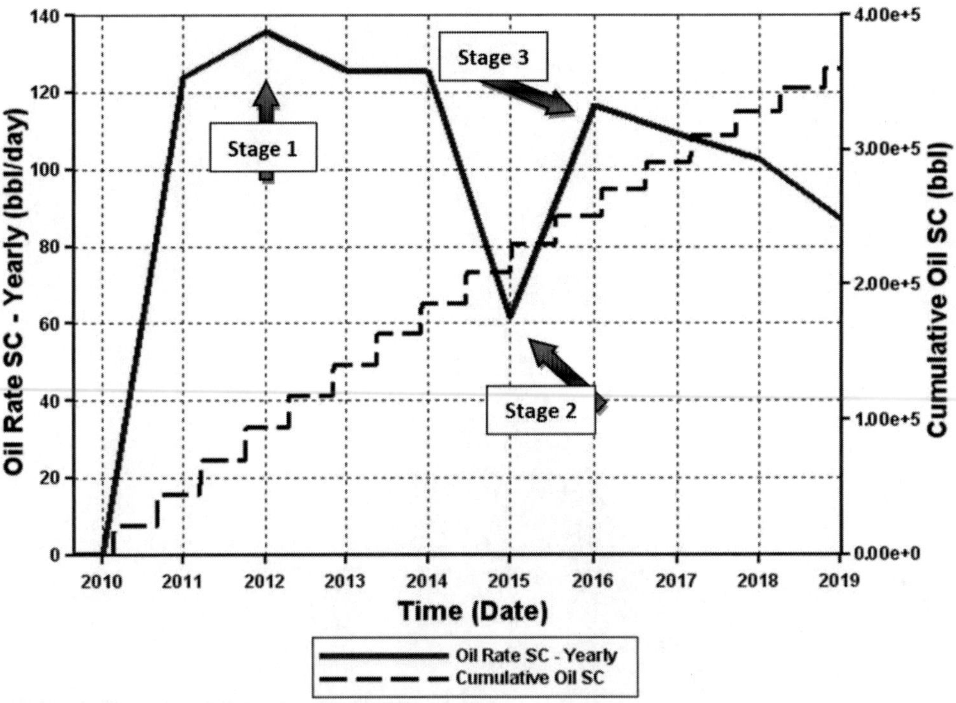

Figure 8. Yearly oil production rate and cumulative oil rate in the base case model.

Next, by increasing the steam injection over time steam escapes through vertical fractures and consequently rock/reservoir cannot redresses depletion of fractures and a reduction in oil rate occurs (Stage 2). After this stage, steam more penetrates into upper blocks and heat losses and therefore a pulse in oil production rate seems as steam rises into blocks in Above Well Region (AWR).

Additionally, the existence of vertical fractures can be another reason for creating that pulse in oil production trend (Stage 3). How to growth and develop the steam chamber of basic model is shown in Figure 10.

In contrast to conventional oil reservoirs, this Figure displays that the steam chamber in NFR during CSS process rises just vertically till it reaches the top of the reservoir and then starts a lateral expansion due to the existence of fractures.

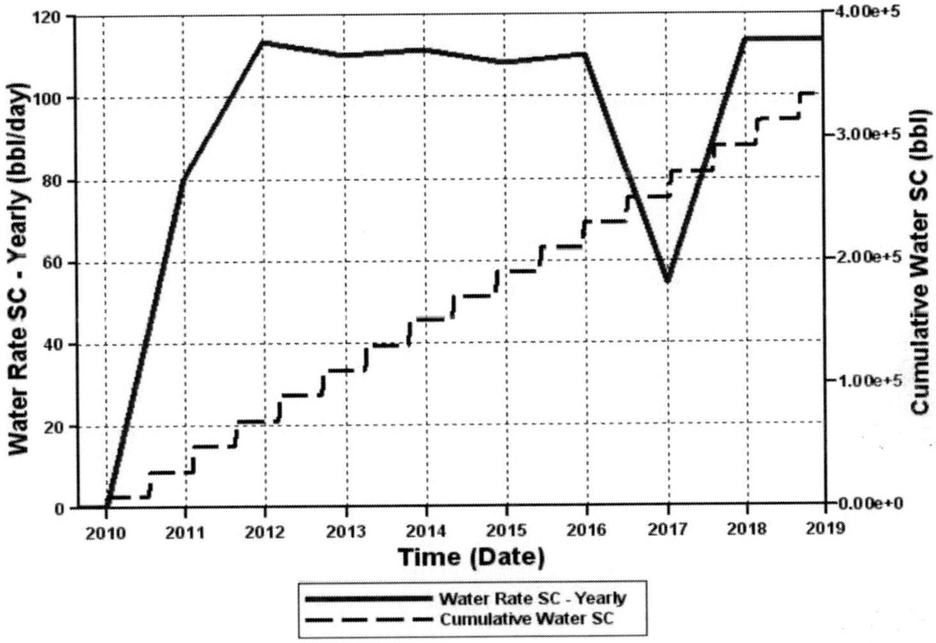

Figure 9. Yearly water injection rate and cumulative oil rate in the base case model.

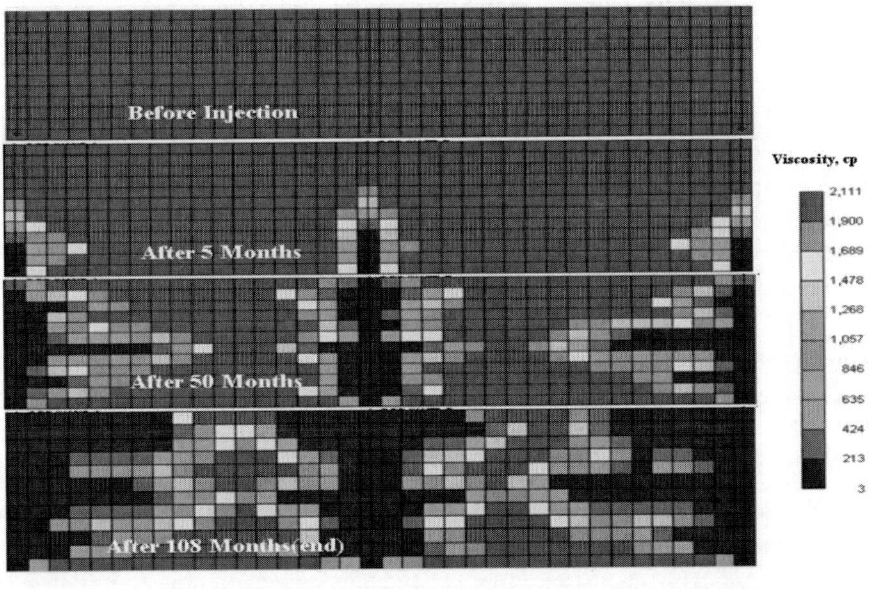

Figure 10. IK 2-D view of steam chamber growth during CSS process in the NFR.

4.2. Fracture Properties Impact

To investigate the impact of fracture properties on the performance of CSS process, we considered properties such as direction of fracture permeability (K_f), fracture spacing and fracture porosity (Φ_f,). In order to study the effect of increasing fracture permeability in ***i,j*** and ***k*** directions, three scenarios have been selected consists of 2500-2000-2000 mD, 2000-2500-2000 mD and 2000-2000-2500 mD for ***i***, ***j*** and ***k*** directions, respectively. The results indicate that the steam moves faster and more freely, and covers more upper blocks in the higher permeability of fractures. However, by increasing fracture permeability in ***k***-direction RF decreases compared to ***i*** and ***j*** directions and even the basic model. This may have two reasons. First, steam quickly escapes through vertical fractures and causes the heat to be wasted and lost. Second, steam moving through fractures in ***i*** and ***j*** directions or horizontal fractures causes the steam chamber to be expanded and therefore the steam will have more contact area with the reservoir. Figure 11 shows that the first and second stage of the production is almost the same for all cases due to appropriate preheating period, but in next stage, the oil rate trend of higher fracture permeability becomes more fluctuated. This inconstancy in the oil rate is encouraged when the permeability of fractures increases as high permeable fractures have key role in the steam transforming (Kiasari and Sedaee Sola, 2012).

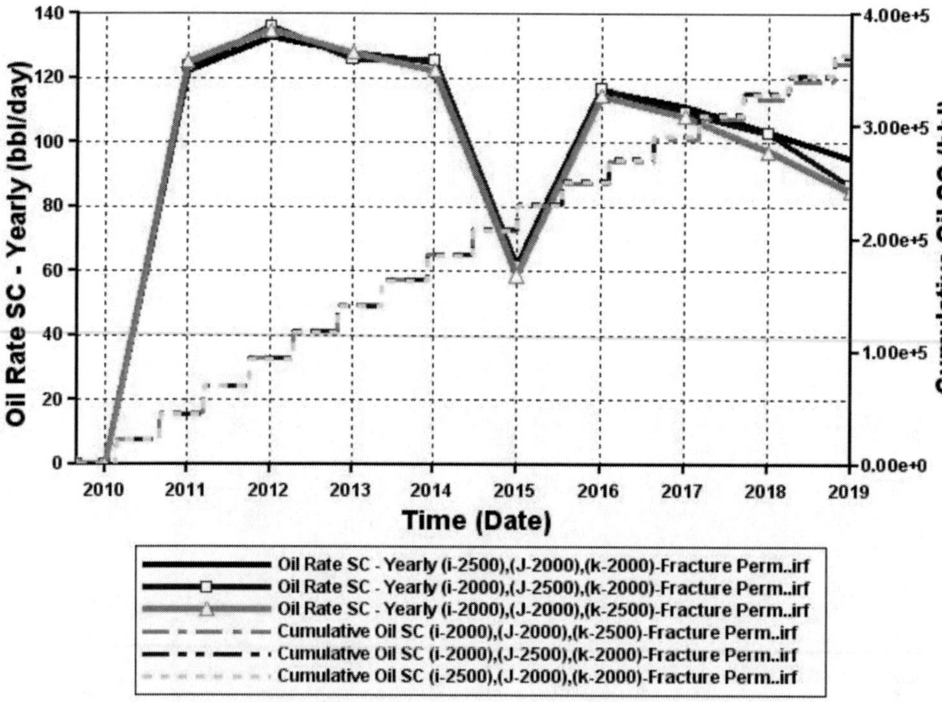

Figure 11. Effect of fracture permeability on production profile in the NFR.

Lower fracture densities are in higher fracture spacing systems and therefore reservoir oil becomes hot and movable because steam heats the matrix more efficiently. Consequently, the temperature of the model is decreased in higher fracture spacing systems but higher fracture density increases reduces steam rising speed (Kiasari and Sedaee Sola, 2012). With this definition, to observe the influence of fracture distance on various stages of oil production profile, three values for fracture spacing such as 1, 10 (base case), and 20 ft were selected. By increasing the fracture spacing value, the initial oil rate increases and also oil rate pulses seem earlier with greater picks value in the next stages (Figure 12).

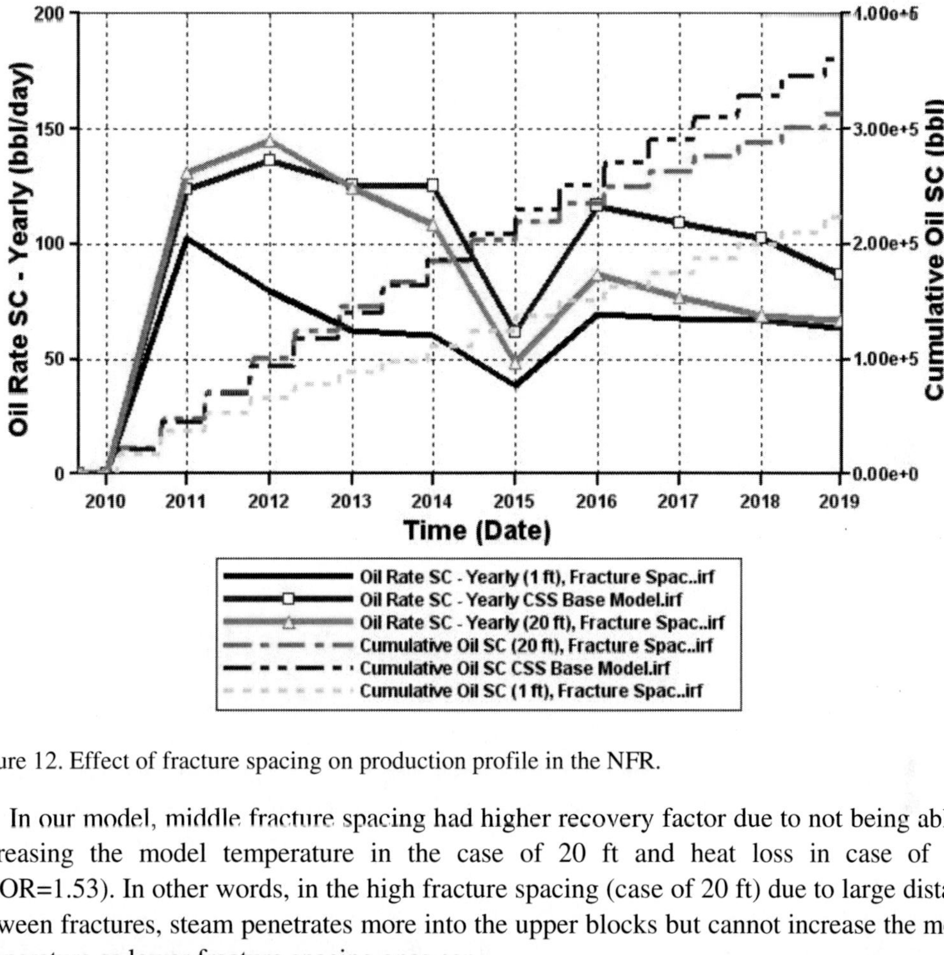

Figure 12. Effect of fracture spacing on production profile in the NFR.

In our model, middle fracture spacing had higher recovery factor due to not being able to increasing the model temperature in the case of 20 ft and heat loss in case of 1 ft (cSOR=1.53). In other words, in the high fracture spacing (case of 20 ft) due to large distance between fractures, steam penetrates more into the upper blocks but cannot increase the model temperature as lower fracture spacing ones can.

The volume of rock contacted with the steam injected to the reservoir is controlled by the porosity of fracture and in other words fracture porosity has a fundamental and large impact on the quantity of heat transported. As a result, more volume of fluid is extended from the well in systems including low fractures porosity (Briggs, 1989). Hence, the heat capacity of the fractures is directly associated with the porosity of fractures as more heat will transfer to the upper matrix blocks at larger values of Φ_f. Therefore, to study the effect of fractures porosity on cSOR and RF during CSS process in Kuh-e Mond NFR, we considered case of 0.01 in addition to base case (0.006).

Figure 13 shows that case of 0.006 has a higher oil rate at ultimate stage of production. The RF and cSOR for case of 0.01 are 0.96 and 25.3 %, and for case of 0.006 are reported 0.92 and 26.7 %. This little difference can be due to steam penetration into upper fractures and heat loss at ultimate stage of production. Table 5 summarized performance results for all fracture properties cases.

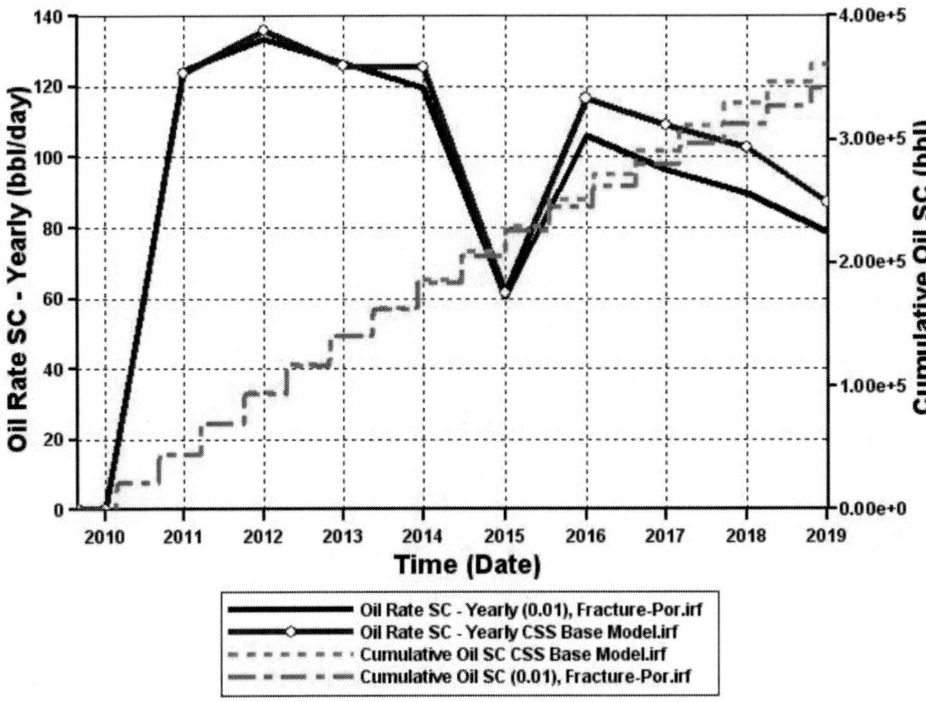

Figure 13. Effect of fracture porosity on production profile in the NFR.

4.3. Matrix Properties Impact

In order to study the effect of matrix properties on the performance of CSS process, we considered properties such as matrix permeability (K_m), matrix porosity (Φ_m) and matrix wettability. According to Darcy equation, reservoir quality and oil flow rate are associated to matrix porosity and permeability. Matrix porosity controls the capacitance of formation/reservoir or its storage capacity; therefore, decreasing matrix porosity should cause a decrease in oil recovery.

Moreover, by both vertical and horizontal steam escape and move through matrix the growth of the steam chamber is controlled (Gotawala and Gates, 2010). This points out that both the vertical and horizontal permeabilities as well as porosity affect the steam chamber growth as the variability of the matrix properties such as porosity and permeability controls the variability of the steam chamber size. In order to study the impact of matrix porosity on cSOR and ultimate RF, we considered three cases of 8 %, 19.5 % (base case) and 31 %. In addition, to investigate the influence of matrix permeability, three scenarios have been selected consists of 0.1, 50 (base case) and 500 mD. Figure 14 shows the yearly oil rate and cumulative oil production for three cases of porosity. This figure confirms that efficiency of CSS process in NFRs is strongly associated to matrix porosity. Figure 15 displays the oil production curves for three considered cases of matrix permeability. The response of our NFR model versus matrix permeability changes is that the model RF significantly decreases in low matrix permeability. On the other hand, in high matrix permeabilities the model RF is almost constant. This means that an appropriate matrix permeability value is necessary for applying CSS process in NFRs.

Table 5. Results of CSS performance against fracture properties changes

Scenario Name	Cumulative Oil Production, bbl	Cumulative Water Injection, bbl	cSOR	Ultimate RF, %
(i-2500),(j=2000),(k=2000)-K_f	363340	330749	0.910	26.911
(i-2000),(j=2500),(k=2000)-K_f	360691	334122	0.926	26.715
(i-2000),(j=2000),(k=2500)-K_f	355566	331371	0.931	26.335
1 ft-Fracture Spacing	223119	341395	1.530	16.525
20 ft-Fracture Spacing	313008	348759	1.114	23.183
10 ft (Base Case)-Fracture Spacing	360691	334122	0.926	26.715
0.01-Fracture Porosity	341600	330828	0.968	25.301
0.0006 (Base Case)-Fracture Porosity	360691	334122	0.926339	26.71521

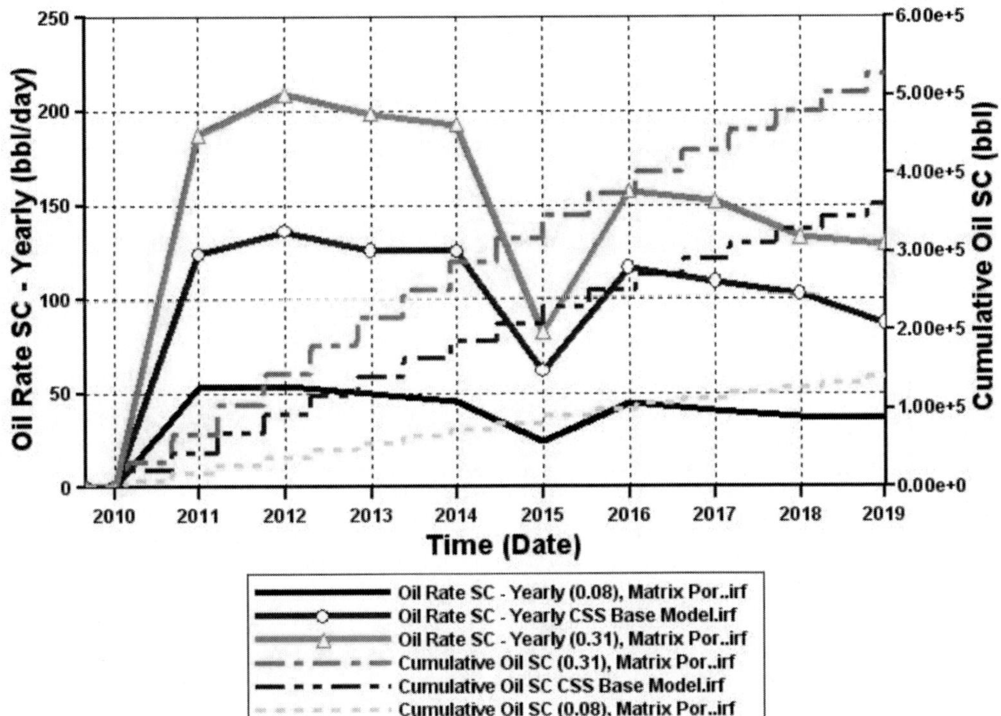

Figure 14. Effect of matrix porosity on production profile in the NFR.

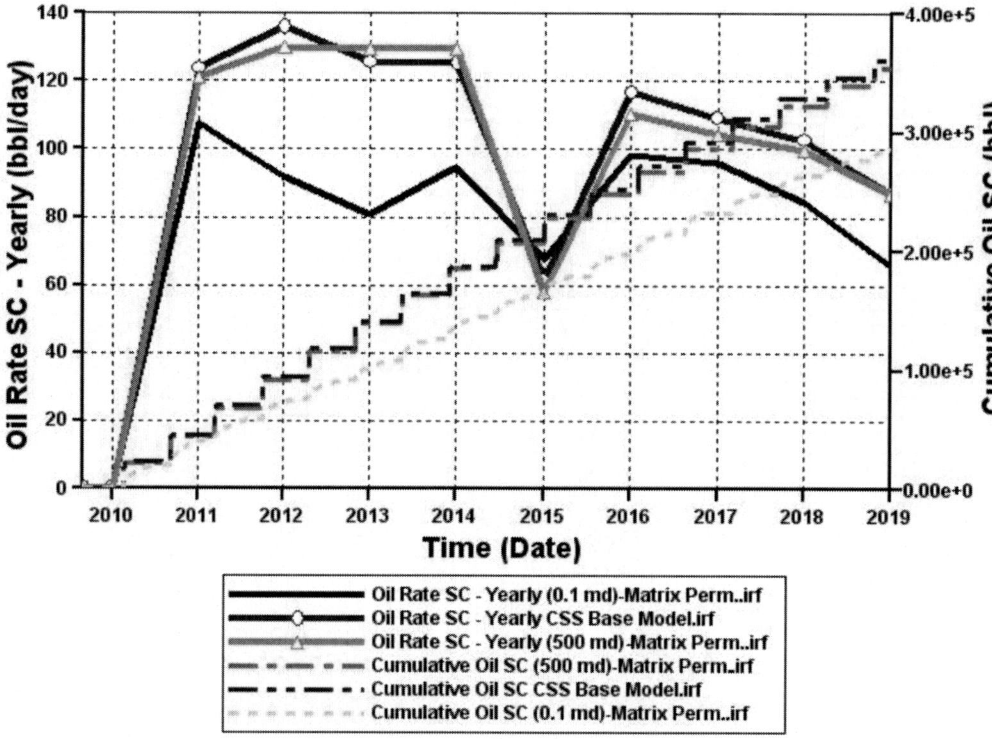

Figure 15. Effect of matrix permability on production profile in the NFR.

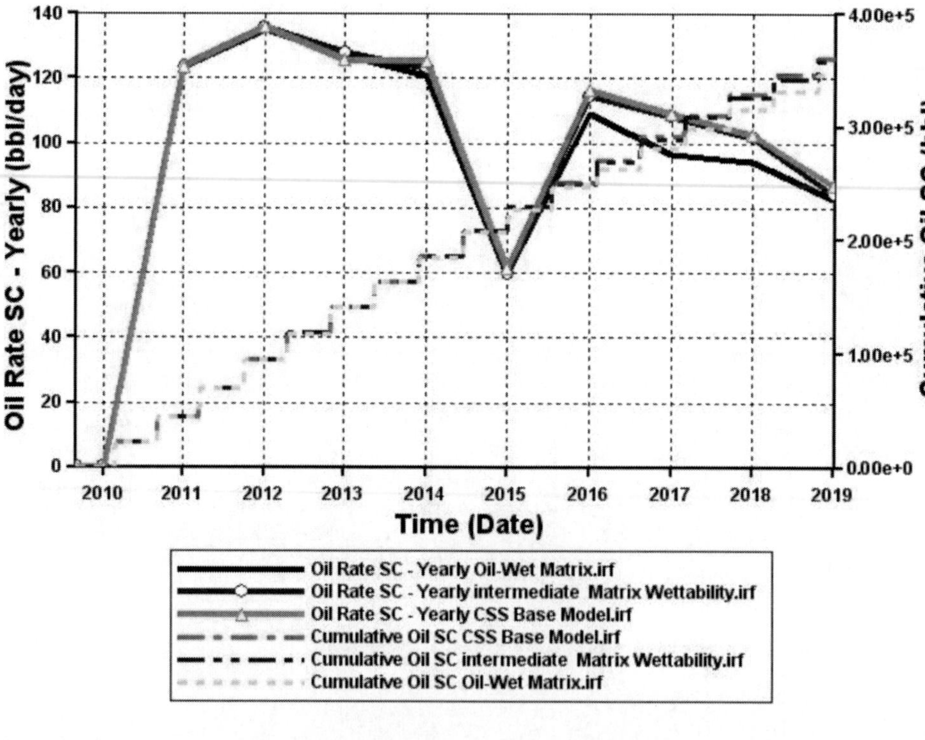

Figure 16. Effect of matrix wettability on production profile in the NFR.

In NFRs, spontaneous imbibitions has significant role in recovery of oil, especially when the reservoir rock is low permeable; hence much attention has been considered to reservoir wettability issues and its impact on imbibitions process (Fernø et al., 2011). In this study, to observe the effect of matrix wettability on CSS performance three cases were chosen; water-wet (base case), intermediate (½ oil-wet pores + ½ water-wet pores) and oil- wet. In oil-wet matrix condition, capillary pressure serves as a barrier and oil depletion stage from matrix blocks to fractures becomes linear, but in the water-wet matrix condition, capillary pressure leads to better steam moving from fractures into matrix blocks (Kiasari et al., 2010).

In water-wet matrix condition, both fracture and matrix collaborate in early production; consequently oil production rate immediately increases particularly at the first stage of production profile. In other words, capillary pressure has a positive impact on steam moving and holding heat through downstream layers. On the other hand, in the oil-wet matrix condition only preheated fractures affect early production and capillary pressure decreases the steam moving into matrix blocks and spreads out steam more into upstream fractures and therefore the number of upstream warmed layers increases and heat loss occurs (Kiasari et al., 2010), as shown in Figure 16. Obviously, temperature of NWR in oil-wet matrix condition is lower than water-wet matrix condition (Figure 17). Table 6 lists the obtained results versus matrix properties changes.

Table 6. Results of CSS performance against matrix properties changes

Scenario Name	Cumulative Oil Production, bbl	Cumulative Water Injection, bbl	cSOR	Ultimate RF, %
0.1 md-Matrix Permeability	287358	353509	1.230	21.283
50 md (Base Case)-Matrix Permeability	360691	334122	0.926	26.715
500 md-Matrix Permeability	354175	328103	0.926	26.232
8%-Matrix Porosity	140728	146850	1.043	10.423
19.5% (Base Case)-Matrix Porosity	360691	334122	0.926	26.715
31%-Matrix Porosity	526410	517924	0.983	38.989
Water-Wet Matrix (Base Case)	360691	334122	0.926	26.715
Intermediate-Wet Matrix	357611	333924	0.933	26.487
Oil-Wet Matrix	346397	333555	0.962	25.656

CONCLUSION

This work is a first start to understand how the fracture/matrix impacts CSS performance in an Iranian naturally fractured heavy oil reservoir. The results indicate that CSS performance within Kuh-e Mond reservoir is impacted by the fracture geometrical properties and how the fluid flow through matrix blocks of Kuh-e Mond reservoir. Specially, due to steam escape through vertical fractures by increasing fracture permeability in ***k***-direction RF decreased. Moreover, higher fracture density increases temperature of the model but decreases steam rising speed, and also at larger fracture porosity values more heat will transfer to the upstream matrix blocks. Matrix porosity and permeability control connectivity and capacity of formation in order to fluid flow through porous media. This implies that permeability and porosity effect on the steam chamber growth and that the variability of the matrix properties controls the variability of the steam chamber size. In addition, the water-wet matrix has a positive impact on reservoir ultimate oil recovery and temperature profile of NWR.

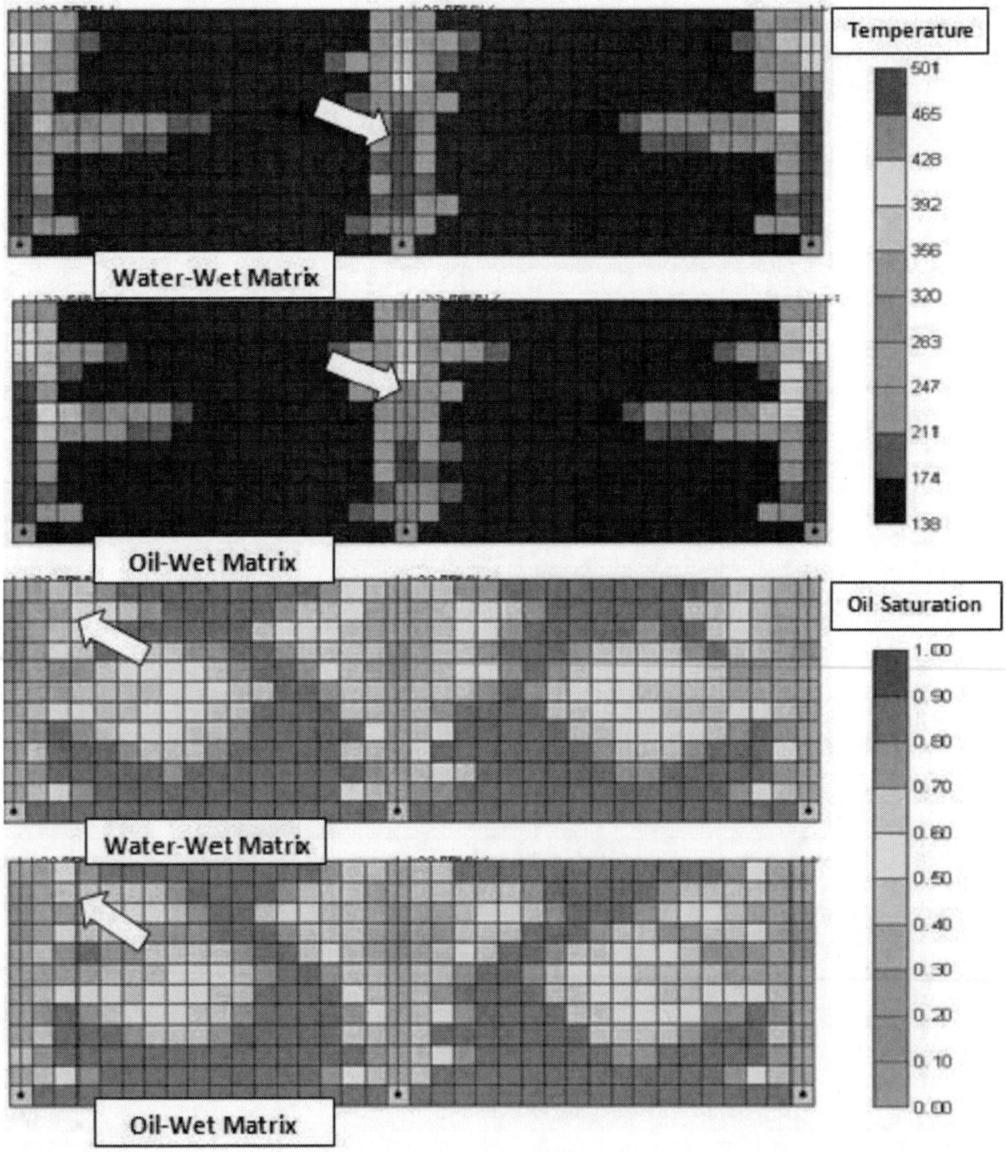

Figure 17. Effect of matrix wettability on temperature profile and oil saturation distribution in the NFR.

NOMENCLATURE

EOR	Enhanced Oil Recovery
RF	Recovery Factor
SOR	Steam Oil Ratio
cSOR	Cumulative Steam Oil Ratio
NFR	Naturally Fractured Reservoir
BHP	Bottom-Hole Pressure
PVT	Pressure-Volume-Temperature
CSS	Cyclic Steam Stimulation
SAGD	Steam Assisted Gravity Drainage
OOIP	Original Oil in Place
NWR	Near Well Region
AWR	Above Well Region
CMG	Computer Modeling Group
F_U	Upper Facies
F_L	Lower Facies
Φ_m	Matrix Porosity
Φ_f	Fracture Porosity
K_m	Matrix Permeability
K_f	Fracture Permeability
K_{ro}	Oil Relative Permeability
K_{rw}	Water Relative Permeability
P_c	Capillary Pressure
S	Saturation

REFERENCES

Alajmi, H., King, P. R., 2011. Assessment of Thermal EOR Methods for a Heavy-Oil Sandstone Reservoir in Middle East. In: SPE Enhanced Oil Recovery Conference, Kuala Lumpur, Malaysia, 19-21 July.

Azin, R., Kharrat, R., Ghotbi, C., 2008. Effect of Fracture Spacing on VAPEX Performance in Heavy Oil Fracture Systems. Iran. *J. Chem. Chem. Eng.,* 27, 35-45.

Babadagli, T., Al-Bemani, A., 2007. Investigations on matrix recovery during steam injection into heavy-oil containing carbonate rocks. *Journal of Petroleum Science and Engineering,* 58, 259–274.

Bagci, S., 2004. The Effect of Fractures on the Steam-Assisted Gravity Drainage Process. *Energy Fuels,* 18, 1656-1664.

Butler, R. M., 1997. Thermal Recovery of Oil and Bitumen, GravDrain Inc., Calgary, Alberta.

Beattie, C. I., Boberg, T. C., McNab, G. S., 1991. Reservoir Simulation of Cyclic Steam Stimulation in the Cold Lake Oil Sands. *SPE Res. Eng.,* 6, 200-206.

Briggs, P. J., 1989. A Simulator for the Recovery of Heavy Oil from Naturally Fractured Reservoirs Using Cyclic Steam Injection. In: The SPE Middle East Oil Technical Conference and Exhibition, Manama, Bahrain, 11-14 March.

Escobar, E., Valkó, P., Lee, W. J., Rodríguez, M. G., 2000. Optimization Methodology for Cyclic Steam Injection with Horizontal Wells. In: The 2000 SPE/Petroleum Society of CIM International Conference on Horizontal Well Technology, Calgary, Alberta, Canada, 6-8 November.

Fatemi, S. M., Kharrat, R., Ghotbi, C., 2011a. Toe-to-Heel Air Injection: Investigation of the Effect of Fractures Geometrical Properties on Process Performance. *Energy Sources,* Part A, 33,2067–2077.

Fatemi, S. M., Kharrat, R., Vossoughi, S., 2011b. Investigation of the Effect of Geometrical Properties of Networked Fractures on the Efficiency of Steam-Assisted Gravity Drainage Process. *Pet. Sci. Tech.,* 29, 1625–1636.

Fattahi, A., Akhondzadeh, H., Derahman M. N., Yunan M. H., Namazi, S., 2012. Impact of reservoir heterogeneity on Steam Assisted Gravity Drainage in heavy oil fractured reservoirs. *Energy Explor. Exploit.*, 30, 553–566.

Farzaneh, S. A., Dehghan, A. A., Kharrat, R., Ghazanfari, M. H., 2011. An Experimental Investigation of Fracture Physical Properties on Heavy Oil Displacement Efficiency during Solvent Flooding, *Energy Sources*, Part A, 33, 1993–2004.

Fernø, M. A., Haugen, Å., Graue, A., 2011. Wettability effects on the matrix–fracture fluid transfer in fractured carbonate rocks. *J. Petrol. Sci. Eng.,* 77, 146–153.

Gotawala, Dh. R., Gates, I. D., 2010. On the Impact of Permeability Heterogeneity on SAGD Steam Chamber Growth. *Nat. Resources Research*, 19, 151-164.

Gozde, S., Chhina, H. S., Best, D. A., 1989. An Analytical Cyclic Steam Stimulation Model for Heavy Oil Reservoirs. In: SPE California Regional Meeting, Bakersfield, California, 5-7 April.

Gunadi, B., 1999. Experimental and analytical studies of cyclic steam injection using horizontal wells. PhD Thesis, Texas A&M University, USA.

Hemmati-Sarapardeh, A., Hashemi Kiasari H., Alizadeh, N., Mighani, S., Kamari, A., 2013. Application of Fast-SAGD in Naturally Fractured Heavy Oil Reservoirs: A Case Study. In: The 18th Middle East Oil & Gas Show and Conference, Bahrain, 10-13 March.

Hascakir B., Kovscek A. R., 2010. Reservoir simulation of cyclic steam injection including the effect of temperature induced wettability alteration. In: The SPE Western Regional Meeting, California, USA, 27-19 May.

Kamari, A., Nikookar, M., Sahranavard, L., Mohammadi, A. H., 2014. Efficient screening of enhanced oil recovery methods and predictive economic analysis. *Neural Comput. Appl.*, 815-824, 25.

Kamari, A., 2013. Part B: Evaluation of Impact of Wettability Alteration and Oil Relative Permeability Changes with Temperature during Cyclic Steam Injection in Naturally Fractured Reservoirs. M.Sc. Thesis, Islamic Azad University, Omidiyeh Branch, Iran.

Kamari, E., Shadizadeh, S. R., Rashtchian, D., 2012. The Effect of Fracture Geometrics on Breakthrough Time in the Immiscible Displacement Process Through Strongly Oil Wet Fractured Porous Media: Experimental Investigation. *Energy Sources*, Part A, 34, 867–876.

Kiasari, H. H., Mortazavi-Kiasari, M., Sola, B. S., 2010. Investigation on effects of rock-fluid properties on the SAGD production profile in naturally fractured reservoir (NFR), *Pet. Coal,* 52, 243-248.

Kiasari, H. H., Sedaee Sola, B., 2012. The Investigation of SAGD Production Profile in a Naturally Fractured Reservoir (NFR). *Pet. Sci. Tech.,* 30, 1343–1352.

Liu, Sh., Gao, Y., Liu, Sh., 2006. Efficient Technology Following Cyclic Steam Stimulation for thick Massive Heavy Oil Reservoirs. In: International Oil & Gas Conference and Exhibition, Beijing, China, 5-7 December.

Owens W. D., Suter V. E., 1965. Steam Stimulation – Newest Form of Secondary Petroleum Recovery, *Oil Gas J.,* 82-87, 90.

Peng D. Y., and Robinson D. B., 1976. A New Two-Constant Equation of State. *Ind. Eng. Chem. Fund.*, 59–64, 15.

Razavi, S. D., Kharrat, R., 2009. Application of Cyclic Steam Stimulation by Horizontal Wells in Iranian Heavy Oil Reservoirs. *Scientia Iranica*, 16, 125-139.

Reis, J. C., 1990. Studies of Fractures Induced During Cyclic Steam Injection. In: The 60th California Regional Meeting, California, USA, 4-6 April.

Revana, K., Erdogan, H. M., 2007. Optimization of cyclic steam stimulation under uncertainty. In: The 2007 SPE Hydrocarbon Economics and Evaluation Symposium, Dallas, Texas, USA, 1-3 April.

Rodrigues, M., 1999. Bachaquero-01 reservoir, Venezuela - increasing oil production by switching from cyclic steam injection to steamooding using horizontal wells. MSc Thesis, Texas A&M University, USA.

Saidi, A. M., 1988. Reservoir Engineering of Fractured Reservoirs: Fundamental and Practical Aspects, Total Ed. Press, Paris.

Tamim, M., Farough Ali, S. M., 1998. A New Analytical Cyclic Steam Stimulation Model Including Formation Fracturing. *J. Can. Pet. Tech*., 37, 31-40.

Thomas, S., 2007. Enhanced Oil Recovery–An Overview. *Oil Gas Sci. Tech.,* 63, 9-19.

Van Golf-Racht, T. D., 1982. Fundamental of Fractured Reservoir Engineering, 2 ed. Elsevier Scientific Publication Company, Netherlands.

Watkins, D. R., Kalfayan, L. J., Blaser, S. M., 1987. Cyclic Steam Stimulation in a Tight Clay-Rich Reservoir. In: SPE California Regional Meeting, California, USA, 8-10 April.

Yitang, Z., Bingyu, J., Xiuluan, L., Guangzhi, L., Dehuang, S., Yunting, X., Lianjie, S., 2005. Study on Developing Low Permeability Reservoir with Waxy Oil by Cyclic Steam Stimulation. In: SPE/PS-CIM/CHOA International Thermal Operations and Heavy Oil Symposium, Calgary, Alberta, Canada, 1-3 November.

Zellou, A. M., Ouenes, A., 2001. Integrated fractured reservoir characterization using neural networks and fuzzy logic: three case studies. *J. Pet. Geol.,* 24, 459–476.

In: Enhanced Oil Recovery
Editor: Alicia Knight
ISBN: 978-1-63463-917-0

Chapter 13

EVALUATING THE RESERVOIR QUALITY PARAMETER (PERMEABILITY) FOR EFFICIENT OIL RECOVERY

***Arash Kamari*[1], *Mohammad Nikookar*[*2], *Leili Sahranavard*[2] *and Amir H. Mohammadi*[*,1,3]**

[1]Thermodynamics Research Unit, School of Engineering, University of KwaZulu-Natal, Howard College Campus, Durban, South Africa
[2]IOR Research Institute, National Iranian Oil Company (NIOC) - R&T, Tehran, Iran
[3]Institut de Recherche en Génie Chimique et Pétrolier (IRGCP), Paris Cedex, France

ABSTRACT

Reservoir permeability plays a key role in determining the quality of reservoir for efficient recovery of oil, in particular during enhance oil recovery (EOR) methods. The permeability can be obtained from experimental measurements and or by in-place testing. Both experimental measurement and well test analysis are expensive and time-consuming. In order to find the impact of well log data on permeability, a more powerful method than the traditional ones is necessary. As a result, artificial intelligence technique is one of the most important and efficient soft-computing approaches to approximate any function with finite number of discontinuities by learning the relationships between input and output vectors. In this communication, several nets with various inputs and hidden neurons are developed, and then the best net with optimum input parameters and hidden neuron were selected in order to investigate the effect of input parameters on permeability. Afterward, to confirm the obtained results and study the impact of well log data on permeability, a sensitivity analysis was conducted using sensitivity about the mean (SAM) feature. The results show that permeability has lowest sensitivity to parameter of thermal neutron apparent porosity (NPHI). On the other hand, parameter of corrected log porosity has the highest effect on estimation process of permeability.

[*] Corresponding authors email addresses: M. Nikookar: mohamad_ni@yahoo.com & A.H. Mohammadi: a.h.m@irgcp.fr and amir_h_mohammadi@yahoo.com.

Keywords: artificial neural network (ANN), permeability estimation, sensitivity analysis, input, output, parameter

NOMENCLATURE

AI	artificial intelligence
ANN	artificial neural network
FL	fuzzy logic
GA	genetic algorithm
MLP	multi-layer perceptron
BP	back-propagation
SAM	sensitivity about the mean
PHIT	corrected log porosity
GR	gamma ray
DT	sonic compression transit time
NPHI	thermal neutron apparent porosity
RHOB	bulk density
ILD	deep induction log
D	Darcy
MSE	mean square error
MAE	mean absolute error

1. INTRODUCTION

As a result, petroleum reservoirs, in particular naturally fractured reservoirs, are complex heterogeneous geological systems. Consequently, the problems related to reservoir characterization are difficult due to complexity, natural heterogeneity, non-uniformity and non-linearity in their effective parameters such as permeability. Permeability is characteristic of hydrocarbon reservoirs which is typically distributed in a spatially non-uniform and non-linear manner (Nikravesh and Aminzadeh, 2001; Nikravesh et al., 2003). As a matter of fact, permeability parameter plays a significant role in reservoir characterizations so that represents and controls the relationship between pores and production rate, respectively.

The permeability of a porous medium can be determined by experimental measurements and well testing analysis (Karimpouli et al., 2010; Wong et al., 2000). Both experimental measurement and well test analysis are normally tedious, expensive and time-consuming compared to the electric techniques. Generally, core measurements are used as the primary method for the calculation of permeability (Arpat et al., 1998). However, accessing the representative core samples from a large number of drilled wells may be considered as a very expensive operation. On the other hand, well test analyses are time-consuming, and also includes high operation costs. Therefore, it is important to develop a reliable model for the determination of permeability in terms of well log data.

Artificial neural networks (ANN), fuzzy logic (FL), and genetic algorithms (GA) are most commonly-used artificial intelligence (AI) methods which are increasingly applied in oil and natural gas reservoirs simulation, production, enhanced oil recovery(EOR), optimization, drilling automation, process control, and data mining (Al-Bulushi et al., 2009; Aminian and Ameri, 2005; Aminzadeh, 2005; Asadisaghandi and Tahmasebi, 2011; Darabi et al., 2010; El-Sebakhy, 2009; Huang et al., 2003; Irani and Nasimi, 2011; Jafari Kenari and Mashohor, 2013; Jahanandish et al., 2011; Kamari et al., 2014; Kamyab et al., 2010).

In the recent years, artificial intelligence techniques, in particular neural networks, have gained popularity in solving complex nonlinear problems (Al-Fattah and Al-Naim, 2009). ANN is a fast and accurate method for the prediction of reservoir properties, so that can be applied in reservoir modeling and characterization. As a consequence, previous studies show that permeabilities of heterogeneous reservoir rocks can be predicted with high accuracy by implementing artificial neural network models (Jeirani and Mohebbi, 2006; Karimpouli et al., 2010; Kaydani et al., 2011; Saemi et al., 2007; Tahmasebi and Hezarkhani, 2012).

The first aim of this paper is to predict the permeability of heterogeneous reservoirs. Then, we tried to recognize the most important geophysical parameters for the prediction of permeability. Therefore, several nets with various inputs and hidden neurons were developed to find the optimum ANN model for predicting the permeability. Moreover, to investigate the impact of well log data on permeability, a sensitivity analysis was performed by sensitivity about the mean (SAM) feature.

After selection of the optimum network for prediction of permeability, effect of threshold on the net efficiency is studied. Finally, permeability was estimated by using the most important input parameters (well log data).

2. Model Development

2.1. Artificial Neural Network Background

Artificial intelligence is the knowledge of making intelligent machines and tools. AI approach is interested in designing ways to make computers which were previously thought to require human brain. Expert systems or intelligent-knowledge-based systems are computer programs that understand and apply the thinking methodologies of human experts (Gharbi, 2005). Expert system utilizes its intelligent-knowledge-based to offer a consultation, and solves complex problems.

As a result, a simple artificial neural network is a kind of artificial intelligence technologies with various applications in electronics, financial, robotic, medical and economic, and finally oil and gas industry (MATLAB, 2000; Tahmasebi and Hezarkhani, 2012).

ANNs mimic the human's brain and nervous system; in other words it has been originated by the biological nervous system of humans (Mohaghegh et al., 1996). ANN is an especially efficient algorithm to approximate any function with finite number of discontinuities by learning the relationships between input and output vectors (Ganguly, 2003; Laugier and Richon, 2003).

2.2. Neuron Model and Network Architectures

ANNs were made of a large number of simple computing components, called nodes or neurons which is arranged to form an input layer, one or more hidden layers and an output layer (Hagan et al., 1996). A simple neuron with a single scalar input and no bias is shown in Figure 1. The scalar input *p* is transmitted through a connection that multiplies its strength by the scalar weight *w*, to form the product *wp*, again a scalar. Here, the weighted input *wp* is the only argument of the transfer function *f*, which produces the scalar output *a*.

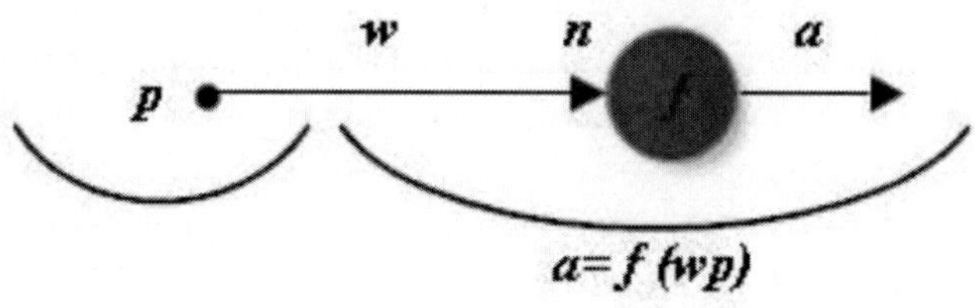

Figure 1. Graphical performance of a simple neuron without bias.

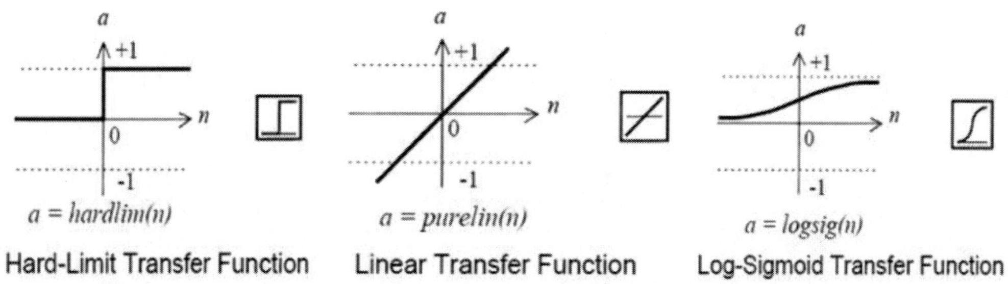

Figure 2. Graphical performance of the most commonly used functions (MATLAB, 2000).

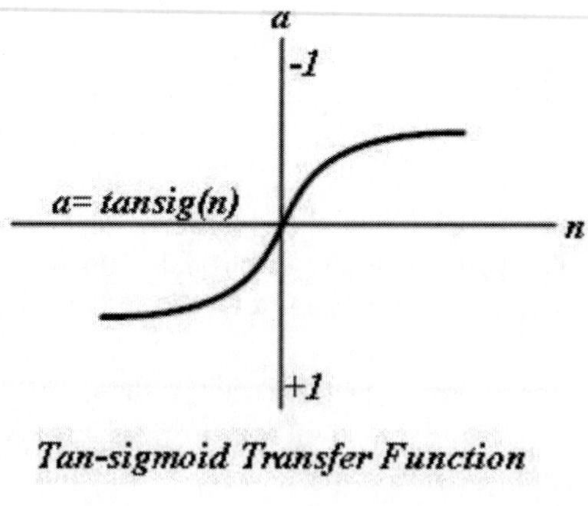

Figure 3. Graphical performance of the used transfer function in this study.

Input variables of the network have different values. These values are defined by weights. Weights are used in calculations before hidden and output layers. By using weights of each input variable, the outputs are calculated (Ramgulam, 2006; Zabihi et al., 2011). Transfer functions are used in the output and hidden layers. As a result, there are different types of transfer function that can be selected by the user. Three of the most commonly-used functions (MATLAB, 2000) are shown in Figure 2. The symbol in the square to the right of each transfer function graph is shown in Figure 2, which represents the associated transfer function. The transfer function used in this study is a tangent sigmoid function as shown in Figure 3.

2.3. Multi-Layer Perception (MLP) Network

A multilayer ANN consists of different layers. The first and the last layers are the input and output layers, respectively. The intermediate layer which is exact between these two layers, connects them indirectly. This layer is called the hidden layer. As a consequence, the number of hidden layer in an ANN system can consist of one or more hidden layers. The number of the neurons in the hidden layer may vary from one to the optimum one. The optimum number of neurons in the hidden layer is depends on problem, the complexity of the input and output mapping, the amount of noise in the data and the amount of training data available (Ramgulam, 2006). How to connect neurons in a neural network makes the type of network. There are different types of neural network. The two general types are feed-forward and backward neural networks. Multi-layer perception is the most popular feed-forward network (Nowroozi et al., 2009; Zuluaga, 2000).

In order to design artificial neural network, the related experimental data should be used. Therefore, experimental data are divided into three groups of data including training, validating and testing. Training data are used for the determination of optimum amounts of the model, testing data are used for the checking of training, and finally validating data are used for study of model accuracy. The MLP uses a back-propagation (BP) method for training the network which is a type of supervised learning methods (Tahmasebi and Hezarkhani, 2012). As a result, there are three criteria for stopping the training: maximum number of epochs, training time, and target mean square error (MSE). In this study, the mean absolute error (MAE), in addition to MSE, is considered as follows:

$$MSE = \frac{\sum_{i=1}^{n}(K_{pred_i} - K_{\exp_i})^2}{n} \quad (1)$$

$$MAE = \frac{\sum_{i=1}^{n}(\frac{K_{pred_i} - K_{\exp_i}}{K_{\exp_i}})^2}{n} \quad (2)$$

where K_{pred} is the estimated permeability value for a training vector of data i by network, the K_{exp} is the real permeability value of data i and the n is the total number of available data for training. Flowchart of procedure of the ANN model proposed in this study is illustrated in Figure 4.

2.4. Sensitivity about the Mean Feature

As already mentioned, testing phase is employed to check the performance of training phase. During testing, the learning is turned off and the chosen data set is fed through the network. The network output is collected and a report is then generated showing the testing results. SAM feature is a method of testing data in order to find the relationship between inputs and output(s). This testing process provides a measure of the relative importance among the inputs of the neural model and illustrates how the model output varies in response to variation of an input. By default the first input is varied between its mean +/- a user-defined number of standard deviations while all other inputs are fixed at their respective means. The network output is computed for a user-defined number of steps above and below the mean. This process is repeated for each input. An alternate variation of this process is to vary the input of interest between its minimum value and its maximum value. This option is particularly useful for binary inputs or inputs which have a non-Gaussian distribution.

3. Case Study and Data Preparation

The current case study is located in the one of offshore reservoirs located in Saudi Arabia. According to the obtained geological setting and the available data, this reservoir can be classified as a structural and heterogonous reservoir. The high variation in the data, due to heterogeneity of the reservoir, indicates that the traditional methods are not able to estimate the petrophysical parameters effectively. In this oil field, a lot of wells have been drilled and consequently several petrophysical parameters are acquired. But, all of them are not suitable for this study. Therefore, among all of the available data, we selected the following parameters: corrected log porosity, gamma ray (GR), sonic compression transit time (DT), thermal neutron apparent porosity (NPHI), bulk density (RHOB), and deep Induction log (ILD).

The collected database includes a wide range of permeability from 0.0001 to 9.929 D. Furthermore, this indicates the high degree of heterogeneity of reservoir formations. About 701 core measurements for permeability and their corresponding well logging responses were available for network training and testing. Ranges of the applied parameters for training and testing phases are given in Table 1 and Table 2, respectively. Due to high variation in some parameters, mean values are calculated for all parameters shown in Table 1 and 2.

Table 1. Range of input parameters for training phase

Parameter	Min.	Average	Max.
Porosity	5.7	26.45782	36.3
GR	7.643	28.13008	80.6332
DT	64.4	93.04267	109.3
NPHI	0.213	0.297864	0.484
RHOB	1.873	2.237579	2.623
ILD	0.1702	71.30138	2000

Table 2. Range of input parameters for testing phase

Parameter	Min.	Average	Max.
Porosity	5.7	26.63125	36.2
GR	9.309	28.42411	83.0234
DT	64.4	93.08769	104.675
NPHI	0.213	0.294892	0.484
RHOB	1.873	2.232628	2.6822
ILD	0.1682	87.39667	2000

4. Results and Discussion

In this work, the process of the network developing was performed by NeuroSolutions for Excel Release 5 software (NeuroSolutions, 2005).

4.1. Selection of the Optimal Network

In this study, a multi-layer perceptron neural network was developed to recognize the effective parameters for the estimation of permeability. The tanh-axon was selected as transfer function, and Levenberg–Marquardt back-propagation was used in all training steps. The default MSE and MAE were used to measure the performance of the model. The threshold is initially adjusted over 0.01. The maximum number of epochs per each training case according to the MSE is set over 1000.

In order to analyze the effect of input parameters on networks performance, we considered seven sets of input data so that each set includes three networks with 6, 8 and 10 hidden neurons.

The available dataset was divided randomly into three distinct subsets consisting of training, validation and testing. The proportions of these subsets were 60%, 15% and 25% of whole of the data, respectively. In the all of made nets, one hidden layer is considered. After training and testing of developed networks, various sets of inputs parameters, MSE, MAE and accuracy of the models has been evaluated to reach the optimum performance. The obtained results indicate that the best network is the net with inputs of porosity, GR, DT, RHOB and ILD, whose number of neurons was found by trial and error (8 neurons).

As a result, for the mentioned net, MSE and MAE are 0.16 and 0.28, respectively. These results show that the parameter of NPHI has a negative impact on permeability during estimation process, at least in this case study. Performance summaries of 21 nets developed in this study have been tabulated in Table 3. By comparing the above results, it can be concluded that the neural network of no. of 14 has minimum deviation. Therefore, this network can be used for prediction of permeability with high accuracy.

Table 3. Various sets of input data and hidden neurons for choosing optimal ANN model. Model no. 14 is the best model for predicting permeability

No.	Neuron Input Parameters	Training		Testing	
		MSE	MAE	MSE	MAE
1	6 Porosity, GR, DT, NPHI, RHOB, ILD	0.434516	0.418102	0.531856	0.453126
2	8 Porosity, GR, DT, NPHI, RHOB, ILD	0.268454	0.336536	0.261281	0.355307
3	10 Porosity, GR, DT, NPHI, RHOB, ILD	0.251227	0.320949	0.226787	0.317024
4	6 GR, DT, NPHI, RHOB, ILD	0.2337	0.315103	0.221131	0.33439
5	8 GR, DT, NPHI, RHOB, ILD	0.262202	0.331554	0.248097	0.345622
6	10 GR, DT, NPHI, RHOB, ILD	0.304605	0.360875	0.351924	0.380962
7	6 Porosity, DT, NPHI, RHOB, ILD	0.281534	0.347818	0.23083	0.342222
8	**8 Porosity, DT, NPHI, RHOB, ILD**	**0.245772**	**0.326662**	**0.262742**	**0.348573**
9	10 Porosity, DT, NPHI, RHOB, ILD	0.222485	0.312489	0.388836	0.343489
10	6 Porosity, GR, NPHI, RHOB, ILD	0.260779	0.334302	0.262741	0.348721
11	8 Porosity, GR, NPHI, RHOB, ILD	0.312989	0.373818	0.288956	0.378707
12	10 Porosity, GR, NPHI, RHOB, ILD	0.26874	0.330066	0.270544	0.323942
13	6 Porosity, GR, DT, RHOB, ILD	0.299845	0.362973	0.340894	0.389071
14	8 Porosity, GR, DT, RHOB, ILD	0.240387	0.311356	0.160770	0.288441
15	10 Porosity, GR, DT, RHOB, ILD	0.461427	0.449843	1.797553	0.564504
16	6 Porosity, GR, DT, NPHI, ILD	0.496994	0.433013	0.650132	0.451124
17	8 Porosity, GR, DT, NPHI, ILD	0.288318	0.356762	0.240154	0.3495
18	10 Porosity, GR, DT, NPHI, ILD	0.222325	0.300426	0.241225	0.338049
19	6 Porosity, GR, DT, NPHI, RHOB	0.313943	0.365206	0.313185	0.369557
20	8 Porosity, GR, DT, NPHI, RHOB	0.270442	0.335099	0.233678	0.350636
21	10 Porosity, GR, DT, NPHI, RHOB	0.290567	0.355694	0.281892	0.360885

4.2. Validation, Efficiency Improving and Prediction

To confirm the results obtained in the previous section, a sensitivity analysis was performed. To this end, the input parameters was tested by using sensitivity about the mean method. Hence, standard deviation was adjusted over 1, and a network which involves the input parameters of porosity, GR, DT, NPHI, RHOB and ILD with 8 hidden neurons was designed. After training and sensitivity testing, the results show that the sensitivity of permeability to porosity, GR, DT, NPHI, RHOB and ILD are 0.76, 0.4, 0.29, 0.12, 0.36 and 0.34, respectively (Figure 5). This means that permeability has lowest sensitivity to parameter of thermal neutron apparent porosity. Also, parameter of corrected log porosity has the highest effect on estimation process of permeability. Figure 6 reveals that by increasing the network output, porosity and ILD increase and GR, DT and RHOB decrease. As can be seen in Figure 6, the NPHI parameter has lowest changes against increasing the value of network output.

Consequently, we can conclude that obtained results in Section 4.1 are reliable, because the obtained optimal network in the previous section has input parameters without existing NPHI parameter. Also, sensitivity analysis shows that NPHI has lowest impact on permeability during estimation process. Hence, the network without existing NPHI as input parameter can be used for the prediction of permeability with high accuracy. The structure of the selected optimal net with eight hidden layer neurons is shown in Figure 7.

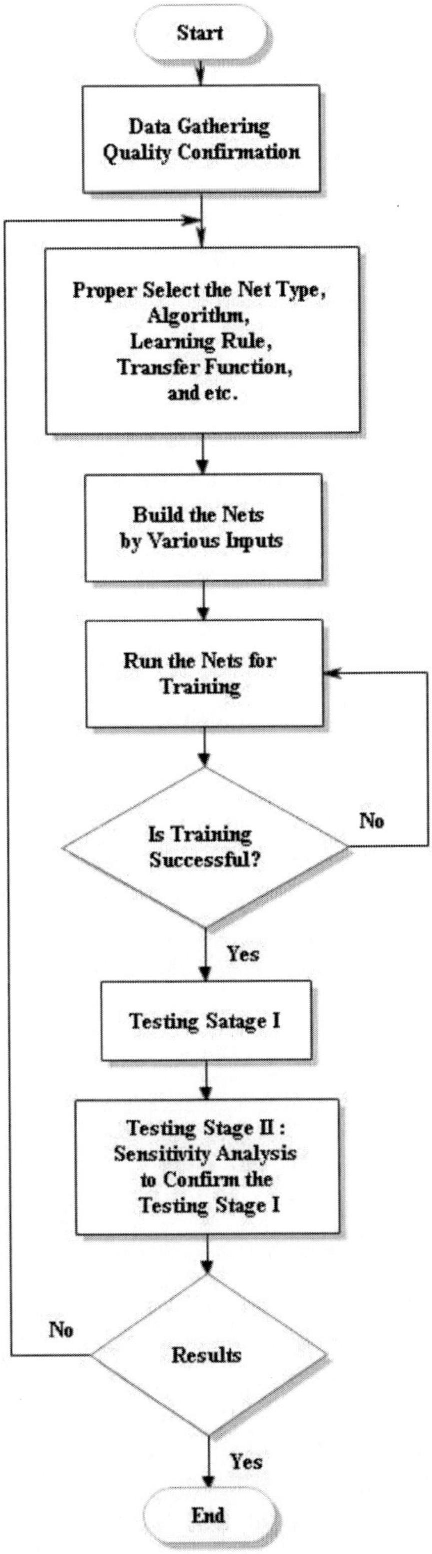

Figure 4. Flowchart of procedure of ANNs design and development proposed in this study.

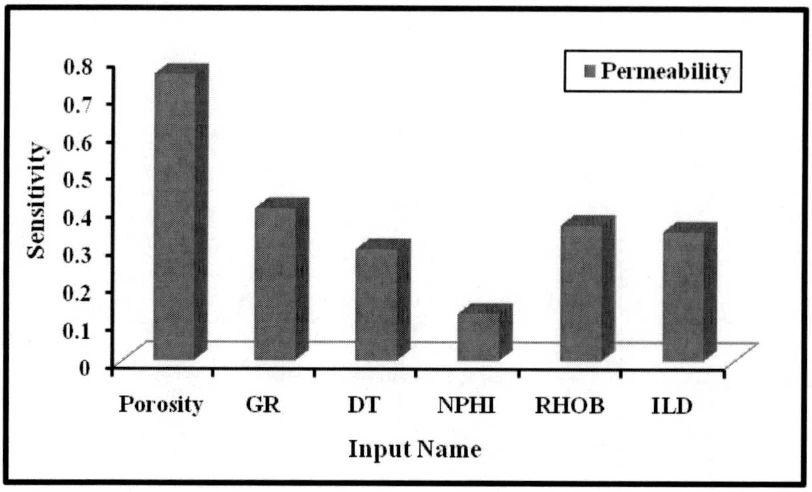

Figure 5. Results of sensitivity analysis for the six input parameters.

Testing of various thresholds during training stage is due to avoid the over- or under-training (Lee et al., 2011). In this study, a number of thresholds are tested with the model during the training to improve the efficiency of the selected network. The results indicate that the ANN model has good performance with all given threshold (Table 4). The best performance is shown in the initial model trained with a threshold equals to 0.01, which is eventually selected as the final model in order to predict the permeability.

Table 4. Performance prediction of optimal ANN model (one hidden layer with 8 neurons) according to the threshold

Threshold	0.1			0.01			0.001			0.0001		
	Train.	Valid.	Test.	Train.	Valid.	Test.	Train.	Valid.	Test.	Train.	Valid.	Test.
# of Rows	421	105	175	421	105	175	421	105	175	421	105	175
MSE	0.37	0.47	0.30	0.24	0.24	0.16	0.21	0.24	0.17	0.16	0.22	0.19
MAE	0.41	0.47	0.38	0.31	0.35	0.28	0.30	0.36	0.28	0.27	0.34	0.31

After selection, validation and improving stages, permeability will be predicted using a net with the aforementioned input parameters, eight hidden neuron and 0.01 threshold. The variation of the experimental data of permeability against the ANN model predictions with their corresponded correlation coefficient is shown in Figure 8 and Figure 9 for training and testing phases, respectively. These figures show that the results of the ANN are in agreement with the experimental data. The the predicted permeability for the network with one hidden layer, 8 hidden neurons, 0.01 threshold and input parameters of porosity, GR, DT, RHOB and ILD is shown in Figures 10 and Figure 11 for training and testing phases, respectively. According to both figures, there are acceptable agreements between the real and estimated permeabilities for the final obtained ANN network.

Regarding the results obtained in this paper, it was found that correct and exact select of the input parameters is very important during estimation process of permeability using artificial neural network technology. On the other hand, optimal selection of hidden layer neurons number and threshold during training phase has a key role in improving the efficiency of ANN model and increasing the accuracy of predicted values. In actual, ANN technology is an efficient and useful tool, if it is used properly and constructively.

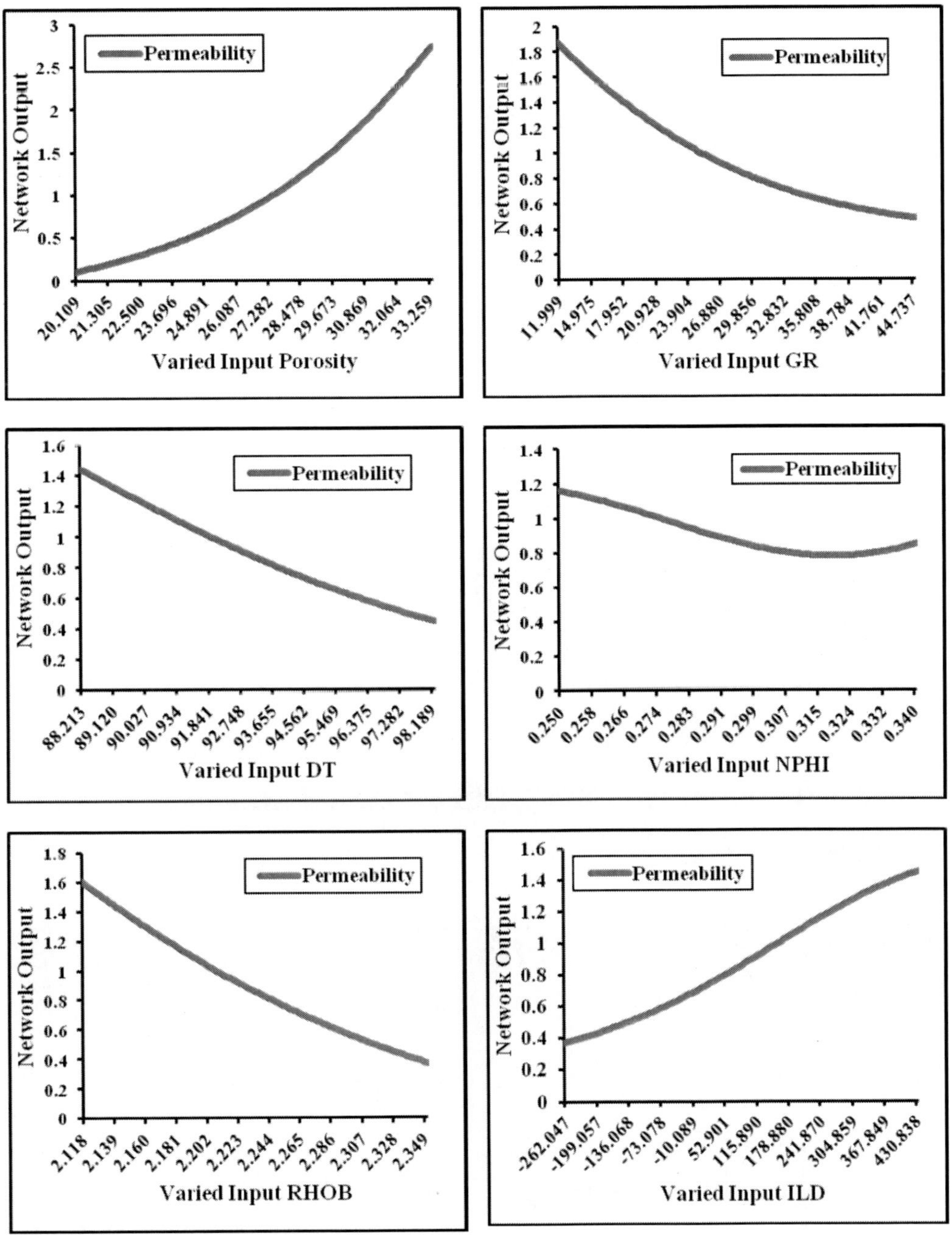

Figure 6. Behavior of the network output against the varied inputs.

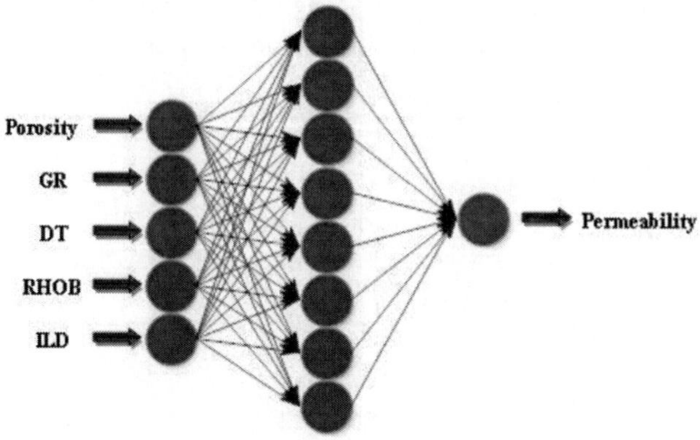

Figure 7. Structure of the optimal network with five input parameters, on hidden layer and eight hidden layer neurons.

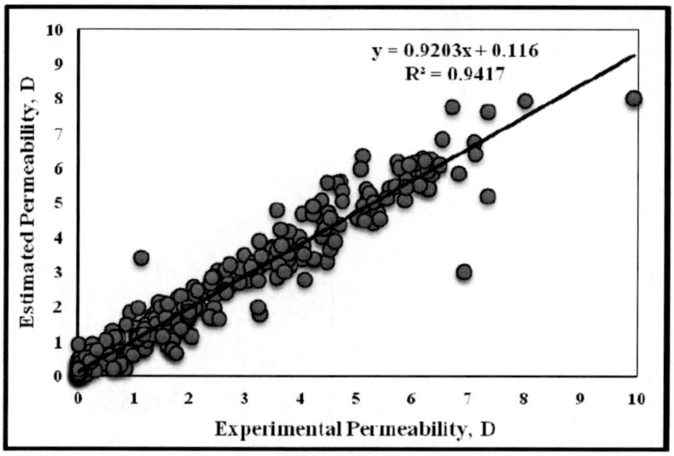

Figure 8. Comparison of the experimental and estimated permeability with their corresponded correlation coefficient for the optimal network in training phase.

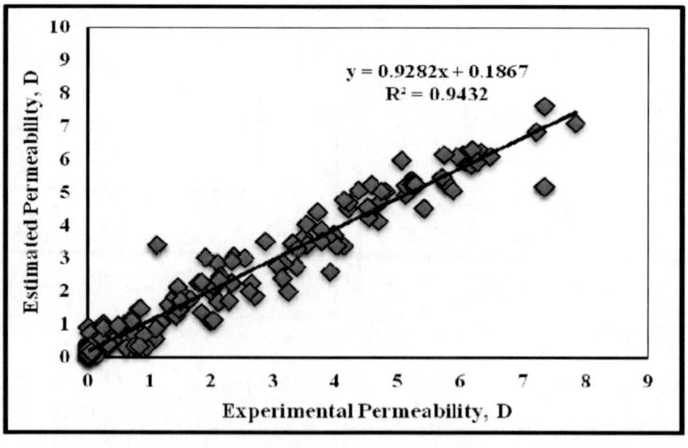

Figure 9. Comparison of the experimental and estimated permeability with their corresponded correlation coefficient for the optimal network in testing phase.

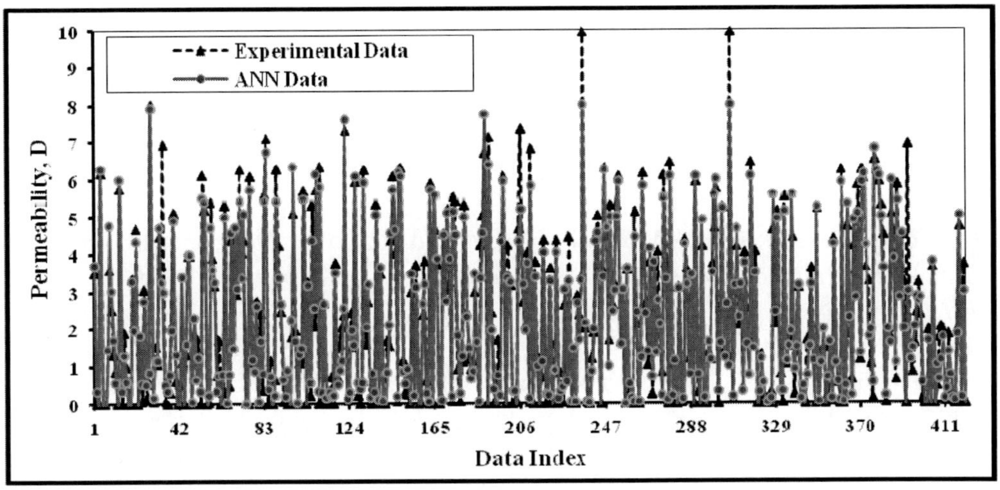

Figure 10. Comparison of the estimated values of the optimal network and the real permeability in training phase.

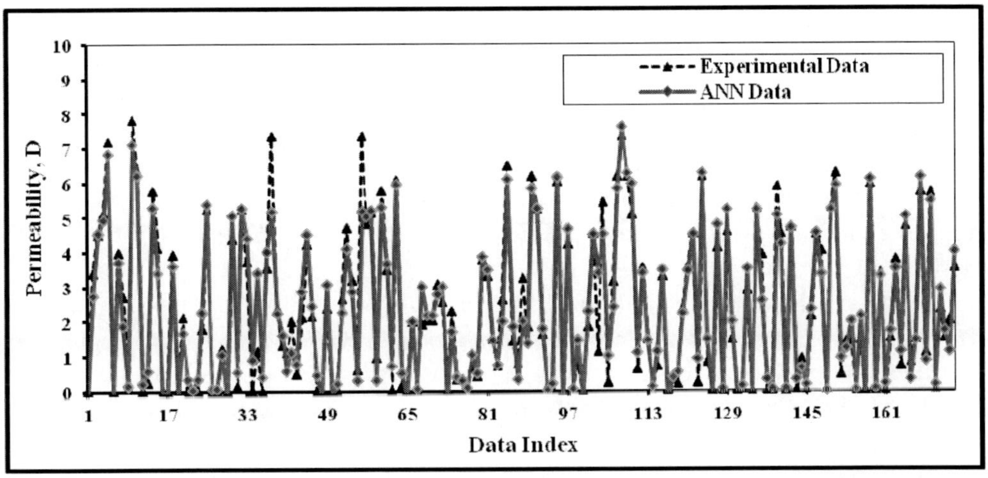

Figure 11. Comparison of the estimated values of the optimal network and the real permeability in testing phase.

CONCLUSION

Due to importance and the fundamental role of permeability in reservoir characterizations, in this paper we developed several neural networks for the estimation of permeability considering importance of well log data on permeability. Therefore, 21 neural networks with various hidden neurons and input data have been developed. By testing the developed neural networks against field data, it was shown that a designed neural network is able to predict permeability. The approach presented in this study reveals the importance of correct selection of input parameters in order to predict the permeability by ANN. According to the results obtained in this study, NPHI and porosity have lowest and highest impact on permeability during estimation process, respectively.

REFERENCES

Al-Bulushi, N., King, P.R., Blunt, M.J. and Kraaijveld, M., 2009. Development of artificial neural network models for predicting water saturation and fluid distribution. *Journal of Petroleum Science and Engineering*, 68(3): 197-208.

Al-Fattah, S.M. and Al-Naim, H.A., 2009. Artificial-intelligence technology predicts relative permeability of giant carbonate reservoirs. SPE Reservoir Evaluation & Engineering, 12(01): 96-103.

Aminian, K. and Ameri, S., 2005. Application of artificial neural networks for reservoir characterization with limited data. *Journal of Petroleum Science and Engineering*, 49(3): 212-222.

Aminzadeh, F., 2005. Applications of AI and soft computing for challenging problems in the oil industry. *Journal of Petroleum Science and Engineering*, 47(1): 5-14.

Arpat, G., Gümrah, F. and Yeten, B., 1998. The neighborhood approach to prediction of permeability from wireline logs and limited core plug analysis data using backpropagation artificial neural networks. *Journal of Petroleum Science and Engineering,* 20(1): 1-8.

Asadisaghandi, J. and Tahmasebi, P., 2011. Comparative evaluation of back-propagation neural network learning algorithms and empirical correlations for prediction of oil PVT properties in Iran oilfields. *Journal of Petroleum Science and Engineering*, 78(2): 464-475.

Darabi, H., Kavousi, A., Moraveji, M. and Masihi, M., 2010. 3D fracture modeling in Parsi oil field using artificial intelligence tools. *Journal of Petroleum Science and Engineering*, 71(1): 67-76.

El-Sebakhy, E.A., 2009. Forecasting PVT properties of crude oil systems based on support vector machines modeling scheme. *Journal of Petroleum Science and Engineering*, 64(1): 25-34.

Ganguly, S., 2003. Prediction of VLE data using radial basis function network. *Computers & chemical engineering,* 27(10): 1445-1454.

Gharbi, R., 2005. Application of an expert system to optimize reservoir performance. *Journal of Petroleum Science and Engineering*, 49(3): 261-273.

Hagan, M.T., Demuth, H.B. and Beale, M.H., 1996. Neural network design, 1. Pws Boston.

Huang, Y., Huang, G., Dong, M. and Feng, G., 2003. Development of an artificial neural network model for predicting minimum miscibility pressure in CO_2 flooding. *Journal of Petroleum Science and Engineering*, 37(1): 83-95.

Irani, R. and Nasimi, R., 2011. Application of artificial bee colony-based neural network in bottom hole pressure prediction in underbalanced drilling. *Journal of Petroleum Science and Engineering,* 78(1): 6-12.

Jafari Kenari, S.A. and Mashohor, S., 2013. Robust committee machine for water saturation prediction. *Journal of Petroleum Science and Engineering*, 104: 1-10.

Jahanandish, I., Salimifard, B. and Jalalifar, H., 2011. Predicting bottomhole pressure in vertical multiphase flowing wells using artificial neural networks. *Journal of Petroleum Science and Engineering,* 75(3): 336-342.

Jeirani, Z. and Mohebbi, A., 2006. Estimating the initial pressure, permeability and skin factor of oil reservoirs using artificial neural networks. *Journal of petroleum science and engineering,* 50(1): 11-20.

Kamari, A., Nikookar, M., Sahranavard, L. and Mohammadi, A.H., 2014. Efficient screening of enhanced oil recovery methods and predictive economic analysis. *Neural Computing and Applications*, 25: 815-824.

Kamyab, M., Sampaio Jr, J.H., Qanbari, F. and Eustes III, A.W., 2010. Using artificial neural networks to estimate the *z*-factor for natural hydrocarbon gases. *Journal of Petroleum Science and Engineering,* 73(3): 248-257.

Karimpouli, S., Fathianpour, N. and Roohi, J., 2010. A new approach to improve neural networks' algorithm in permeability prediction of petroleum reservoirs using supervised committee machine neural network (SCMNN). *Journal of Petroleum Science and Engineering,* 73(3): 227-232.

Kaydani, H., Mohebbi, A. and Baghaie, A., 2011. Permeability prediction based on reservoir zonation by a hybrid neural genetic algorithm in one of the Iranian heterogeneous oil reservoirs. *Journal of Petroleum Science and Engineering*, 78(2): 497-504.

Laugier, S. and Richon, D., 2003. Use of artificial neural networks for calculating derived thermodynamic quantities from volumetric property data. *Fluid phase equilibria*, 210(2): 247-255.

Lee, J.-Y., Shin, H.-J. and Lim, J.-S., 2011. Selection and evaluation of enhanced oil recovery method using artificial neural network. *Geosystem Engineering*, 14(4): 157-164.

MATLAB, 2000. *MATLAB User's Guide.* Version 4, Neural Network Toolbox. The MathWorks Inc.

Mohaghegh, S., Arefi, R., Ameri, S., Aminiand, K. and Nutter, R., 1996. Petroleum reservoir characterization with the aid of artificial neural networks. *Journal of Petroleum Science and Engineering,* 16(4): 263-274.

NeuroSolutions, 2005. *NeuroSolutions for Excel Release 5 software,* Copyright 2005, NeuroDimension, Inc.

Nikravesh, M. and Aminzadeh, F., 2001. Past, present and future intelligent reservoir characterization trends. *Journal of Petroleum Science and Engineering*, 31(2): 67-79.

Nikravesh, M., Zadeh, L.A. and Aminzadeh, F., 2003. *Soft computing and intelligent data analysis in oil exploration,* 51. Elsevier.

Nowroozi, S., Ranjbar, M., Hashemipour, H. and Schaffie, M., 2009. Development of a neural fuzzy system for advanced prediction of dew point pressure in gas condensate reservoirs. *Fuel Processing Technology*, 90(3): 452-457.

Ramgulam, A., 2006. Utilization of artificial neural networks in the optimization of history matching, The Pennsylvania State University.

Saemi, M., Ahmadi, M. and Varjani, A.Y., 2007. Design of neural networks using genetic algorithm for the permeability estimation of the reservoir. Journal of Petroleum Science and Engineering, 59(1): 97-105.

Tahmasebi, P. and Hezarkhani, A., 2012. A fast and independent architecture of artificial neural network for permeability prediction. *Journal of Petroleum Science and Engineering*, 86: 118-126.

Wong, P.M., Jang, M., Cho, S. and Gedeon, T.D., 2000. Multiple permeability predictions using an observational learning algorithm. *Computers & Geosciences*, 26(8): 907-913.

Zabihi, R., Schaffie, M., Nezamabadi-Pour, H. and Ranjbar, M., 2011. Artificial neural network for permeability damage prediction due to sulfate scaling. *Journal of Petroleum Science and Engineering,* 78(3): 575-581.

Zuluaga, E., 2000. *Prediction of permeability reduction by external particle invasion using artificial neural networks and fuzzy models*, Canadian International Petroleum Conference. Petroleum Society of Canada.

INDEX

D

H

I

K

L

M

N

O

P

R

S

T

U

V

W

Y

Z